Integrated Reservoir Studies

COVER ILLUSTRATION
Reservoir Modelling Cycle
(Courtesy of Beicip-Franlab)

FROM THE SAME PUBLISHER

- Well Production Practical Handbook
  H. CHOLET
- Geophysics of Reservoir and Civil Engineering
  J.L. MARI, G. ARENS, D. CHAPELLIER, P. GAUDIANI
- Well Completion and Servicing
  D. PERRIN
- Well Testing: Interpretation Methods
  G. BOURDAROT
- Rock Mechanics. Vol. 1. Theoretical Fundamentals.
  Vol. 2. Petroleum Applications
  PH. CHARLEZ
- Formation Imaging by Acoustic Logging
  Edited by J.L. MARI
- Drilling J.P. NGUYEN
- Basics of Reservoir Engineering
  R. COSSÉ
- Drilling Mud and Cement Slurry Rheology Manual
- Blowout Prevention and Well Control
- Multiphase Flow in Porous Media
  C.M. MARLE
- The Reservoir Engineering Aspects of Fractured Formations
  L.H. REISS
- Properties of Reservoir Rocks: Core Analysis
  R.P. MONICARD
- Enhanced Oil Recovery
  M. LATIL
- Comprehensive Dictionary of Petroleum Science and Technology
  English-French / French-English
  M. MOUREAU, G. BRACE
- Dictionary of Drilling and Boreholes
  English-French / French-English
  M. MOUREAU, G. BRACE

INSTITUT FRANÇAIS DU PÉTROLE PUBLICATIONS

Luca COSENTINO
Senior Reservoir Engineer
Project Manager
Beicip-Franlab

Integrated Reservoir Studies

Foreword by
Jean-Claude Sabathier

2001

Editions TECHNIP 27 rue Ginoux, 75737 PARIS Cedex 15, FRANCE

ISBN 2-7108-0797-1

A Paola e Michele

The author's proceeds of this book will be used by

World Vision

one of the world's leading aid and development agencies, to help transform the lives of children and families around the world in their struggle against poverty, hunger and injustice.

Foreword

Ever since the 1986 crisis, the price of crude oil has been fluctuating severely. Oil companies have had to comply with this situation by cutting their costs and putting more effort into estimating as accurately as possible the economics of projects and associated risks.

Technical advances in well design and drilling, now allow the drilling of horizontal wells several kilometres long, as well as extended multilateral wells, with new production-injection architecture. Such wells permit the development of the fields using less wellheads and hence more convenient surface infrastructures. Also, new types of structures have replaced traditional platforms, allowing a reduction in capital expenditure and hence increasing the possibility of deep offshore development.

While drilling less but longer wells allowed for a significant cost reduction, the technical risk involved in such operations is higher. These complex wells are more expensive than normal vertical wells and whenever there is a failure, the impact on the economics of the project is significant. Furthermore, such wells are prone to technical problems in the drilling phase, and also running logging tools is often not straightforward.

In addition, the types of completion commonly utilised for such complex wells do not allow for easy interventions, with the possible exception of horizontal wells completed with cemented liners. Early water or gas breakthrough may cause the well to shut-in prematurely, with a significant decline in the total field production.

Even more than in the past, it therefore becomes essential to carefully plan the development strategy of the reservoir, both in terms of number and type of wells to drill and the recovery process (depletion, injection…). These choices, together with a correct prediction of the field performance, will impact heavily on the surface structure design and hence the global economics of the project. To stay within fixed economic bounds and minimise risk, oil companies make use of reservoir studies. While such studies have always been performed, in the present climate they have to be more accurate and less expensive.

The basic fluid flow equations have been used for more than 50 years and their most recent application is linked to the relatively recent development of reservoir simulators. Current models often work with 10^5 gridblocks, but megacell simulations (10^6 gridblocks) are becoming increasingly common. Compositional simulation allows for a better modelling of variations in fluid composition, while wellbore hydraulics and surface networks are being coupled to the reservoir model.

Nevertheless, such models still represent a simplified approximation of a complex and unknown reality. The main problem is related to the knowledge of the reservoir parameters and their discretisation on a large-scale support grid. In this respect, sophisticated upscaling techniques have been developed in the last years, however no definitive solution is available yet and the information loss which results from any upscaling process has to be taken into account when defining the reservoir model.

In any case, the knowledge of the reservoir is the most critical factor. The parameters governing the dynamic behaviour of the field are essentially:

- Structural parameters (depth and thickness maps, faults…).
- Internal architecture (correlation scheme).
- Petrophysical properties (porosity, permeability, capillary pressure, relative permeability).
- Fluid contacts.
- Thermodynamical properties of fluids.

These data are only partially accessible, given the small number of sampling points (wells) and the difficulty of in-situ measurements. Furthermore, these data are not directly measurable and instead must be inferred from other available measurements (e.g., resistivity, radioactivity, pressure). Also, the drilling of complex wells entails less sampling points, while the interpretation of the available measurements becomes generally more difficult. In all cases, the estimation of the reservoir properties must be performed starting from just a few points.

In recent years, data acquisition has been developing considerably, due to the improvement of existing techniques and the capacity to record new physical parameters that can be related to basic reservoir characteristics. One notable technique is 3D seismic, which completely changed the structural modelling of reservoirs and that, under favourable circumstances, may help in assessing the distribution of some reservoir properties; recent logging tools, which discriminate mineralogy, fluids, porosity, faults, fractures …; permanent gauges, which allow for continuous reservoir monitoring.

At the same time, interpretation techniques have become increasingly sophisticated and allow a better definition of reservoir characteristics. In this respect, the most spectacular progress concerns the spatial modelling of reservoir properties. Sequence stratigraphy represents a rigorous framework for well to well correlation, minimising the errors in difficult sedimentary environments. In addition, the probabilistic approach to the problem of estimation led to the development of Geostatistics. The parallel evolution of the theory, the numerical methods and the computer capabilities formed the basis for the development of statistical methods that generate equiprobable images of the reservoir, starting from a sparse set of data.

Such techniques require a high level of technical expertise, as well as powerful computing resources, but their successful application is still dependent upon the quantity and quality of the available data. Mostly, as it has long been recognised, the cooperation of the various specialists (synergy) and the concept of integrated study are the main issue, as far as the improvement of results is concerned.

Nonetheless, it is obvious that the realisation of such a concept is a difficult task. Companies soon realised that putting a geophysicist, a geologist and a reservoir engineer in the same working room was not enough. While these conditions are favourable to the generation of team work, they do not guarantee in themselves that the resulting study will be really *inte-*

grated. The main problems are in the choice of the methods and the difficulty of managing different tasks in parallel, through a continuous communication among the team members.

Each phase (log interpretation, well testing, spatial analysis...) may be carried out using various techniques, which can differ significantly in terms of time and money involved. An old rule of thumb says that 80% of the work can be achieved in 20% of the time. It is therefore necessary, from the planning phase, to choose the interpretation techniques as a function of the available data and the importance of that extra 20% of results.

The importance of integration is related to the scarcity of the available data, that must be supplemented through hypothesis, analogs and correlations, which in turn may have a significant impact on the final results. These various elements must be validated in terms of coherency through all the phases of the study. For example, the reservoir engineer may suspect the presence of a sealing fault on the basis of production data, but this must be consistent with the geological scheme. The difficulty lies in the fact that the study is divided in tasks that are not independent, the results of each task representing the feedback for the others. If a *downstream* task is not consistent with another *upstream* task, the latter should be revised and this process may imply a delay in the project execution and a cost increase. It is therefore necessary that each specialist, before starting a new task, cross check the coherency of the hypothesis with the other disciplines, which in turn implies that all the tasks should be performed, as much as possible, simultaneously.

As can be appreciated, the planning and the realisation of an integrated study is a considerable challenge. Usually, each specialist tends to propose and perform the best study and to attain the best results, even though this is not relevant to the global objective of the study. Frequently, it is possible to see very sophisticated (and expensive) studies that are not utilised in the subsequent study. An opposite attitude is also frequently encountered: specialists may limit the degree of detail of their work, not being interested in providing more accurate results. For example, a geologist may generate a 3 layers vertical discretisation « since reservoir engineers hardly use more than 3 layers in their models ». Another geologist, on the contrary, might generate 15 layers, when really a 5 layers description would be a good compromise from a static and a dynamic point of view.

The concept of parallel planning, on the other hand, may pose some difficulty to the specialist which does not intend to share intermediate results of his work for feedback with other disciplines. Certainly, most professionals prefer to complete their work before delivering their results.

This quick review of the main hurdles of an integrated reservoir study highlights the importance of the role of the project manager. Together with the team members, he is responsible of the definition of the work plan, as well as of the modifications that inevitably will have to be considered during the realisation. He must check that the results of individual phases be shared among the various disciplines at the right time, always taking into account possible delays and, mostly, the overall objective of the study. A development study for a new field, where few data are available, must be accurate from a global perspective. In a typical infill drilling project, on the other hand, more data are available and a more detailed study is required. Finally, a secondary or tertiary recovery study requires an even more accurate knowledge of the reservoir, in terms for example of fluid distributions.

Generally, the project manager of an integrated study belongs to one of the key disciplines (geology, geophysics, reservoir engineering), however he must be aware of the fundamentals of the other domains. Of course, he cannot be specialist in every branch, since the individual disciplines are becoming increasingly complex.

Excellent textbooks exist in the market that deal with individual reservoir disciplines. However, these manuals are conceived for specialists and they are not suited for the project manager, who must understand the added value of different and alternative techniques that he does not necessarily know in detail. A manual for the reservoir project manager should explain what could be done and at which price, without entering (as far as possible!) into much technical detail. Such book should be a guide for the project manager, by dealing with what can be done to meet the objective of the study, optimising at the same time the cost involved and the degree of technical detail. To my knowledge, such a manual did not exist.

Luca Cosentino has worked to fill this gap. A reservoir engineer with a geological background, he has managed several integrated studies on various type of reservoirs, both for oil and consulting companies. Making use of his experience, he wrote this book, which presents a simple and complete summary of the different methods that can be used in a reservoir study and their utilisation. At the same time, he provides us with an extensive bibliography to go into the technical subjects in more depth.

It is without hesitation that I recommend to every professional to keep this precious manual at hand, as a reference for those decisions that are outside their domain of competence. I only regret I had to wait till the year 2000 to encounter such a manual. The explanation lies perhaps in the large experience, the capacity to summarise and… the amount of work that is necessary to complete such a task. In any case, the results are worth the effort!

Jean-Claude Sabathier

Preface

Integration has probably been the most fashionable word in the Exploration and Production domain in the last decade. Papers, conferences and technical meetings have focussed on this concept and the advantages that it entails when successfully applied.

During these last years professionals belonging to the various reservoir disciplines, i.e., Geology, Geophysics, Petrophysics and Reservoir Engineering, have been taught to work together, to search for some synergy and to integrate their individual pieces of work. Vendors have created integrated databases, shared earth models and interoperable applications.

Managers, on the other hand, have created asset teams, organised common working environments and encouraged cross-disciplinary educational courses. They are searching for integration, believing that the extra value which could be gained is worth the effort. One cannot disagree, of course. Integration is one of those magic words that always has a positive meaning, however it is applied. *Integrated* is always better than *disintegrated*, and it is right to look for some sort of integration.

But when we come to the everyday working reality, what does integration *really* mean? What kind of change does it mean to our way of working? Is integration just a new solution to old problems or it is a completely different approach that also poses new and unexpected questions? And if this is the case, are we able to identify those problems and to propose adequate solutions? What new technical and professional challenges have to be faced? Who is responsible for the correct implementation of such integration in the course of the study?

The objective of this book is to try to give some answers to these questions and to highlight those aspects of a study that become relevant when we want to perform an *Integrated Reservoir Study*.

Acknowledgements

A book is never a one-man effort. In the case of *Integrated Reservoir Studies*, a lot of people have been involved in the project and have helped me in some way throughout the work, friends, colleagues, family.

In the first place I want to thank Jean-Claude Sabathier, for believing in this work and supporting me in the finalisation phase, the most crucial part. His views are always illuminating.

Giuseppe Spotti, who shared so many years of professional life with me in Agip first and Beicip-Franlab later, is the other friend I want to acknowledge especially. Many parts of this book arise from our long discussions on the most diverse aspects of reservoir studies.

I am also grateful to a number of friends who contributed to this work by reviewing all or parts of it and providing useful suggestions and corrections. Philippe Le Bars and Patrick Rouvroy, for their help in the Database issues (Chapter 2), my old companion Christian Ravenne, for his contribution to the geological parts, Pietro Consonni and Patrick Lemonnier for reviewing the simulation part (Chapter 7), Pierre Lemouzy for providing his ideas, always original, about the philosophy of upscaling and many other reservoir matters, Marco Thiele for his help in the streamline section.

I also want to acknowledge other collegues: Jean-Marc Chautru, Paul Van Lingen, Bernard Borbiaux, my venezuelian friends Roberto Munoz and Jose Edmundo Gonzalez, the staff of the IFP school (Gilles Gabolde, Bernard Durand, Jean-Pierre Roy, Jean-Hector de Galard), the staff of Editions Technip (Sylvie Haxaire and Philippe Catinat).

Sabrina Albornoz provided me with invaluable help for the bibliography, digging out the most obscure papers, while German Parra and mostly Gerard Dyot helped me in the long and tedious work of assembling the graphical part.

I also owe a special acknowledgement to Ros Stallard, who patiently revised the English version.

Finally, I am grateful to Beicip-Franlab management and in particular to Jean Burrus, Honoré Le Leuch and Paul Bia for supporting me in bringing this book to fruition.

Last but not least, thanks to my wife Janet for understanding.

Contents

Chapter 3
INTEGRATED GEOLOGICAL MODEL

Chapter 4
ROCK PROPERTIES

Chapter 5

HYDROCARBON IN PLACE DETERMINATION

Chapter 6

BASIC RESERVOIR ENGINEERING

Chapter 7

NUMERICAL RESERVOIR SIMULATION

Chapter 8
PLANNING A STUDY

CHAPTER 1

Integration Issues

1.1 WHAT IS INTEGRATION?

Following Webster's definition, integration is a combination and coordination of separate and diverse elements or units into a more complete or harmonious whole [1]. Since it implies the creation of a more complete or harmonious whole, integration can therefore be considered as a process whereby extra value is produced. How this extra value can be generated is precisely the scope of this book.

In the petroleum industry and in particular in the Exploration and Production (E&P) domain, integration is primarily concerned with the manner in which different disciplines are combined to improve an established (or create a new) analytical process.

Actually, however, integration is a difficult concept to define. E&P processes are highly complex activities, which involve a number of disciplines, each one carrying its own integration problems. Therefore, integration of disciplines actually means integration of all the aspects which constitute those disciplines. There is the aspect of integrating different pieces of work, produced by different professionals, but also the aspect of integrating different geoscientists belonging to different professional cultures. Moreover, there is the need to integrate different human beings, coming from different places and talking different languages. Finally, there is the problem of the integration of the software and hardware platforms, a necessary condition of any integrated study.

To further complicate the issue, the degree of integration within a team or a work process, by definition, is always changing. Consequently, while the study progresses, the integration problem changes and needs to be continuously redefined.

Despite the inherent complexity of defining what integration is, it is possible to try and categorise some basic issues, which relate in particular to an Integrated Reservoir Study:

- ***Vertical vs. horizontal integration***. A basic distinction that could be made refers to *vertical* integration within an E&P discipline and *horizontal* integration across the disciplines. An example of vertical integration is the workflow along the seismic interpretation of a particular reservoir, when different geophysicists work together on the same platform on the different aspects of the geophysical interpretation, e.g. the structural interpretation on the original volume and the attribute analysis on a processed volume of data. The need for vertical integration within each discipline is obvious and has already been implemented in many computing systems. It is also normally easy to achieve, since the professionals belong to the same branch, geophysics in this example, and can easily communicate. In contrast, *horizontal* integration means integration across the different disciplines of an integrated study and it is a far more complex problem. The main hurdles are the relatively reduced interoperability of the software platforms normally available and mostly the tendency of the different professionals to concentrate on their particular study and to have a reduced level of communication with other professionals of the team.

- ***Loose vs. tight integration***. Another useful distinction that could be made refers to *loose* and *tight* integration. As will be discussed in the next chapter, this mainly refers to the degree of interoperability of the different software applications, but loose and tight integration can easily be defined in a general sense also for the work process. When a geologist discusses with a reservoir engineer about the position of the faults that he has identified in his study, we could talk about loose integration. When the same geologist and engineer work together to define the position of the faults, using their respective tools, we can probably talk about tight integration.

In the next sections, it will be shown how the integration problem has been defined and tackled in the completely different field of business administration and how those concepts can be applied to E&P projects. Later, some of the relevant integration issues within an reservoir study will be discussed, with particular emphasis on those aspects that mark the difference between a **integrated** and a **traditional** reservoir study.

1.2 SYSTEMS THINKING

Some years ago, during the crisis of the American cars market, some engineers of a Detroit Company decided to deconstruct a Japanese car, to try to understand some of the secrets that made those cars so cheap and efficient. Looking into the engine they found, among other things, that a particular type of part was used three times to join different components in three different parts of the cylinder block. In the American cars, the same assembling required 3 different types of parts, which needed 3 different keys and 3 inventories of parts, so that the assembling procedure was much slower and more expensive.

Analysing the design organisation of the 2 companies, it was realised that in the Detroit Company there were different departments in charge of the design and production of the 3 parts. The engineers responsible for their product considered that their own components were good and that they fitted perfectly in the assembling procedure. By contrast, in the Japanese Company, one single engineer was supervising the whole assembling procedure of the engine.

This simple example is taken from a *best seller* book on Organisations Management [2] and it shows how, in a complex activity like engine building, joining components of good quality is not enough, when the global efficiency of the project is being considered. The Japanese engineer, while supervising the whole design procedure, was able to focus on the links between the different parts of the assembling process and to optimise the process itself. There is no doubt that the single parts produced by the 3 departments were **good** products in themselves; however, when the general objective of the project is considered, their intrinsic quality becomes less important, and the simpler and more flexible Japanese part proves to be **better**.

The idea of concentrating on the patterns of a particular process, rather then on its constituting parts, is the key concept of Systems Thinking.

Systems Thinking is a discipline whose principles and tools have been developed in the last 50 years in the different fields of Physics, Engineering, Biology and Mathematics and that in recent years has been largely applied in Organisations Management.

Simplifying the issue, Systems Thinking is a way to analyse a process (physical, natural, economic, political and in general any kind of process), which is based on the study of all the factors, internal or external, that have an influence on the development and outcomes of such a process. Leaving the interested reader to the huge existing bibliography on this theme, we can try to summarise here some of the basic principles of Systems Thinking:

- Understand the process of change, instead of focussing on the individual constituent parts of the process itself.
- Understand the interrelationships among all the constituent parts, rather than the linear cause-effect concatenations.
- Concentrate on the *dynamic* complexity of the process, rather than on the *static* complexity of its details.

Any type of process can be considered this way, from the study of the orbit of an electron around its nucleus to the analysis of the global heating of the earth. In Systems Thinking, there is no one simple cause that generates one effect. Rather, there is an interrelation of different elements (the system) that determined the effect we are considering.

As far as reservoir studies are concerned, Systems Thinking provides a useful theoretical framework to describe the different approach required by integration.

An integrated reservoir study is by definition a complex process, which results from the integration of different disciplines and which has a definite objective. Like the engineers of the Detroit Factory, we are faced with the problem of integrating elements that have been produced by different professionals (departments). We also have the same objective, which is producing a study (an engine) that is reliable and accurate. And in the same way, we would not want to fall into the same trap, which is producing good pieces of the study (parts) that are not linked to the overall objective of the project in an effective way.

When performing a study, we need to understand the patterns of interconnections of the different activities and, like the Japanese engineer, we need to plan the work flow by thinking about the optimisation of the process. Systems Thinking may help us to see through the static (intrinsic) complexity of the geological or engineering work and to identify those parameters that have an impact over the global objectives of the reservoir study.

In the course of this chapter, some of those aspects of a Reservoir Study which are critical to the integration process will be discussed. Most of these aspects require a change of focus with respect to the traditional way of working. This change can be often be related to Systems Thinking principles.

1.3 A CHANGE OF FOCUS

Until a few years ago, the workflow of a reservoir study was a far different process than it is today. The approach was a sequential-type one, where the geophysicist, the reservoir geologist, the petrophysicist and the reservoir engineer worked almost independently, while the results of each one were passed to the other without significant feedback.

One of the main consequences of this kind of approach is that each discipline defines its own objectives, which are in general different from the others and possibly only loosely related to the general objectives of the reservoir study. Within each single discipline, professionals tend to think that the more detail is used in the analysis, the better the quality of the results. Therefore, the various professionals deliver a product that is the best they can achieve with the available technology, compatibly with their professional experience and the time available. In this approach, the *best analysis* implicitly brings the *best results*.

The process of integrating different E&P disciplines to perform an integrated reservoir study requires a change of focus. According to Systems Thinking, the approach becomes multidisciplinary and the professionals work in an interrelated way, where the feedback from other disciplines is fundamental to the validation of the work that is being performed (Fig. 1.1).

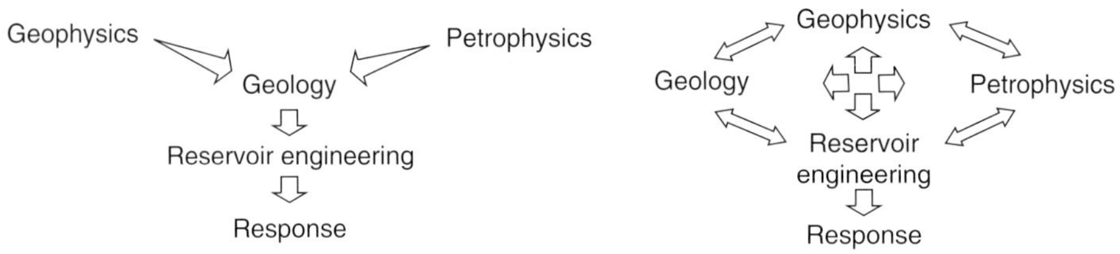

Figure 1.1 Professional links in a reservoir study: traditional (left) and integrated (right) approach.

In this case, the key point becomes the understanding of the global objectives of a particular study, while all the disciplines involved in the project must define sub-objectives which are strongly dependent upon the global ones. This also implies that, in the framework of an integrated reservoir study, the technology available to the different disciplines needs to be compared to the common goal, in order to select the appropriate tools. In fact, as it will be discussed later, a simpler and more traditional approach can sometimes be preferred to leading edge technology.

This change of focus is a difficult task to achieve, because it poses to the professionals and to the project manager unusual questions and problems, which need to be tackled and solved from a new point of view. In the next sections, some of the key points that become important and must be considered when performing an integrated study will be analysed.

1.4 INTEGRATING THE INFORMATION

An integrated study is a challenging task to perform. The reservoir is actually a very complex object in itself, which must be characterised from a variety of viewpoints, with a large number of parameters and with a remarkable degree of accuracy. Furthermore, in addition to the inherent, *natural* complexity of reservoirs, we may also have the man-induced contributions to reservoir unknowns, i.e. fractures, formation damage, cementation problems, etc.

As a consequence of these factors, studies always carry a degree of uncertainty, which in turn can be considered as an (unknown) measure of our imperfect knowledge of the reservoir.

As a matter of fact, geoscientists, unlike other scientists, have a very limited access to the object of their investigations, i.e. the reservoir. The information that is normally available to them is peculiar for at least 3 reasons:

- ***It is mostly indirect***. The only direct access to the reservoir is through coring. Using samples of core we can perform some direct measurements of the reservoir properties, e.g. rock porosity. In all other cases we can derive information on those properties **indirectly**, via some other type of measurement. Then, we correlate those measurements to the reservoir parameters of interest through some type of transfer function. For example, in a geophysical survey we measure travel times and we convert them into depths through a time/depth relationship. Likewise, during a well test we measure pressures and then we convert them into a number of parameters of interest like permeability or skin factors. The transfer function is in this case some type of solution of the diffusivity equation.
- ***It is based on a small support volume***. With the notable exception of seismic and, to a lesser extent, well testing, all the information that we interpret is computed on a small or very small support volume, that we implicitly assume to be representative of the whole reservoir. For example, we measure rock wettability on a few 1 or 1-1/2 inch plugs and then we extend the results to the whole reservoir or to a part of it. Likewise, we observe diagenetic phenomena through the microscope on a thin section of the reservoir rock and we assume that the same phenomena will have acted throughout the producing unit. Fig. 1.2 compares the usual sources of measurement of reservoir data in a dimensional scale, normalised to a reservoir volume unit. It can be appreciated as most of the data refer to what can be defined as macroscale, while typical modelling work will be performed at the megascale. The figure highlights the problem of scale implicit in reservoir studies and the consequent related upscaling issues. Ref. [3] provides a comprehensive discussion about these points.
- ***It is varied***. Information is gathered in a number of different ways, in the core, in the borehole or from the surface. The amount of methodologies used to infer reservoir properties is surprisingly high, from analog geological outcrops studies to axial tomography on a small plug in the laboratory. To further complicate the issue, the same reservoir property can be computed by means of different methodologies, which provide independent estimates at different scale.

From this point of view, it is clear that one of the most relevant problems of a reservoir study is to properly integrate all this information into a consistent model. The challenge is

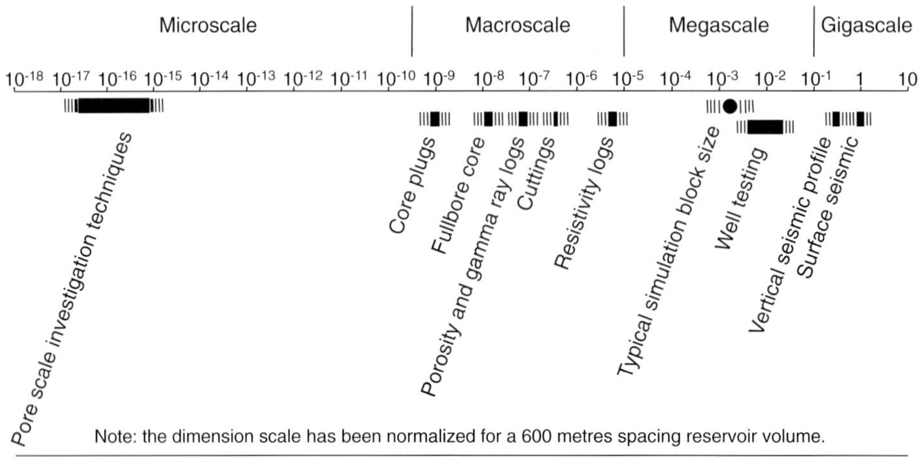

Figure 1.2 Comparison of sampling techniques as a function of reservoir scale.

noteworthy, since integrating the information means to coordinate and combine different kinds of data, coming from different sources, acquired by means of tools which access different portions of the reservoir and with a different resolution.

A typical example is porosity. Normally, we can get information on reservoir porosity by means of 3 types of data: cores, logs and seismic. Porosity coming from cores is a *direct* measurements, which is computed on a very small portion of the reservoir, the plug, which represents normally a fraction of about 10^{-9} of the total reservoir volume (Fig. 1.2).

On the other hand, porosity from logs is an *indirect* measurement. What we actually measure is attenuation of gamma rays, hydrogen content or formation travel times and then we convert these quantities into porosity using different equations (transfer functions). In this case the volume of rock upon which the computation is made is approximately 3 000 times greater with respect to the core plug, but still it is absolutely negligible with respect to the reservoir volume.

Porosity can also be derived from seismic data, again as *indirect* measurement. We measure travel times, we process it to derive amplitude and/or impedance and then we look for a correlation with porosity. In this case the whole volume of the reservoir is sampled, but the resolution is normally very poor.

Eventually, when we need to compute porosity for the final reservoir model, we are faced with the problem of integrating all this different data, which are relevant to the same property but carry different pieces of information as far as accuracy, resolution and areal distribution are concerned.

Another typical and possibly more interesting example is permeability. Like porosity, permeability can be measured directly or indirectly, using in this latter case some type of transfer function. Additionally, its value depends on the saturation conditions of the reservoir rock, so that permeability values can be compared only when those saturation conditions are identical.

From a general viewpoint, we can compute permeability on cores, in a borehole or by means of well testing.

In the laboratory, we can measure **gas absolute permeability** in a core by means of a minipermeameter or flowing core plugs with gas. In this case, the transfer function is Darcy's law and the fraction of reservoir that we sample, as in the case of porosity, is infinitesimal [1].

We can also compute absolute permeability in the borehole from NMR logs, sampling in this case a much bigger volume of rock, at reservoir (pressure and temperature) conditions and using a different transfer function. Still in the borehole environment, through WFT measurements, we can derive **effective permeability to oil** at a scale possibly similar to that of core plugs.

Finally, in a well test, by means of another different transfer function (the diffusivity equation), we can again derive **effective permeability to oil** at reservoir saturation, sampling a portion of reservoir which is normally millions times bigger than the **gas absolute permeability** that we have measured on core plugs.

This short digression highlights the complexity of the problem. However, this is not limited to porosity and permeability. If we listed every variable of a reservoir study, we would realise that almost all of them can be sampled at different scale levels and often in an indirect way.

Sure enough, methodologies have been developed to deal with these kind of problems. Petrophysicists for example, know very well how to reconcile Density and Sonic porosity with core data. However, when the same property has been computed in different professional domains, then the correct integration of data is no longer straightforward.

In its broadest sense, integrating the information means to highlight the differences, to understand the relative contribution of each piece of information and to investigate all the possible ways of reconciling the data. This will allow the choice of the best way of representing the reservoir, in relation to the objectives of the study.

Integrating the information is probably the most challenging task in an integrated reservoir study, because there is no one single way to tackle this problem. As we will see, it is the Project Manager's responsibility to guarantee its implementation.

1.5 ACCURACY VS. PRECISION

The terms accuracy and precision, applied to reservoir measurements, are often used interchangeably in the geoscience practice. In fact, the two words are related to different concepts: accuracy refers **to the extent to which a measurement approaches the true value** of the measured quantity, while precision refers to the **degree of refinement** of a measure. In other words, precision is not related to the true value of the quantity being measured, but rather to the **repeatability** of the measurement process. Fig. 1.3 graphically illustrates the concept.

---

1. It could be argued that, being based on the application of a transfer function, even laboratory measurement on core samples are indirect estimations. In this context however, they will be considered as direct measurements.

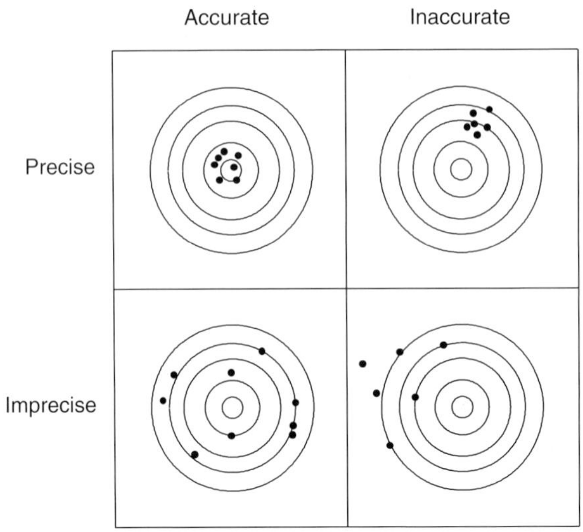

Figure 1.3 Accuracy vs. precision.

Typical measurements available in a reservoir study span all possible combinations of accuracy and precision and it is important to understand, case by case, what kinds of data are being handled and which are their characteristics.

In many instances, reservoir parameters of interest can be estimated by means of different techniques, which provide complementary information about the parameters themselves. Saturation pressure, for example, can be measured with a good degree of precision in the laboratory on fluid samples, but these measurements are often inaccurate, because of the poor representativeness of the fluid samples. By contrast, saturation pressure can in some cases be accurately (but not precisely) estimated by means of field production data.

Understanding the characteristics of the information that is being handled is a critical factor in any estimation problem. In general, the use of a single type of data will not provide the best results. On the other hand, the possibility of integrating different sources of independent data, each one related to a certain degree of accuracy and precision, may help in providing unbiased estimates of the reservoir parameters.

1.6 COMPLEXITY VS. ACCURACY

A model is by definition a simplification. The degree of such a simplification, or conversely, the degree of complexity of a given study, depends on the information available and the human and technological resources allocated to the project.

An important problem that makes integration difficult is the increasing *complexification* which the individual E&P disciplines are currently undergoing. Complexification is the process of adding incremental levels of detail to a study to represent its complexity more rigorously [4]. To a greater or lesser degree, complexification is a process that affects every E&P discipline. A typical cycle of complexification is shown in Fig. 1.4.

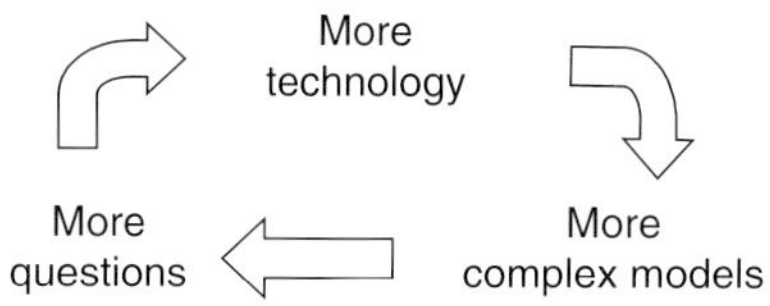

Figure 1.4 Cycle of complexification [4].

New technologies offer today to the geoscientist powerful tools to investigate the details of a particular problem. However such detail can be more difficult (and sometimes less relevant) to integrate in the study work flow. In other words, new technologies pose new problems of integration, unknown until a few years ago.

The analysis of a cycle of complexification raises 2 important points related to the integration process:

- Increasing the level of complexity of a particular work does not necessarily ensure improved accuracy in overall results.
- Improved accuracy does not automatically guarantee compliance with the objectives of the study.

When performing an integrated study, we need to make sure that we are not allocating human and technological resources in searching for some false detail or for an accuracy which does not add anything but complexity to the study itself. The degree of accuracy must always be measured against the overall objective of the study. This is why, where an integrated study is concerned, the concept of *good* or *best* work changes.

An interesting example of complexification is presented in Fig. 1.5, which shows the interpreted fault pattern for a North African oil field. The original 3D seismic interpretation (left figure) has been revised recently through the analysis of seismic attributes, like dip azimuth, dip magnitude, coherency cube and amplitude, providing a much more detailed image of the fault pattern (right figure). The revised interpretation could delineate faults with throws less then 25 ft or tight flexure zones likely to be heavily fractured. A total of more than 4 000 faults have been picked in this interpretation.

The level of detail attained by this study is certainly remarkable and the right image is possibly a better representation of the actual fault pattern of the reservoir, compared with the original. However, this level of detail could hardly be maintained in a simulation study, with the technology available today. It is impossible to reproduce all these small scale faults in a normal simulation model, where the practical cell size is bigger than most of such features. The most likely reaction of the reservoir engineer would be a simplification of the proposed fault pattern, whereby only the most important faults would be retained. The effort of detailing a fault pattern of the reservoir would be followed by the effort of simplifying the same pattern, the net result being probably a loss of time.

Another obvious example of complexification is illustrated in Fig. 1.6, taken from a recently published paper [4]. The figure shows a set of 263 imbibition relative permeability

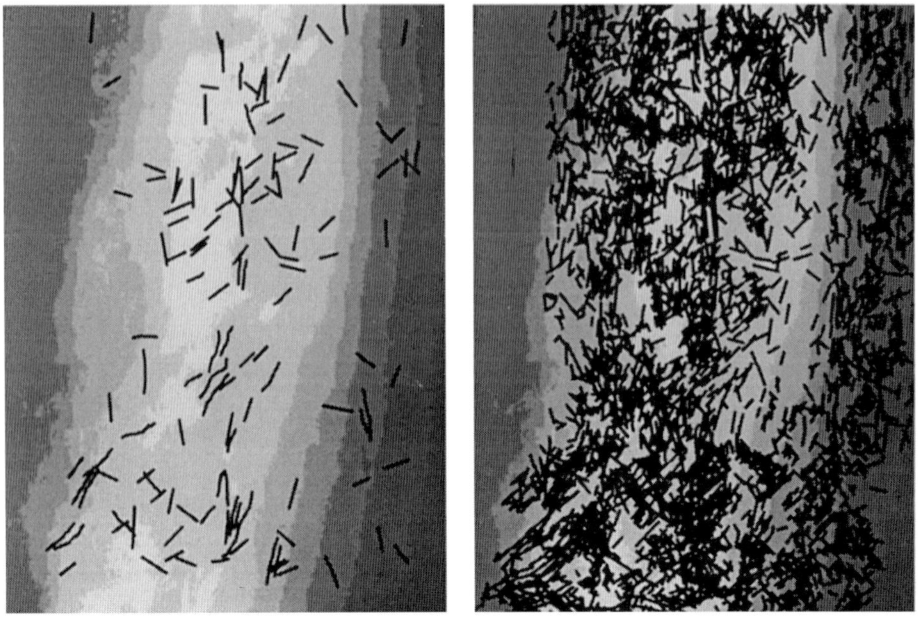

Figure 1.5 Fault pattern for a North African field, original (left) and revised interpretation.

curves used to describe a given reservoir. When considering the particular saturation functions study that led to the generation of those curves, it could be said that the degree of detail that has been attained certainly captures the rock property variability within the reservoir and, from this viewpoint, the study is certainly accurate. However, from another point of view, some questions should be considered: what does that degree of accuracy add to the study? And what is the impact in the following phases of the project? If the results of such a study have to be used in a reservoir simulation study, this complexity becomes overwhelming for the engineer to manage. In addition to that, such complexity may also prove to be totally unpractical (ironically perhaps) in exploring the impact of relative permeability variations in model results. Is looking for such detail really justified?

The same paper also quotes that the simulation model results obtained with a single average relative permeability curve were comparable with those obtained with the complete set of 263 curves. This raises another key point: what is the degree of complexity that we can cope with in a reservoir study? The problem of course is subtle, and while in this particular relative permeability example the answer is clear, in the majority of cases it would not be so obvious.

Another aspect of the problem of complexification is related to the misallocation of technology between the various phases of the study. As a matter of fact, leading edge technologies are frequently applied in domains which have a second-order effect on reservoir performance, while most relevant reservoir characteristics are given less or little attention.

While digging into the existing reservoir studies and reports, it is not uncommon, for example, to find exhaustive sedimentological studies and facies stochastic modelling

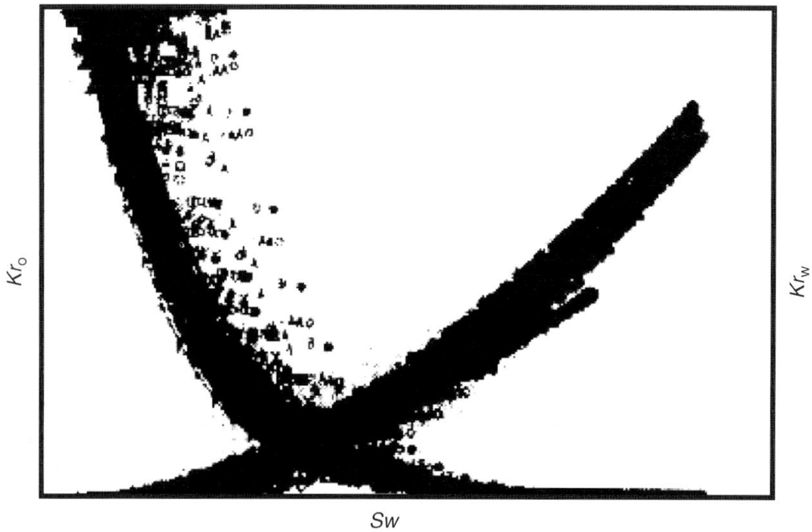

Figure 1.6 Relative permeability curves for reservoir characterization [4].

applied to fields where reservoir compartmentalisation is the main driving parameter in reservoir performance, as could be the case for many pre-cretaceous reservoirs in the North Sea. In such cases, when the objective of the Reservoir Study is to define the (re)development plan of the field, it is obvious that the structural model of the reservoir is far more important than the lithological and/or petrophysical descriptions, since the number of wells to be drilled (and consequently the capital cost of the project) basically depends upon the connectivity of the reservoir. Simulation models run using different stochastic realisations of the sedimentological model will not capture the critical features linked to the structural model of the reservoir and will provide a biased image of the reservoir performance. This, in turn, could open the way for unwise management decisions.

This simple example shows that, when planning a Reservoir Study, we must be able to identify the critical factors with respect to the general objective of the study, and concentrate our efforts accordingly. In the above example, it would have been much better to dedicate more time and more technology to the definition of the structural setting of the reservoir, while modelling lithology and petrophysics through faster and more conventional methodologies.

These concepts can be generalised and the following points should be considered, when starting an integrated study:

- Identify the critical features of the field, with respect to the overall objectives of the study. This in turn allows for the identification of the main requirements of the project, in terms of human and technical resources.
- Rank the critical parameters. This process aims at identifying those parts of the study that do not have a considerable impact in the final results, i.e. the parts that **do not make a difference**.

- Define the degree of complexity of the critical phases of the study, compatibly with the project constraints (availability in terms of budget, time, professionals and technology).

Managing the degree of complexity of an integrated study and its constituent parts is a very important issue and should be faced early in the project life. It represents one of the main aspects of the planning of the study, since it defines how and where the bulk of the resources will be allocated. These issues will be discussed in more detail in Chapter 8.

1.7 OTHER INTEGRATION ISSUES

Properly integrating the available information is certainly the most difficult stage of an integrated study. However a number of other integration issues can frequently be faced in a reservoir study project.

One aspect that is often encountered is the difficulty of integrating professionals belonging to different disciplines. Different backgrounds, different cultures and possibly different languages are the main factors that make communication difficult within a workteam. In most cases, the global objectives of the integrated study are hardly understood by the various geoscientists, who tend to apply methodologies that proved to work in the past and that are more familiar to them. Additionally, the lack of knowledge of other professionals' work is a factor that can induce diffidence in the relationships among the team members. Behind this attitude, which unfortunately is very common among E&P professionals, there is probably the fact that working independently is somehow easier than working with other people, who often discuss our methodology or question our conclusions. It is the Project Manager's task to smooth over any existing tensions among the team members and to find a compromise among different technical positions. The key factor in this case is always the comprehension of the workflow and the objectives of the project by the various team members.

Another aspect which is frequently encountered within oil or consulting companies is the physical distance among the various constituents of the team. This distance can be measured in meters or thousand of kilometres, but often the impact over the study is the same: lack of communication and hence of integration. There is no doubt that today's communication technologies are powerful and allow the different team members to share information of any type in a fast and effective way. World wide computer networks, video conferences and virtual teams are only a few examples of the available means to minimise the effect of the physical distance. A recently published paper [5] illustrates the example of a study successfully performed through the co-operation of State and Federal Agencies, an industry partner, a national laboratory, one university and 2 independent consulting companies. A **virtual enterprise** was established among these participants, where the professionals interacted in a common workspace everyday, using the worldwide web.

However, despite examples of this type, the physical distance among team members does not facilitate in itself the integration. More often than not, this can be taken as a good justification for the most reticent professionals to perform the work in an independent way. When no alternatives exist for the physical distance among the team members, the Project Manager has to watch over this problem, which always constitutes a danger to the successful development of the study.

Last but not least, problems of integration can derive also from the differences that exist in the databases used by the various groups of professionals and in the limited interoperability of the applications software. More often than not, Geophysicists, Geologists, Petrophysicists and Reservoir Engineers have their own interpretation packages and do not share a common database. This issue will be the object of Chapter 2.

1.8 THE ROLE OF THE PROJECT MANAGER

An integrated reservoir study must have a Project Manager. Simply stated, the responsibility of the Project Manager is to successfully achieve the objectives of the study, within the allocated budget and timeframe. More specifically, a number of points can be identified, where the Project Manager is directly responsible: working environment, team effectiveness, budgeting and reporting, project commissioning and decommissioning, links with higher level management and so on. However, as far as integration is concerned, there are some specific points that can be identified, where the responsibility of the Project Manager is direct and where the role proves to be very important:

- Define the general objectives of the project, identifying those phases of the work that are critical to the final results.
- Allocate the human and technological resources to the project, according to the prioritisation of work phases mentioned above.
- Guarantee the correct integration of the different sources of information.
- Make sure that the correct level of technology is applied within each discipline, avoiding the use of expensive and time-consuming methodologies in non-critical tasks.
- Ensure that factors like lack of communication among team members, physical distance and low interoperability of the different software application, have a minimal impact over the development of the project.

The importance of the role of the Project Manager cannot be overstated. Projects often fail because of bad management, and in many cases the problem can be identified to be a lack of understanding of all the different aspects of an integrated study. Typical examples refer to studies where the Project Leader has a strong background in one of the disciplines and knows little about the others. Sometimes for example, when the project manager is a Reservoir Engineer, the integrity of a sound geological model can be sacrificed to meet history matching in the reservoir simulation phase. This attitude not only can be dangerous from a technical viewpoint, but can also generate frustration in the geologists who worked hard to produce the best image of the subsurface. Conversely, when the Project Leader is a geoscientist, the bulk of the work can be allocated to the *static* part of the study and very often too little attention is paid to the production performance of the field and to the potentially huge amount of information that can be derived from dynamic data.

For these reasons, the Project Manager must have a good understanding of all the phases of the project and must be able to balance the activities accordingly to the type of study and its objectives.

He has to distinguish between sound application of high level technology and misuse of technology in every discipline of the project, even though he does not need to be (and indeed he cannot be) aware of all the details.

Most importantly, he must look for integration within his team. He has to understand all the benefits that could be gained when integration is achieved, but also he has to be aware of all the problems that may be encountered during the study, trying to integrate people and tools.

In any case, if we accept that integration is the key to the realisation of high quality reservoir studies, then the Project Manager is directly responsible for its application.

There is no single solution to integration management. However, talking Systems Thinking, a **holistic** attitude by the Project Leader is the best guarantee for the integration to be a success.

References

1 Webster's Third New International Dictionary (1981). Merriam Webster Inc.

2 Senge PM (1990) The Fifth Discipline. Doubleday.

3 Haldorsen HH (1986) Simulator parameter assignment and the problem of scale in reservoir engineering. *In*: Reservoir Characterization. Academic Press.

4 Saleri NG (1998) Re-Engineering Simulation: Managing complexity and complexification in reservoir projects. SPE-REE, Febr.

5 Withney EM, Brickey MR, Coombs SE, Duda CA, Duda VK, Integrated reservoir management for the long term - The Carpinteria offshore field. SPE paper 38284.

CHAPTER 2

The Integrated Database

2.1 DEFINITIONS

This chapter deals with the construction and management of an Integrated Database for reservoir study purposes. Before beginning the discussion, it is useful to define a few basic concepts that will be used throughout this text.

- An **Integrated database** is a data repository system to interactively store, retrieve and share E&P data, within a controlled and secure environment.
- A **Data Warehouse** or **Data Storage** can be defined as an integrated, non-volatile, time variant collection of data to support management needs. From this viewpoint, it implies a reduced degree of interaction with the end user (the geoscientist, in this case).
- **Data Management** is the process of storing, organising, retrieving and delivering information from a database or a Data Warehouse.

It is also worth mentioning that the terms **data** and **information** will be largely used interchangeably in this text, despite the fact that there is an inherent relationship between the two (data *generate* information).

2.2 THE PROBLEM OF THE INTEGRATED DATABASE

The integrated database is one of the key issues in an integrated reservoir study. The availability of high quality data, both static and dynamic, and the rapidity of access to this data, is always a necessary condition in the workflow of a study.

Unfortunately, in the majority of cases, the condition of the Database at the moment of starting the study is far from optimal. The experienced geologist or engineer knows the problem very well: data are frequently spread around in different places and are accessed by different software. Sometimes data is duplicated and we do not know which is the **right** version to use. Also, it is common to find out that some or most of the information has not been transferred in digital format (or it has been incorrectly transferred), which means a tedious and lengthy work of digitalisation has to be done prior to the beginning of the study. In many cases, even the information stored in the existing database proves to be of poor quality, difficult to access or partially or totally corrupted. In the worst cases, data have been lost.

From a general viewpoint, the relevance of this problem varies from study to study, because it depends on the quantity of information being handled, its age and the effectiveness of the data-warehousing policy of the company. Additionally, the problem is often difficult to detect and evaluate *a priori*, i.e., before actually starting the study. However, the little published information is impressive: according to the available reports, reservoir geologists and engineers spend 50 to 80% of their time searching for data and only 15 to 30% for their interpretation, i.e. supposedly the real work. Fig. 2.1 shows these results in a graphical form [1]. If we translate this into money, we will realise that we are spending the biggest part of our budget on something that we probably had not considered at the beginning: data management.

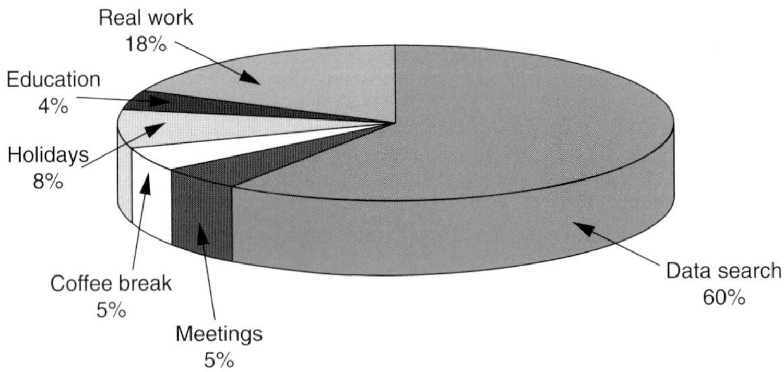

Figure 2.1 Working time of a E&P engineer [1].

This explains why the database is often **a problem** to a project. An integrated study always begins with the collection of all the available information from the existing discipline-oriented databases and from other sources, but this phase can prove to be a difficult and very long process, whose impact on the overall planning of the study can be very large. More often than not, delays in the project execution are related to unforeseen amounts of work spent in the construction of a complete and reliable database.

In the next paragraphs, we will see how data is generally organised in an oil company and will focus on some of the characteristics of the main object we are concerned: the Project Database.

2.3 THE THREE LEVELS OF E&P DATABASES

Historically, data management has not been considered a critical issue within oil companies. Indeed, it was a far simpler task than today. The data was generally stored in a central archive as a single (the original) piece of information, i.e., a well log. Geologists simply borrowed that original and used it until their work was finished; the result of their interpretation, for example a structure map, was then filed and referenced for future use.

The situation is much more complex today. On one hand, the huge amount of reservoir data that have been acquired in the last years, especially 3D seismic surveys, require large storage capabilities and efficient retrieval software. On the other hand, the relatively new concepts of multidisciplinary teams and asset management require a new approach to data management: while before each group of professionals had their particular storage information software, now information needs to be shared with other groups of professionals, working on the same asset.

Most companies, however, experienced the difficulty of re-organising the data management system and in most cases hybrid solutions were envisaged to store the data. These solutions, often obscure to the user, generated a proliferation of the data. It was most professionals' perception that data were sparse in a number of different locations, spanning from individual PC's to remote and somehow inaccessible central archive systems. The users did not know the quality of these data, often could not distinguish the original from the copies and consequently tended to generate their own database, the only one in which they could trust. This phenomenon started in the late eighties, with the proliferation of Personal Computers and spreadsheets, but it is still fairly common today.

At the beginning of this decade, oil companies finally realised that data management was a critical issue and started to organise new and more efficient data warehouses, while vendors made available new integrated software for data interpretation and storage.

Today, a tendency is emerging among the oil companies to organise their data in three different levels of databases (Fig. 2.2, [2]):

- **Corporate database**.It stores the official data of the company. Data quality is high and the rate of change (**volatility**) is low. No new data is created within the Corporate database, and it does not feed any application, except its own set of utilities for browsing, selecting and exporting. Data are delivered in a format compatible with the Project database. Although the database can be accessed by anyone, changes in content are controlled by an administrator. It usually resides in a mainframe and is characterised by the many controls that are placed around it.
- **Project database**. It contains data relevant to a particular project or asset. It is made up of information withdrawn from the Corporate database and is accessed using software from different vendors. Its size is highly variable, from few to thousands of wells, and it may contain multiple versions of the same data. All the professionals working on the team can access and modify the database, so that the volatility is high. New data is generated through the interpretation stages. When the project has been completed, the interpreted data is returned to the Corporate database and becomes the new reference information.
- **Application database**. It contains data relevant to a single application. It is normally accessed only by the geoscientist working on that particular application and the information is therefore highly volatile. Also, the information may not be easily shared with other application databases, when vendors are different, unless a dedicated interface software is available. When the interpretation is completed, the data is stored in the Project database.

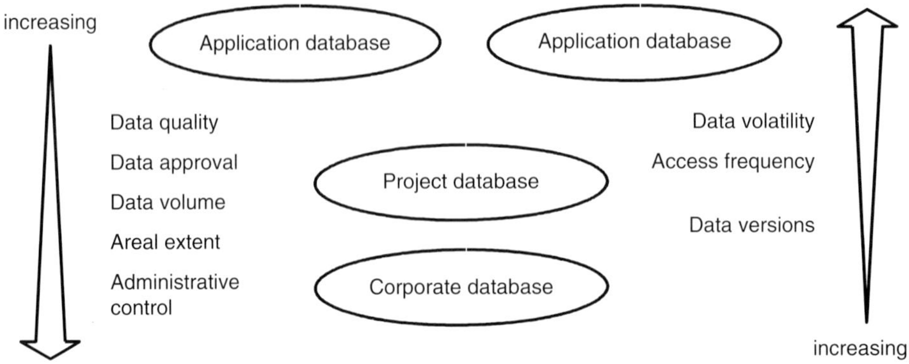

Figure 2.2 The 3 levels architecture of E&P databases.

As data management concepts in Exploration and Production are rapidly changing, simplified alternatives to the 3-level architecture described above could be envisaged in the future. The Application database, for example, might soon disappear, since vendors are more and more evolving towards industry standards that should improve software interoperability (see paragraph 2.5).

Another level of data storage that might have a bright future is Data Banking. In recent years several multi-company initiatives have been started, with the objective of building a shared repository system, which allows for bulk data to be managed in archival form. Warehousing is managed by a third party, the information is treated like money or stocks, and its access can be bought or sold. The main advantage of such a system is that it allows fast retrieval on line of any kind of high quality information.

In the future, outsourcing the data management activities should help oil companies to cut costs and to concentrate on core activities, while the banking system could represent one of the most interesting and viable options.

2.4 THE PROJECT DATABASE

The Project database is the actual interface for the geoscientist. It is here that much of the action takes place and where the final interpretation resides. It is generated at the beginning of the project by downloading the relevant data from the Corporate database and it normally resides in a multi-user UNIX server.

The generation of a Project database has a lot of advantages compared to direct access to the Corporate database: it permits faster access speed, it works with formats compatible with common E&P applications and it allows for the creation of multiple versions of data.

The key characteristics of the Project database is that it is (or it should be) integrated. That is, it contains all the information pertinent to the study and it is accessed by all the applications relevant to different E&P disciplines. Modifications generated by one application are recorded in the database and made available to the other applications.

From a structural viewpoint, a typical Project database should contain 4 main types of data. These are Definitive Data, Reference Information, Project Data and Personal Data. Each type can be characterised in terms of attributes such as confidence, access audience and life span (see Table 2.1, [3]).

Table 2.1 Types of data in a Project database.

	Confidence	Audience	Life Span	Example
Definitive data	High	Large	Forever	Well logs
Reference information	Moderate	Moderate	Moderate	Official maps
Project data	Low	Low	Short	Working map
Personal files	–	Individual	Indeterminate	Preference files

In actual practice, an integrated database should typically contain the following raw information:

- Lease and Cultural data
- Well data (name, location, RT, deviation data …)
- Open Hole and Cased Hole well logs
- Seismic data (2D and 3D surveys, navigation data, borehole seismic …)
- Completion data (perforations, workovers …)
- Production and injection data

The list is probably not exhaustive, since the different types of data that can be collected in a reservoir study project is surprisingly high, but it shows the information that is normally available in all cases and that should always be present in the Project database. Of course, in addition to the raw information, the Project database will contain the results of the interpretative stages (quantitative logs interpretation, maps and surfaces …).

Figure 2.3 displays the simplified structure of a Project database and its relationship with the Corporate database.

It is important to stress that the Project database should contain not only geophysical and geological information, as is very often the case, but also production and completion data. The importance of this concept cannot be overemphasised. In fact, while performing an integrated study the geological model is cyclically updated and this in turn generates changes in the dynamic model. For example, if a well top is changed, this may also have an impact in the production database, since the new interpretation may cause the perforated interval to be partially or totally in a different geological unit. This in turn requires a re-allocation of the production volumes according to the modified geological framework (see paragraph 6.5.1). In an integrated environment, these modifications are transparent to the user, however if dynamic data are not tightly connected to the Project database, the engineer must run some interface software to update the production history on a unit by unit basis.

Unfortunately, as discussed in paragraph 2.6, such tight integration of static and dynamic data does not exist yet. While vendors are starting to propose fully integrated solutions, in

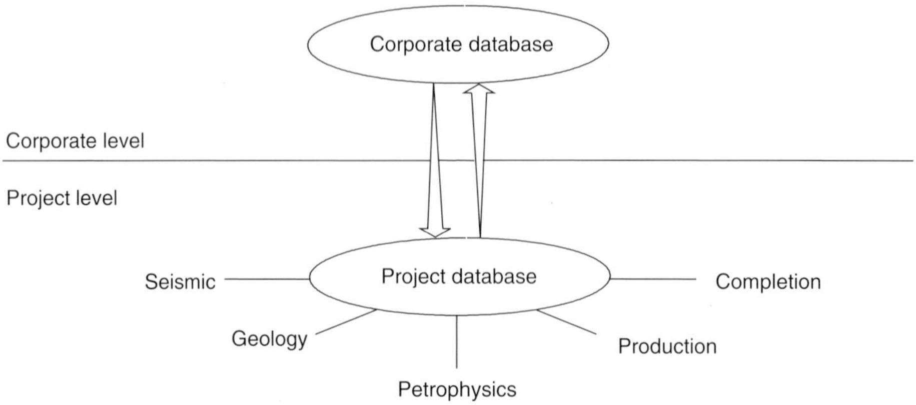

Figure 2.3 Ideal structure of a Project database.

the real E&P world such solutions will not probably be effective for a few years. In the meanwhile, geoscientists must rely on provisional links between the different applications.

Another key issue of the Project database is that it should contain, as much as possible, all the information relevant to the asset under study. This is very difficult in practice, because part of the data is always lacking, maybe simply because we do not know of its existence, especially when dealing with large and old fields. However, the completeness of the database is an essential requisite of any study, which should be given maximum attention.

Another aspect of the completeness of the database is relevant to particular types of data, whose nature make it difficult to store them in digital format and which have no obvious interaction with the data contained in the Project database: previous studies and reports, data like special core and PVT analyses and so on. This means that the computer database must be integrated with a physical database, including all the information that cannot be transferred into digital format.

The importance of the completeness of the database can be appreciated by considering the simple example of duplicated wells (side-track, re-drill, re-entry ...). These wells have been drilled at some stage of the field life because of mechanical problems of the original borehole or other reasons. Very often, the geologist tends to ignore this information, especially if it is not readily available and it is difficult to locate, simply because it does not add anything to the geological or petrophysical model of the reservoir. However, these duplicated wells can hold invaluable information on the saturation changes of the reservoir, which can be used in the fluid saturation distribution and monitoring study (paragraph 6.5).

2.5 PROJECT DATABASE MANAGEMENT

The Project database is not just something that is consructed at the beginning of the study and transferred to the Corporate database when the work has been completed. It is a living part of the project, which changes and evolves with the study itself and, as such, needs continuous management.

The management of the Project database is an important task in the integrated study. Experience shows that many of the sticky points in a reservoir study are related to the quality and/or the completeness of the database. From this viewpoint, the presence of a Project database Manager is essential to assure minimal drawbacks during the course of the study.

The Database Manager is responsible for loading all the relevant information into the system, making this information easily available to the other team members, updating the database with the new information available and upgrading the hardware and software configurations (however, he will not be responsible for the quality of the information, which needs to be guaranteed by the geoscientists).

In addition to that, he is responsible for the security of the database, and should limit the access to the entitled professionals. The Database Manager should also make sure that a common working area is allocated to the project, thus avoiding the proliferation of home directories that would increase data spreading and multiplication.

There is little doubt that big studies require a Database Manager assigned to the working team. He or she may have an IT background, but in all cases needs to be aware of the various steps of a reservoir study in order to understand the importance of the different pieces of information and their interrelationships.

Another important point in managing the database is controlling its life-cycle. There is a clear tendency in actual projects to keep all the information locked in the Project database until the study has been completed. However, many projects have a life span of several years and in some cases, when the asset is critical, may be almost permanent. This obviously generates a problem, because value is gained during the study and this value remains locked in the Project database, accessible only to the team members who performed the study. At the same time, the Corporate database from which the original information has been copied, becomes outdated. In other words, the longer the life span of a project, the more out of date becomes the information upon which it is based. From this point of view, the Database Manager has to regularly transfer to the Corporate database the information that can be considered as the final product of some important step of the project. For example, the corrected suite of well logs and their quantitative interpretation when the petrophysical study is complete, or the final geological maps when the dynamic simulation study has already been successfully history matched.

The final transfer of the study results to the Corporate database can be considered as the Database **decommissioning**. This often neglected aspect of an integrated study requires the participation of the whole team, especially concerning the data reduction phase. It can be estimated that at the end of a project as much as 95% of the information contained in the Database can be discarded. Referring to the types of data illustrated in Table 2.1, every personal file must be deleted, unless properly referenced, while the project data must be **promoted** to be used as the new reference information. All the intermediate interpretations must be deleted, since no duplicate reference data can be admitted. The geoscientists that participated in the study should indicate exactly to the Database Manager which is the final and unique information to be transferred to the Corporate database.

2.6 SOFTWARE INTEGRATION

The integrated team needs a common working platform, if an integrated study has to be performed. The creation of a such a platform has been the major concern of E&P software vendors in the last few years and undoubtedly giant steps have been made, when we compare the way we normally work today with what used to be our way of working only a decade ago.

Open software and integrated applications are already a reality which is on our desktop. However, not all the problems have been solved. These platforms have initially been developed for geoscience applications and they are able to integrate fairly well all the different disciplines relevant to the *static* part of an integrated study. As has already been mentioned, when the integration of the dynamic data is considered, the situation is far less exciting. In addition to that, while these platforms offer truly integrated solutions in themselves, they normally prove to be much less integrated with each other. From a user viewpoint, it becomes very difficult to transfer data from one platform to another, without having to reformat the relevant files in some way.

The present situation is what we could refer to as **loose integration** [4]. The transfer of data from one application to another needs to be performed by means of ad hoc solutions, which normally imply some software interface. Loose integration today exists not only between applications of different vendors but also between one integrated platform and other applications (company proprietary or other vendors' software) that perform tasks not available in that platform. To give an example of loose integration, when a geostatistical package is to be used, which is external to the working platform, data need to be extracted from the Project Database and reformatted through some kind of software, in order to be loaded into the geostatistical software. The inverse operation has to be performed when reloading the interpreted data to the database. Geologists who have already performed these transfers know how painful and time consuming such a simple operation can be.

Loose integration basically means that data reside in different databases, accessible by different software. Extractions and reloading operations are necessary steps, if activities are to be integrated. The great majority of the reservoir asset team in oil and consulting companies are working today in a loosely integrated environment, where only islands of proper integration exist.

In contrast with that, **tight integration** could be defined. Tight integration implies that many applications coexist with one another and data are shared among them without having to transfer or reformat the data, no matter the task or the vendor. The situation depicted in Fig. 2.3 actually illustrates a tight integrated environment.

Ideally, tight integration would require a unique database, accessible to all the different applications. In recognition of such problem, some multi-company initiatives started at the beginning of the last decade. The Petrotechnical Open Software Corporation (POSC) and the Public Petroleum Data Model (PPDM) are examples of non-profit making organisations sponsored by world-wide oil companies and vendors, that had the objective of establishing a common data model for the oil industry.

These initiatives, however, did not generate the expected results, because of commercial conflicts among the participants, and as of today the idea of the unique platform still remains a dream.

At the moment the idea of having to deal with different and partially communicating systems has been accepted as a kind of necessary evil in most petroleum and service companies. In this respect, two tendencies have been emerging in recent years:

- Choosing to use only one vendor's platform, offering the whole chain of reservoir applications (Shared Earth Model, or SEM). These systems offer in themselves tight integration, but individual applications are not necessarily the best on the market, while communication with other, external systems becomes difficult.
- Choosing to work with the **best** applications in the market, regardless of the platform. The problem in this case is to establish a viable communication system among the different databases, through standard interfaces. Recent multicompany projects like OPENSPIRIT address these problems.

The problem of systems interoperability keeps being the focus of the attention of both petroleum and service companies. New perspectives are being explored, one of the most interesting being the possibility of communicating through different shared earth systems and different locations through the world wide web (Fig. 2.4). Such opportunities may have a significant impact on the way integrated studies will be performed in the future.

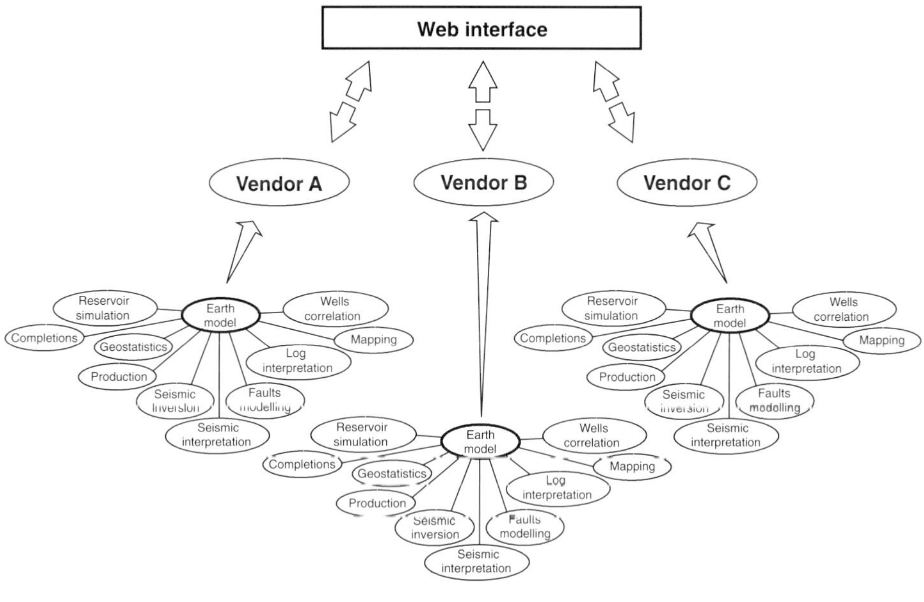

Figure 2.4 Interoperability among different Shared Earth Models through the WEB.

However, the definitive solution to the interoperability problem is still to come. In the meanwhile, it is the project manager's responsibility to pay attention to this problem and to assure that no time and resources are lost because of interoperability issues.

References

1 Lesslar PC, Van der Berg FG, Managing data assets to improve business performance. SPE paper 39728.

2 Managing oilfield data management, Oilfield Review, June 1994.

3 Wild J (1995) Integrating People: Realising the value of Systems and Data. EXPRO Transactions.

4 Sidney S (1997) Defining the relationship between integration and data management, Hart's Oil and Gas World, May.

CHAPTER 3

Integrated Geological Model

The definition of the geological model of the reservoir represents one of the most important phases in the workflow of a typical reservoir study, both concerning the volume of work involved and the impact on the final results.

The relevance of a sound geological model in the overall reliability of a reservoir study has been repeatedly emphasised in the technical literature, being long recognised that the static description of the reservoir, both in terms of geometry and petrophysical properties, is one of the main controlling factors in determining the field production performance.

It should be emphasised that, in most operational studies, such relationship between static description and field performance is exploited as a measure of the accuracy of the geological model, while little direct use is made of the dynamic information to improve the model itself.

In other words, the geological study is often performed making use of static information alone, i.e., seismic, log and core data, while the dynamic information is used only to check the consistency of the model and its ability to reproduce the observed reservoir performance. This could be referred to as an *a posteriori* control.

In a truly integrated geological model, however, the dynamic information should be better used as a direct input, i.e., an *a priori* constraint. From a practical standpoint, this means that all the professionals involved in the study, working at the same time in the same working environment, contribute to the identification of those data which are deemed useful in the geological characterization. In this way, all the available data can be explored during the execution of the geological study and it becomes possible to build a model that will satisfy both static and dynamic constraints. Such model will have a higher degree of consistency and it has a better chance of being able to reproduce the observed field performance.

The objective of this chapter is precisely to underline some of the integration aspects of a typical geological modelling work and to discuss how, from a practical standpoint, the dynamic information can be utilised to improve the static reservoir description.

The chapter has been divided into 4 separate sections, which detail typical stages of a reservoir study project:

- **Structural model**. In this section, we discuss how the different available information (seismic interpretation, geological evidence and well data) can be used in the definition of the structural top map of the reservoir and associate fault pattern. The uncertainty related to this phase of the study, as well as some issues related to a 3D approach to geological modelling, will also be addressed.

- **Stratigraphic model**. Starting with a short presentation of Sequence Stratigraphy principles and applications, we describe how correlative (deterministic) surfaces can be drawn across the reservoir. The integration of other techniques, which can improve or validate the stratigraphic framework, will also be discussed. Finally, we review some of the issues related to the building of a 3D stratigraphic grid.
- **Lithological model**. This is an essential (yet not mandatory) phase of the study. Here, we discuss the advantages related to the subdivision of the reservoir into a number of elementary facies and we analyse how a sound characterization process can be performed. In the last section, we discuss some of the available stochastic techniques to obtain a detailed 3D facies distribution, while the problem of uncertainty assessment will also be addressed.
- **Reservoir heterogeneity**. In this section, we analyse the presence, extension and importance of internal heterogeneities within a hydrocarbon reservoir. We provide a scale-based classification scheme and we comment on the impact of the different types of heterogeneity on fluid flow. Finally, we illustrate the application and integration of different sources of static, quasi-static and dynamic data in identifying reservoir heterogeneities.

3.1 THE STRUCTURAL MODEL

Building the reservoir structural model refers to the combined work of defining the structural top map of the hydrocarbon accumulation and interpreting the fault pattern that affects the reservoir.

Traditionally, this phase of the study is the domain of Geophysics. Seismic surveys actually offer the only direct means to visualise the subsurface structures and to infer a geometrical model of the reservoir. While other techniques can provide useful information about the structural setting of the reservoir under study, e.g., regional tectonic studies, there is little doubt that reservoir geophysics, either 2D or 3D, still represents in practically all cases the reference source of large scale information.

This section is divided in 4 main parts. The first two parts deal with the definition of the basic structural framework and the identification of the fault patterns, which are typically based on seismic and well data. Later, a discussion will be presented on the uncertainty related to this phase of the study, while in the last section some issues will be discussed which are relevant to 3D geological modelling of reservoirs.

3.1.1 Reservoir Architecture Definition

The definition of reservoir architecture amounts to identifying the basic geometrical framework of the hydrocarbon trap. We will refer here to the definition of the external boundaries of the reservoir, in particular to the structural top map, while the internal framework is considered related to the stratigraphic model of the reservoir and will be discussed in detail in paragraph 3.2.

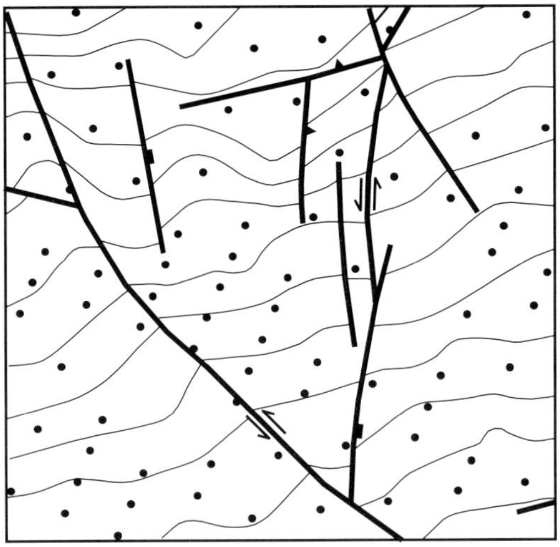

Figure 3.1 Structural top map of a 150 metres well spacing reservoir.

In the majority of cases, the structural top map of the reservoir is defined on the basis of post-stack 2D or 3D seismic data. The interpreter picks significant time horizons in a time seismic block and generates a set of (x, y, t) data which represents the two-way travel time to the picked horizon. These data can then be gridded and result in a time map of the reservoir structure.

In a later stage, this time map is converted to depth by means of a velocity model of the overburden formation. Several approaches exist for this time-depth conversion, the choice being dependent upon the available data and the overall complexity of the geological setting. We refer the reader to the specialised literature for more information about the issue.

In some cases, seismic data are not available or their quality is too low for a reliable interpretation to be performed. This is the case, for example, of fields where surface production infrastructures disturb the acquisition of the seismic surveys and induce noise in the recorded data. Alternatively, bad quality data may be related to the presence of gas in the overburden formation or simply to the existence of strong reflectors above the horizons of interest, that limit the energy of the seismic waves that travel in deeper zones.

In these cases, the structural top map of the reservoir has to be based, mostly or completely, on well data. This generates some uncertainty in the interwell zones where information is lacking, the problem being emphasised of course when few wells are available, as is the case in newly discovered fields.

On the other hand, when a large number of closely infill wells are available, the definition of the structural top map has little uncertainty and the same seismic information gives little advantage (if any) to the interpretation.

Figure 3.1 shows an example relative to a field where wells have been drilled with an average well spacing of 150 metres (small dots in the figure). Seismic information in this case has been used only to define the fault pattern, while the structural map itself has been adequately drawn on the basis of wells data only. The integration of well and seismic data would prove to be a cumbersome process because of calibration problems, while the gain in the accuracy of the resulting map would probably be negligible.

3.1.2 Faults Modelling

Take a pane of window glass and throw it on the ground: this is how my reservoir looks. This description provides a colourful image of the structural complexity of some reservoirs. Actually, nature sometimes is unfair with geoscientists. Highly faulted reservoirs are not uncommon and probably many of us have experienced this *broken glass* feeling at least once.

The impact of structural complexity over the development strategy and consequently on the economics of a given field is immediate. The same volume of oil in place may require a number of wells that is considerably different in the two extreme cases of a continuous and a highly compartmentalised reservoir. In a harsh environment like the North Sea, where drilling and completing wells can be extremely expensive, properly assessing the number of wells necessary to drain a reservoir has an essential impact over the field economics.

There is no waste of time or technology when reservoir compartmentalisation is concerned. On the subject of integration issues (Chapter 1) we already mentioned how important it is to properly address each aspect of an integrated reservoir study with the right degree of detail and technology. In fact, in the case of fields which are highly faulted and likely to be compartmentalised, the study of the structural model should always be given maximum attention.

In most operational studies, faults are identified on the basis of 3 main types of information, whose integration eventually provides the characteristic fault pattern for the reservoir under study.

1. **Geological evidence**. The technique refers to the identification of suspected faults by means of inconsistency in the stratigraphic correlation scheme. Typically, for example, well markers are too high or too low with respect to the *expected* depth, i.e., the depth which the well should have on the basis of geological (geometrical) considerations. In the past, when full 3D seismic acquisitions were not available, this was the main technique to locate faults in the reservoir. Nowadays, faults responsible for such geological evidence are easily picked in typical seismic volumes and this method is certainly no longer the most relevant for modelling faults.
2. **Well evidence**. Faults intersected by wells are, in most cases, easily identified. It is well known that missing sections can be related to normal faults, while repeated sections can be related to reverse faulting (Fig. 3.2). Additionally, some of the commonly available wireline logs may show evidence of the existence of a fault, since faults are often related to anomalous zones, in terms of resistivity and/or density values. Dipmeter logs also offer useful clues about the existence of a faulted zone. Finally, it should be noted that vertical wells have a much lower probability of crossing a fault compared

to horizontal wells, since in the large majority of cases faults are vertical or subvertical. This implies that vertical well data can hardly be used to infer a fault model, nor to generate statistics that could be used in a stochastic approach, since they bring biased information. However, they can be useful to locally validate the seismic interpretation.

3. **Seismic data**. Surface seismic information is gathered by generating elastic waves at the surface by means of a source, vibrating or explosive, and recording the reflected wavetrains at some specific surface locations, by means of geophones. As is known, faults can be detected from discontinuities in the reflection patterns, once this has been processed to reduce the noise and to place the events in their appropriate position in space (migration).

Seismic data make up the basic information, as far as fault modelling is concerned, since they provide a complete coverage of the area under consideration. From a practical standpoint, the seismic volume is loaded into an interactive workstation, and the interpretation is carried out on time sections and slices. Also, a number of seismic attributes can be utilised (amplitude, dip, azimuth ...), that allow the interpreter to make full use of the huge amount of information that seismic data inherently carry. When a good quality 3D survey is available, an accurate description of the reservoir architecture can usually be obtained.

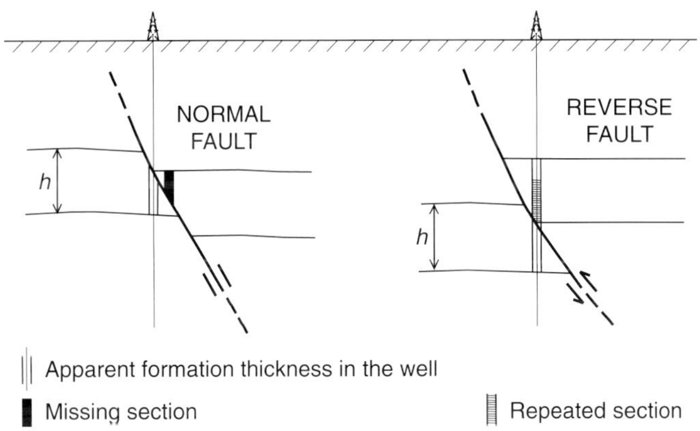

Figure 3.2 Normal and reverse fault evidence in a well.

3.1.2.1 Accuracy of the Fault Model

Petroleum reservoirs tend to become more complex with time. This paradox can be appreciated when different structural maps, produced in different times, are compared. In the large majority of cases the more recent the map, the more complex the structure.

This is obviously a consequence of the increased degree of detail accessible through recent acquisition and interpretation technologies. Actually, as already discussed in Chapter 1, the availability of powerful interpretative tools allows the geophysicist to push the detail of the interpretation very far, to the extent that in many cases a representation of the reservoir is generated, which may be difficult to transfer to the reservoir simulation model (Fig. 1.5). Small scale structural features may represent unnecessary detail and should not appear in the structural map provided to the engineer.

Which is then the degree of detail we are interested in? In the framework of an integrated reservoir study, we are basically interested in identifying the structural features **that have an impact on fluid flow**.

Faults shorter than half of the average well spacing, for example, are likely to have a minor impact on reservoir dynamics (Fig. 3.3, left). Likewise, faults shorter than the simulation gridblock dimension cannot be explicitly represented in the model, therefore their presence in the structural map leave the engineer with the doubt of arbitrarily suppressing these faults or extending them to the minimum gridblock size. The situation is also illustrated in Fig. 3.3 (right): while the longer F1 fault can be approximated with a discrete trace across the simulation grid, there is no way to satisfactorily take into account small faults like F2 or F3.

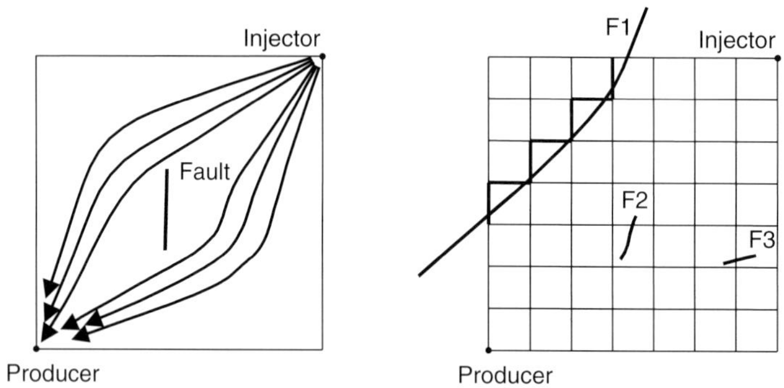

Figure 3.3 Impact of small faults in fluid flow modelling.

The above discussion implies an important point, i.e., the fault modelling phase must be related to fluid flow dynamics. This evidence, in turn, raises two interesting points of discussion regarding seismic data interpretation:

- **Fault Seal Potential**. Geophysics allows us to identify in space the presence of faults, but it does not provide us with information on whether or not the interpreted faults are actual barriers to fluid flow (with the possible exception of large fault throws). In other words, the seismic interpretation allows us to determine **where** a possible reservoir discontinuity is located, but it will not help us in defining its sealing potential. Sometimes this important point is overlooked. We should always use geophysics together with other sources of information, as it will be discussed in paragraph 3.4.2. When used alone, geophysics may only allow us to *guess* the degree of compartmentalisation of a reservoir.
- **Seismic data resolution**. Every seismic survey has a given resolution, which depends on the acquisition technique and the particular field under study. The degree of detail of the relevant interpretation is obviously linked to this resolution. In general, there is little chance that such an interpretative detail will be representative of the actual degree of reservoir compartmentalisation, i.e., the internal structuring of the reservoir that has an impact on fluid flow. In general, the interpretation is more likely to overestimate or

underestimate this true degree. For example, when dealing with shallow reservoirs, where recent tectonics have generated a widespread network of small scale faults, the resolution of the 3D seismic survey may lead us to generate a very detailed reservoir description, which could prove to be irrelevant when the dynamic behaviour of the field is considered. On the other hand, in the case of deep reservoirs and/or when the quality of the seismic survey is inadequate, we may not be able to identify important structural features that could be essential in governing flow paths. Again, in the framework of a reservoir study, we should always compare and integrate the seismic interpretation with other independent data, in order to assess how representative the available interpretation is with respect to the actual reservoir compartmentalisation.

These issues highlight an important point of discussion, i.e., geophysics alone is not sufficient to establish a structural pattern that is relevant to fluid flow. Even when the seismic interpretation is complemented with well observations and other geological evidence, the resulting map does not necessarily respond to the needs posed by an integrated study. Other techniques should be used to integrate the geophysical interpretation, like well testing, fluid characteristics and production information (see paragraph 3.4.2).

The work of establishing a representative fault pattern for the reservoir under study is therefore a complex activity, which involves the combination of data coming from different disciplines, both static and dynamic. Geoscientists involved in the study must be aware of this integration need, in order to generate a reliable faults model and to simultaneously optimise the workflow of the project.

An interesting example of use and limitation of seismic data in the interpretation of a reservoir structure is given in Fig. 3.4. This illustrates the fault pattern of a North Sea oil field of Jurassic/Triassic age, as it has been interpreted using surface seismic data. The field is highly compartmentalised and during the appraisal and early development phase, practically all the wells were drilled in different fault blocks. Also, horizontal wells were drilled to try to drain different blocks and access a larger hydrocarbon volume.

A quick glance at the structural setting of this field is enough to understand that reservoir compartmentalisation is the most important parameter that influences oil recovery and hence economics. As a matter of fact, practically all the wells drilled in this field have found hydrocarbons, but on the other hand it is very difficult to predict how much oil is accessed by each well. There is a chance that not all faults are sealing and also that not all faults have been picked in the seismic interpretation. Therefore, the reserves we calculate are dependent on the volume of rock that we believe is in communication with each well. On the other hand, while it is probably possible to calculate a reliable number of total oil in place, we could scarcely estimate with the same accuracy how the oil is distributed among different fault blocks and eventually how many wells will be needed to produce that oil.

Such examples are not uncommon among the pre-Cretaceous fields of the North Sea. This reservoir is quite deep (13 000 ft) and is overlain by 2 younger reservoirs containing light oil. A complex Jurassic salt withdrawal structure is also present above the producing formation and all these factors make the geophysical interpretation a difficult task. This example shows that the uncertainty related to the actual reservoir continuity is very high when geophysics is not integrated with other types of information. This topic will be discussed in more detail in the next section.

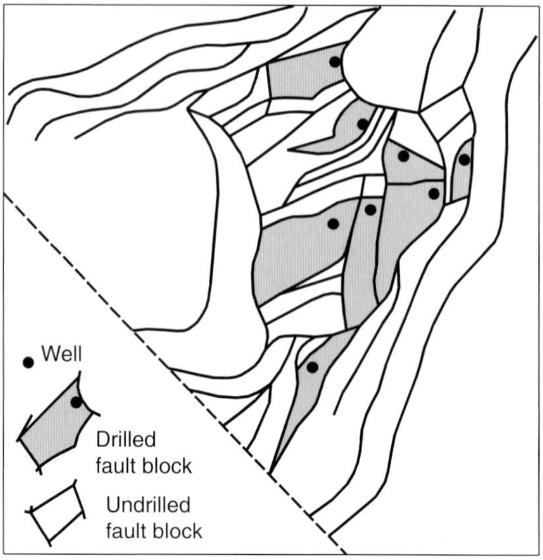

Figure 3.4 Structural top map of a Pre-Cretaceous North Sea reservoir.

3.1.3 Structural Model Uncertainty

The structural model of reservoirs carries an inherent degree of uncertainty that is related to the partial knowledge of the reservoir and the limitation of the techniques that are commonly utilised.

In general, this uncertainty is greater in the cases of fields with a limited amount of wells. On the contrary, mature fields, with closely spaced wells, will have a lower degree of structural uncertainty.

In the seismic interpretation procedure, errors can be basically related to one, or both, of the following factors:

- **Errors in picking**. A number of parameters can actually be included in this category: problems in the processing and migration phases, well-seismic mismatch, interpretation problems and so on. Globally, errors in picking may represent an important source of uncertainty in the structural interpretation.
- **Depth conversion problems**. The uncertainty in the velocity field to use in the time-depth conversion may be another major source of error. Lateral variation in the overburden lithology, presence of gas, limited or low quality well velocity surveys are only some of the problems that can be encountered. The impact on the overall uncertainty may be relevant, especially when poor control exists on the flanks of the structure, as is often the case, since small variation in the velocity field may generate significant fluctuations of the reservoir volume.

A measure of the uncertainty related to the reservoir structural model can be evaluated in a deterministic way, using alternative interpretations and velocity models. A more thorough and rigorous exploration of the uncertainty domain can be done through a stochastic approach [1] [1].

In general, the potential uncertainty existing in the structural modelling phase of a reservoir study is significant. A recent paper discusses the results of the application of probability fields to the evaluation of the structural uncertainty [2]. The OOIP computed by means of 200 realisations of the stochastic model showed a considerable dispersion of the values, with the 5th quantile (Q5) being about half of the 95th quantile (Q95).

Interestingly, similar results were obtained in the framework of the Great Reservoir Uncertainty Study, performed on a North Sea Brent reservoir by a consortium of Norwegian companies. The results of this project showed that the structural geological uncertainty, including fault description and reservoir top and base maps, accounted for three quarters of the total reserves uncertainty [3].

3.1.4 Building a 3D Structural Framework

Three-dimensional (3D) geological modelling is a recent matter. Despite the fact that 3D numerical simulation has been performed for more than 30 years, the 3D approach to geological studies proved to be a much more problematic task because of difficulties in modelling and visualising complex geometrical structures at fine scale.

In recent years, however, geologic software platforms have improved rapidly and for most of the geoscientists the possibility of working in 3 dimensions is an everyday reality. The main advantage of this technique, compared to the traditional 2D approach, is the ability to deal with complex geologic structures with a considerable degree of detail.

Several published papers deal with procedures for building 3D geological models [4], however in general terms the following steps need to be performed:

1. **Defining the main faults**. Main faults are considered to be those who limit major reservoir blocks. Fault planes are explicitly modelled as complex surfaces and determine the overall geometrical framework of the reservoir.
2. **Building the geologic surfaces**. Within each reservoir block, the main geologic horizons (top, bottom, major correlatable events) are modelled by means of mathematical (parametric) surfaces, which interpolate the available data points.
3. **Modelling the minor faults**. The main geological horizons are cut and offset by minor faults, i.e., faults that have a negligible impact over the global geometry of the reservoir.

Figure 3.5 shows a typical 3D reservoir framework, where the main faults and the parametric surfaces are clearly visible. In particular, the presence of reverse faults can be noted: this kind of structural features could not be modelled using the traditional, 2D approach.

The 3D structural model represents the basic geometrical framework of the reservoir. Later, this model will be completed with internal correlatable surfaces, usually defined through a chronostratigraphic approach based on sequence stratigraphy (see paragraph 3.2). Eventually, the model will be populated with facies and/or petrophysical properties, in order to generate what has been referred to as a lithologic model of the reservoir (paragraph 3.3).

---

1. Stochastic models will be discussed in more detail in paragraph 3.3.3.1.

Figure 3.5 3D modelling of complex geological structures (Courtesy of Beicip-Franlab).

3.2 THE STRATIGRAPHIC MODEL

Since the beginning of the oil industry, building the stratigraphic framework of a reservoir has been possibly the most traditional among the reservoir geologists' tasks. It ultimately refers to correlating all the wells, in order to define the surfaces that bound the main reservoir units. In particular, when numerical simulation is concerned, the overall objective of the stratigraphic model is to define the main reservoir **flow units**.

The importance of this stage cannot be overemphasised. Every professional knows the impact of properly correlating wells, and possibly even better the impact of *not* properly correlating wells. The whole importance of this stage is related to the fact that fluid flow takes place largely along the stratigraphic units of geological formations. Consequently, a correct description of the geometry of the sedimentological bodies that make up a reservoir, as well as their interrelations, is an essential requisite for the simulation of the production/injection performance of a field.

Building a reliable stratigraphic framework may require a considerable effort. In fact, the inherent difficulty of this work mainly depends on the sedimentological setting of the particular reservoir under study. In some cases, the depositional bodies of the reservoir formation may exhibit a wide areal extension, which make them easily correlatable between well, even when the well spacing is significant. This is the case of most platform depositional areas, where the lateral continuity of the sedimentary units is often significant. An extreme example of correlatable units is given by the distal facies of deep-sea turbidite complexes: in some fields of the North Adriatic off-shore, individual turbidites events few centimetres thick may be correlated for distances of kilometres throughout the basin.

However, in most of the reservoirs world-wide, the typical correlation length of the reservoir units is much shorter than that and unfortunately it is often shorter than the well spacing distance. This is typical for example in most continental formations, like alluvial, fluvial and deltaic complexes. Under these conditions, the definition of the internal reservoir architecture becomes a very complex issue and may represent the biggest challenge for the reservoir geologist. It is here that the integration of other disciplines may prove to be decisive. The correlation work potentially involves a considerable number of geologically-related disciplines, such as seismic and sequence stratigraphy, sedimentology, well log interpretation, palinology, biostratigraphy, geochemistry, mineralogy, outcrop studies and so on. Additionally, dynamic data can often be used to assess the reliability of the correlation scheme.

Providing general rules for building the reservoir stratigraphic model would be pretentious. In this chapter we will try instead to highlight some of the key points that should be considered when an integrated reservoir study is being performed.

First, an introductory discussion is dedicated to Sequence Stratigraphy, which is considered to be the reference methodology to build the reservoir stratigraphic framework. Later, some of the alternative techniques that may be used to refine and improve the well correlation scheme will be presented. The last section is dedicated to the building of a 3D stratigraphic framework for stochastic simulation purposes.

3.2.1 Sequence Stratigraphy

Sequence stratigraphy is a relatively new discipline. Even though its principles have been in circulation for a number of years, its official appearance can be fixed to 1977, with the publication of an AAPG memoir titled *Seismic Stratigraphy Application to Hydrocarbon Exploration* [5]. Now familiarly known as *AAPG memoir 26*, this breakthrough work contained the basic principles of this new, chronostratigraphic-based approach to pattern deposition analysis. Other milestone publications soon followed, and nowadays the technical literature on this theme and its application to exploration and production issues is huge. Leaving the interested reader to this specialised literature, we will concentrate, as usual, on those aspects that make Sequence Stratigraphy most relevant to an integrated reservoir study.

Sequence stratigraphy can be defined as *the study of genetically related facies within a framework of chronostratigraphically significant surfaces* [6]. The basic principle behind this statement is that the deposition of sedimentation patterns is controlled by changes in relative sea level, and this is in turn controlled by eustasy, subsidence, tectonics and sedimentation rate. The influence of these factors changes in different geological contexts: in passive margin shelves, eustasy is normally the predominant phenomena, while tectonics seems to have a more important impact in active margins.

The interaction of these elements determines the space made available for the potential sediment accumulation (accomodation space) and the resulting geometry of the sedimentation patterns.

In sequence stratigraphy, a hierarchy of depositional patterns can be defined, in relation to the scale of observation. The lamina is the smallest megascopic layer: it is uniform in composition and texture and it is not internally layered.

The sequence is the basic stratal unit for sequence stratigratigraphic analysis. It can be defined as a relatively conformable, genetically related succession of strata, bounded by an unconformity or their correlative conformities [6]. These unconformities, also called

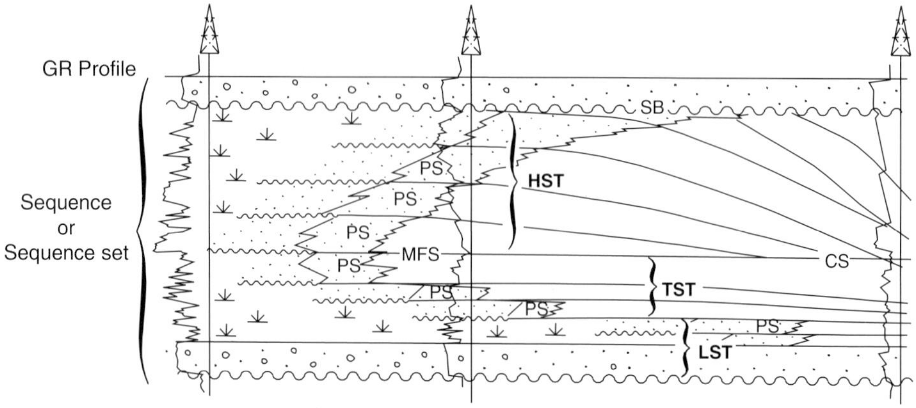

Figure 3.6 Illustration of a sequence unit.

sequence boundaries, record a relative fall in the sea level, and represent the most important surfaces that a reservoir geologist can pick [2]. The sequence boundary is a laterally continuous, widespread surface covering at least an entire basin and seems to occur synchronously in many basins around the world [7]. It has a chronostratigraphic significance, since it is formed in a timeframe of few hundreds or thousands of years, a period that can be considered synchronous from a geological point of view. Fig. 3.6 shows a schematic diagram that illustrates the main components of a sequence, while Table 3.1 summarises the main characteristics of these stratal units, together with the tools that can be used in their characterization.

The correct identification of depositional genetic units provides a high-resolution, chronostratigraphic-based reservoir architecture, which is particularly suited for reservoir studies, as in most cases there is a strong link between chronostratigraphic facies architecture and fluid flow.

There are a number of reasons why sequence stratigraphy can be considered an ideal tool to an integrated reservoir study:

- The application of sequence stratigraphy to the reservoir scale provides a detailed stratigraphic framework that may reduce the risk of miscorrelations between different genetic units.
- Sequence stratigraphy can be studied and identified at different scales and in this sense it is fractal in nature. This allows for the utilisation and the integration of data collected at different scale and with different tools (Table 3.1).
- Within a sequence it is possible to predict, the continuity, connectivity and extension of sandbodies and to establish representative parameters for stochastic modelling.
- It allows for the prediction of the presence and the extension of the reservoir facies outside the developed areas of a mature field.
- Its principles can be applied both to siliciclastic and carbonate systems.

2. An alternative sequence model considers the recognition of *Maximum Flooding Surfaces* as sequence boundaries [9]. Concepts and basic principles of sequence stratigraphy can be found in [8].

Table 3.1 Main characteristic of sequence stratigraphy units
(Van Wagoner et al., AAPG Methods in Exploration Series, No. 7. AAPG © 1990. Reprinted by permission of the AAPG whose permission is required for further use) [6].

Stratal Units	*Definitions*
SEQUENCE	A relatively conformable succession of genetically related strata bounded by unconformities and their correlative conformities
PARASEQUENCE SET	A succession of genetically related parasequences forming a distinctive stacking pattern and bounded by major marine flooding surfaces
PARASEQUENCE	A relatively conformable succession of genetically related beds or bedsets bounded by major marine flooding surfaces or their correlative surfaces
BEDSET	A relatively conformable succession of genetically related beds bounded by surfaces of erosion, non deposition or their relative conformities
BED	A relatively conformable succession of genetically related laminæ or laminasets bounded by surface of erosion, non deposition or their relative conformities
LAMINA SET	A relatively conformable succession of genetically related laminæ bounded by surface of erosion, non deposition or their relative conformities
LAMINA	The smallest megascopic layer

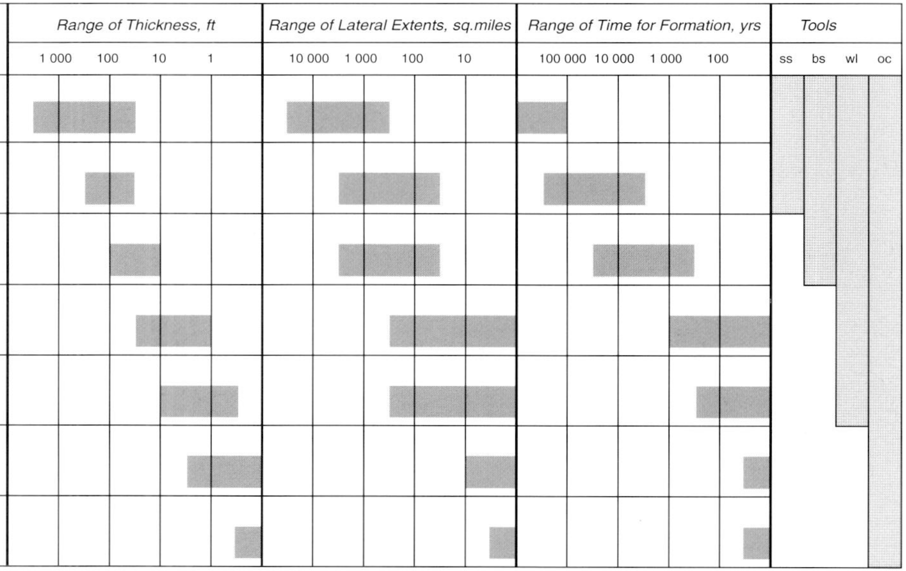

ss = surface seismic bs = borehole seismic wl = wireline logs oc = outcrop and cores

The importance of a correct identification of the reservoir facies architecture can be appreciated observing Fig. 3.7 (from [6]), that shows two different interpretations of the same set of well logs data. In the upper figure, wells have been correlated within a sequence stratigraphic framework: when flattening the cross section using the sequence boundary as a datum, we can interpret a retrogradational sequence of shallow marine sandstones. In the lower figure, wells have been correlated using a more traditional lithostratigraphic approach, where the top of the sands have been picked as a reference surface.

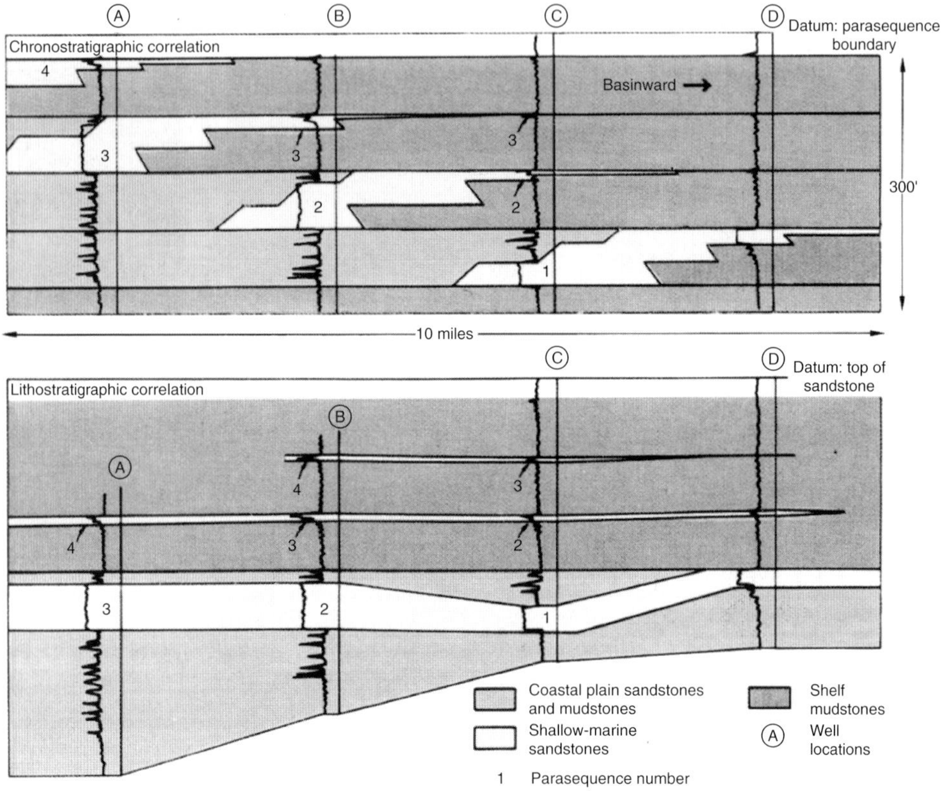

Figure 3.7 Chronostratigraphic vs. Lithostratigraphic correlation schemes (Van Wagoner et al., AAPG Methods in Exploration Series, No. 7. AAPG © 1990. Reprinted by permission of the AAPG whose permission is required for further use) [6].

The resulting reservoir architecture in the two cases is very different: in the upper figure, we recognise a series of progressively younger parasequences that step upwards and landwards, poorly connected and hydraulically independent of each other. In the lithostratigraphic interpretation, genetically different sandbodies are linked together, showing a much greater reservoir continuity. It is evident that, if a simulation study is performed using the two alternative reservoir descriptions, the resulting production performances will be substantially different. In particular, the lithostratigraphic scheme will lead to optimistic results, including to a wrong assessment of the number of wells needed to develop the field.

3.2.2 Other Techniques

As discussed in the previous section, sequence stratigraphy is the reference technique to well correlations since it allows the geoscientist to integrate different types of data (seismic, log and core) into the definition of a coherent stratigraphic framework.

However, the geoscientist should make use of all possible information to corroborate the stratigraphic model. In particular:

- **Biostratigraphy and Palinology**. Rock samples collected on cuttings can be analysed for the presence of particular micropaleontological and/or palinological associations. This data in turn may provide useful information concerning the assumed well correlation. Attention must be paid to the poor vertical resolution of such data and to the possible lack of a unique relationship between chronostratigraphy and biostratigraphy.
- **Drilling data**. Under favourable circumstances, the Rate of Penetration (ROP) is a parameter that provides useful information about the stratigraphic position of the unit that is being drilled. When the vertical sequence of depositional units exhibits a distinct lithological hardness trend, this data can be used to validate the correlation scheme. In any case, care must be taken in this exercise, since the resistance offered by the formation to the drilling bit is not necessarily related to the chronostratigraphic correlation scheme.
- **Pressure data**. As it will be discussed in more detail in following sections (paragraph 3.4.2.4), pressure data measured through static surveys or, better, with WFT (Wireline Formation Tester) tools, may help in validating the correlation scheme. When the existence of a major reservoir heterogeneity (e.g., a fault) is ruled out, different pressures measured in contiguous wells in the same stratigraphic unit may be related to a possible correlation problem.
- **Production data**. Consistent trends should be observed in the production data when a sound correlation scheme has been defined. Deviation from these trends (e.g., anomalous GOR or WOR) may be caused by miscorrelation. In all cases, care must be taken in verifying that these anomalies are not related to individual well completion problems.
- **Fluid data**. The same hydrocarbon type is expected to be found in the same stratigraphic unit. When anomalies in the produced fluids are observed for some wells, e.g., lower than normal oil API gravity, this is a warning that a correlation flaw may exist.

Of course, other techniques can be considered, that could be used to assess and validate the stratigraphic correlation scheme defined through the sequence stratigraphy approach. Every study has its typical association of available data and it is ultimately the geoscientist's responsibility to explore that information and to use it in the most proper and fruitful way.

3.2.3 Building a Stratigraphic Grid

In the traditional 2D approach to reservoir geology, the spatial complexity of the reservoir is represented by means of a series of cross sections and stacked maps of the various geometrical and petrophysical properties.

In recent years, the advent of 3D geological and stochastic modelling has completely changed the traditional perspective. In a 3D approach, the areal variability of any geological parameter is represented at a much finer scale, while the vertical direction is also explicitly taken into account. The final result is a detailed and more realistic representation of the reservoir architecture and internal heterogeneity.

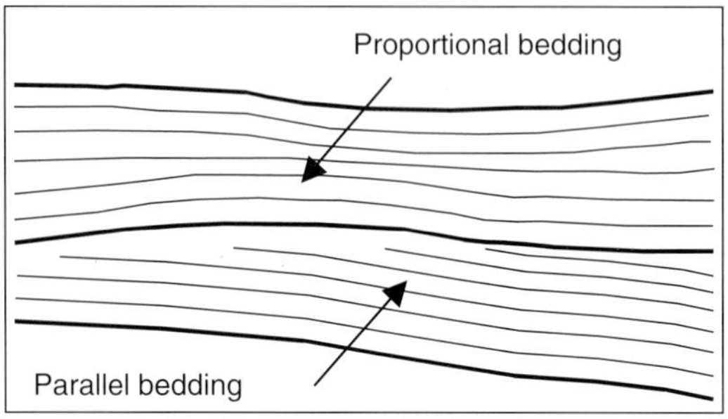

Figure 3.8 Depositional framework for stratigraphic grid building.

The basic geometric framework of such geological representation is the 3D structural model, as it has been defined in paragraph 3.1.4. The 3D stratigraphic model, consisting of a set of correlatable surfaces, is created within that geometrical framework.

From a stratigraphic point of view, the main question to face is probably the definition of a sound internal geometry for the architecture of the formation units. In general, 2 possibilities exist, which are contrasted in Fig. 3.8:

- **Proportional bedding**. The lower scale genetic units (laminae, beds) are deposited throughout the area under study, while their individual thickness may change laterally. The total thickness of the units is also variable, but the vertical sequence is preserved in any point.
- **Parallel bedding**. The thickness of individual lower scale genetic units does not change laterally. Since the total unit thickness may vary, the vertical sequence is not preserved. The series can be parallel to the base or to the top of the unit. The typical example is a series truncated by an unconformity.

The choice of a correct representation of the stratigraphic framework has a considerable impact on the modelling phase, since it defines the spatial architecture of the depositional units within the reservoir.

Figure 3.9 provides an example, relative to the stochastic modelling of a cross section, simulated with a parallel (above) and proportional (below) stratigraphic framework. Erosion of the upper units is evident in the former, while much more continuity in the sedimentary bodies is observed in the latter. These differences, in turn, will have an impact on the results of flow simulations.

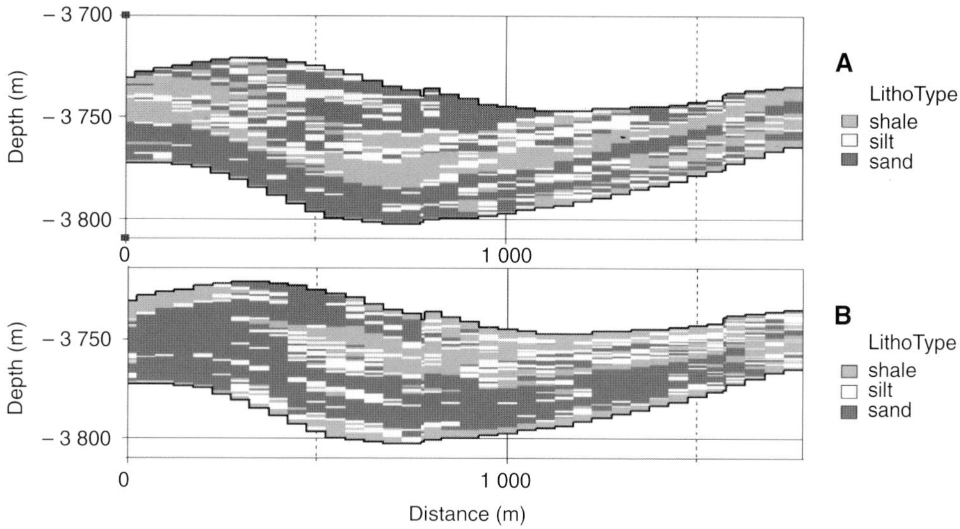

Figure 3.9 Stochastic facies model obtained with a parallel (A) and proportional (B) schemes.

3.3 THE LITHOLOGICAL MODEL

The structural and stratigraphic models of the reservoir, discussed in the previous sections, provide the geologist with a reference geometric framework of the field under consideration.

A subsequent, major phase of a typical study concerns the building of the lithological model of the reservoir, i.e., filling, or populating, that geometrical reference framework with data that describe the lithological characteristics of the reservoir rock and their spatial variability.

It should be appreciated that this phase of the work is not mandatory. Many studies are successfully performed without the explicit modelling of the lithological distribution of the reservoir. Actually, the computation of the hydrocarbon in place and the numerical simulation of any field only require the knowledge of the petrophysical properties of the reservoir, in addition to a simple pay-non pay classification of the reservoir rock. Therefore, there is no need, in principle, for the support of a detailed lithological description.

Nevertheless, a detailed lithological model of the reservoir represents a powerful tool to guide the petrophysical distribution, since, in most reservoirs, the lithological facies and the petrophysical characteristics are intimately related. The approach is based on the assumption that lithology distribution is more predictable than a direct representation of the petrophysical properties.

In the majority of cases, the lithological model of a reservoir is built integrating a conceptual representation (the sedimentological model), a classification phase (facies definition) and a probabilistic approach of the lithological distribution (the stochastic model). In the next section, these topics will be discussed in some detail.

3.3.1 Conceptual Sedimentological Model

The definition of the sedimentological-depositional model of the reservoir is one of the first tasks to be performed in the workflow of an integrated study. Its impact is crucial in all the following phases, since this work provides the conceptual basis of the lithological model of the reservoir. Additionally, the correct description of the sedimentological and depositional systems will provide the geoscientist with a semi-quantitative evaluation of the geometrical parameters to input in the stochastic modelling process (covariance functions, shape and dimension of the reservoir units, etc.).

From a general point of view, the sedimentological study of a reservoir is composed of two main phases:

- **Lithofacies description and classification**. This work is normally carried out on the available core material and has the objective of classifying the reservoir rock from a lithological and depositional viewpoint. The facies identified during this phase will often constitute the elementary building blocks of the reservoir architecture. Related disciplines such as Biostratigraphy, Palinology, Mineralogy, Pore Imaging and Geochemistry will provide additional information concerning the age of the rock, the sedimentological environment, the pore system geometry, the presence and impact of post-depositional processes and so on.
- **Definition of the depositional model**. All the information that has been analysed in the previous phase can be used to define the depositional model of the reservoir. This amounts to identifying the sedimentological setting (fluvial, deltaic, shallow marine ...), as well as the depositional processes (high or low energy currents, debris flows ...) related to the reservoir formation. The analysis of the texture and the internal structure of the rock will also allow for the recognition of deformations and fracturing, linked to possible syn- or post-sedimentary tectonic processes.

The sedimentological model of the reservoir is usually defined through an accurate analysis of the available core material. Other types of data like cuttings from non-cored wells, log interpretation, seismic and outcrop studies of geological analogs can also be used in this process.

Figure 3.10 shows an example of a sedimentological model obtained through the description of cores and synthesised in a classification scheme, whereby each facies number (left column) is representative of a given lithological and depositional character. Note also the relationship with sequence stratigraphy (right column).

However, it should be appreciated that such detail can usually be attained only when working on core material. Since this is not always possible, it is important to establish a procedure to extrapolate this model to uncored sections, using data available in a larger number of wells, i.e., well logs. How to carry out such an extrapolation will be the object of the next sections.

3.3.2 Facies Classification

The facies can be considered to be the basic building-blocks of geological modelling. The importance of the facies concept for reservoir description and characterization has always been appreciated among reservoir geoscientists. In the past, however, the detailed facies

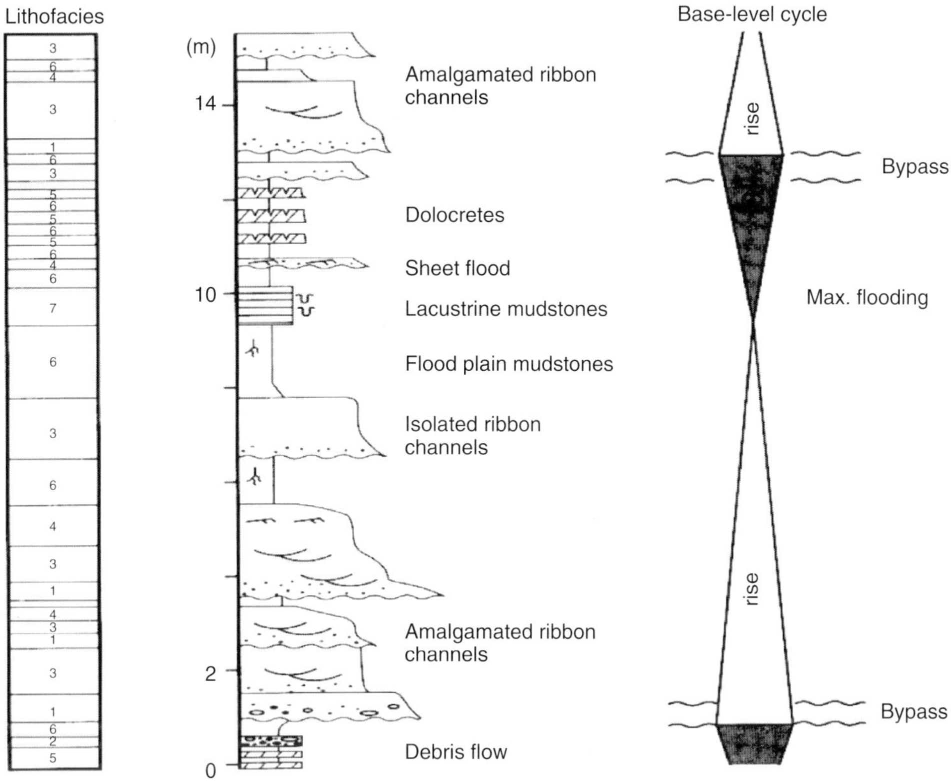

Figure 3.10 Sedimentological model and facies classification [10].

description that could be obtained at the well locations could only with difficulty be extended to the whole reservoir, since no specific tool was available to the geologist for such extrapolation, apart from straight, deterministic well to well correlation. Therefore, the facies classification scheme remained a nice theoretical framework that, in most cases, could offer little advantage to the study.

In recent years, however, with the advent of geocellular 3D modelling and stochastic simulations, the role of the facies as a basic component of the reservoir characterization process has been emphasised. At present, the possibility of creating a detailed 3D architecture of facies allows for a more realistic representation of the lithological complexity of the field, as well as a more reliable calculation of the petrophysical properties distribution across the reservoir.

Before starting the discussion, it should be appreciated that different types of facies have been defined in the technical literature: **lithofacies** or **petrofacies** (defined on cores), **electrofacies** (defined on logs), **seismic facies** (defined on seismic), **rock types** [3] and **lithotypes**

3. Even though the words **facies** and **rock type** are often used as synonyms in the reservoir geology literature, the term rock type is generally used in the simulation model to define zones where different saturation functions are applied. As we will see, however, the two concepts are linked.

(groups of facies). In the remainder of the text, the simple term facies will be used, to avoid reference to a specific classification procedure.

In the next paragraphs, we will expand on the facies concept, trying to show why the facies description can be considered to be an ideal geological characterization tool in the framework of an integrated reservoir study.

3.3.2.1 Facies Identification and Classification

Every time we perform a reservoir study we deal in some way with facies, even though often implicitly. In fact, a geological zonation always implies the generation of some kind of facies classification: in the simplest study, this reduces to the definition of *reservoir* and *non reservoir facies*, based on a pay cut-off criteria. In other, more complex cases, a larger number of facies can be defined on cores or logs and then distributed throughout the field, for example by means of stochastic models.

The simplest way of defining a facies classification scheme is through a more or less simplified process of lithological recognition in logs. The usual procedure in this case is to choose one or more thresholds on the lithological logs to identify different facies, normally spanning from *pay* to *non-pay*. Whenever possible, normalised suites of logs should be used, in order to ensure the objectivity of the facies identification process for all the wells. In simple sand/shale reservoirs, this method allows for a quick and in some cases effective facies discrimination (Fig. 3.11).

When good quality information is available, both in terms of suites of logs and core material, a more sophisticated approach can be attempted, based on a multivariate statistical treatment of the data. In this case, the basic steps of a typical facies classification procedure are as follows (Fig. 3.12):

1. **Definition of key wells**. The basic framework of the facies classification is built using a limited number of key wells, i.e., the wells which have core information, reliable and complete suites of logs and are located in representative areas of the reservoir.

2. **Facies classification**. As discussed in paragraph 3.3.1, facies can be defined on cores, through the description of the lithological, depositional and/or petrophysical features of the rock. This facies classification scheme is then linked to logs, through the recognition of a particular log signature for each facies. Alternatively, facies can be computed from log data and characterised afterwards through an accurate comparison with core data. Statistical algorithms like Cluster Analysis or Principal Component Analysis are commonly used in this kind of approach. A further step that is often performed is the grouping of the basic facies into a reduced number of what could be referred to as **lithotypes**. This has the advantage of providing a simpler description of the overall geological complexity, which can be handled more readily through stochastic modelling.

3. **Aggregation of other wells**. The final classification scheme can be extended to the remaining wells, which typically have older or incomplete suites of logs. The classification of these wells requires some kind of aggregation procedure performed throughout the existing log curves.

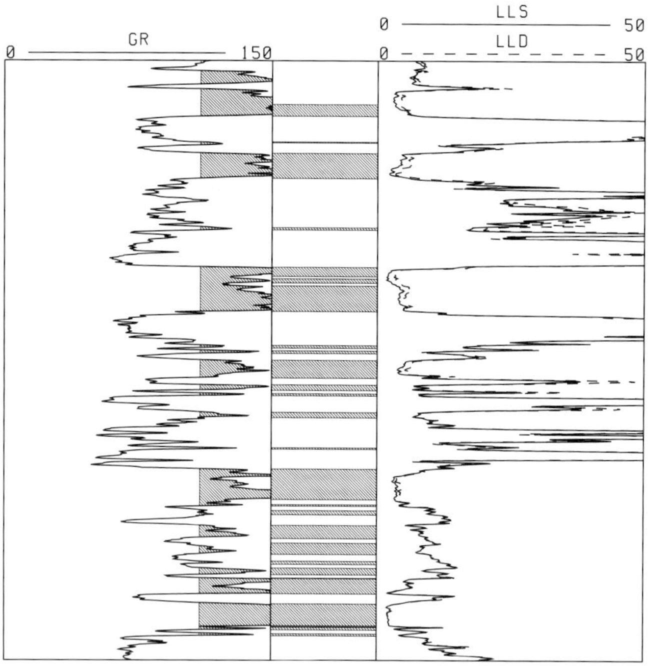

Figure 3.11 Facies identification on the basis of a lithological log cut-off.

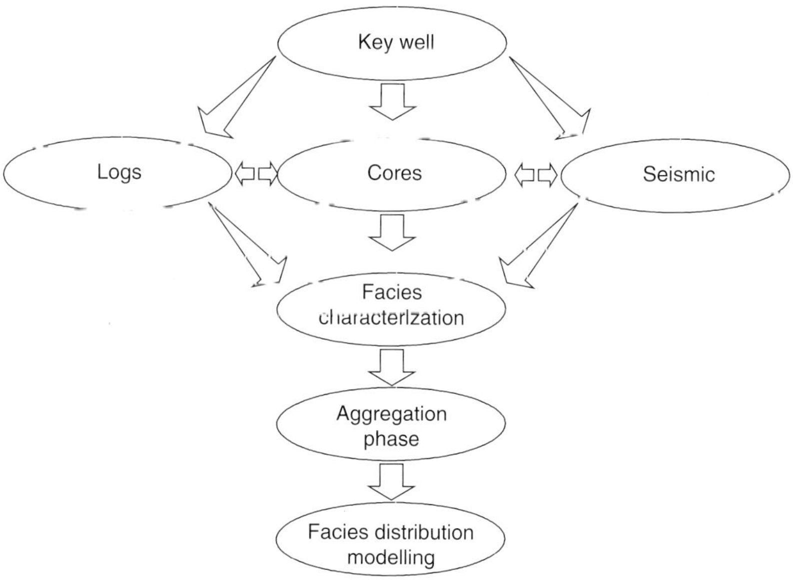

Figure 3.12 Simplified facies classification process.

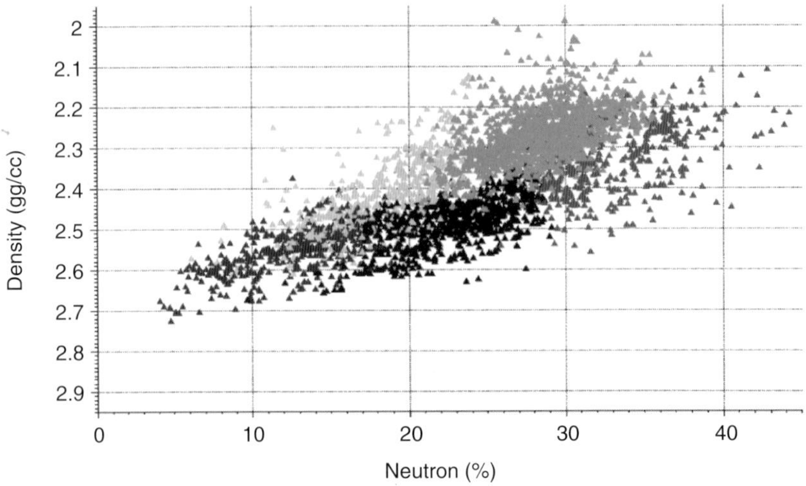

Figure 3.13 Facies identification in a Density-Neutron cross plot.

As an example, Fig. 3.13 shows a typical Density-Neutron cross plot for a given set of key wells, whereby clouds of points have been identified through a cluster analysis algorithm [4]. This methodology allows for the generation of facies profiles for each well, based on log information. In turn, such facies profiles will represent the basis for the subsequent 3D modelling of the reservoir.

3.3.2.2 Facies Characterization

The characterization phase aims at defining typical lithological, depositional and petrophysical parameters for each facies.

In principle, most of the available reservoir data could be used in this characterization process: mineralogical composition, sedimentary structures and textures, diagenetic effects, granulometric distribution, mechanical properties, fracture type and intensity, advanced rock properties like cementation factor and saturation functions, and so on.

However, such a detailed characterization could only be performed in core materials and in most cases when trying to transfer this information to logs an important part of this characterization work is lost. This loss of information is the inevitable price that has to be paid in order to generate a fieldwide facies distribution, no matter how the basic facies scheme has been built (core to logs or vice-versa).

In fact, some of the properties that have been used to characterise the elementary core facies are simply not recognisable on well log signatures, because of tool resolution and limitation. There is no tool that gives direct information, for example, on the wettability of the reservoir rock or its internal sedimentary structure. While these characteristics will influence

4. In reality, such clouds are often defined in a multidimensional space, being the number of dimensions defined by the number of well logs used in the analysis.

the overall tool response, there is no way to deconstruct the bulk signal to recognise each individual component.

In any case, the higher the number of available logs and their quality, the less information will be lost in the characterization phase and the more detailed should be the resulting facies classification scheme. When only a suite of SP and old resistivity curves are available, for example, the only information we can probably transfer from core to logs is a bare lithological sand-shale subdivision. However, when more modern logs exist, like Density/Neutron/PEF and Sonic for example, the characterization phase can be more comprehensive and in this case the facies will synthesise significant lithological and petrophysical properties of the reservoir rock.

Figure 3.14 shows an example of the characterization procedure, where a distinct capillary behaviour has been assigned to each facies (or association of facies). These curves can be used in subsequent phases of the study, typically in the oil in place calculation and in the definition of the capillary pressure curves in the dynamic model building.

It should also be emphasised that a *standard* facies characterization process cannot be defined, since this depends upon the particular reservoir, the available data and the objectives and the constraints of the integrated study. In general, we should focus on the classification process that is most suited to the reservoir under study, keeping in mind that the facies is the basis for a reservoir description that will be used for dynamic purposes.

References [10] through [13] present some recently published case studies, where typical facies classification and characterization procedures have been applied which are particularly suited for the reservoirs under study.

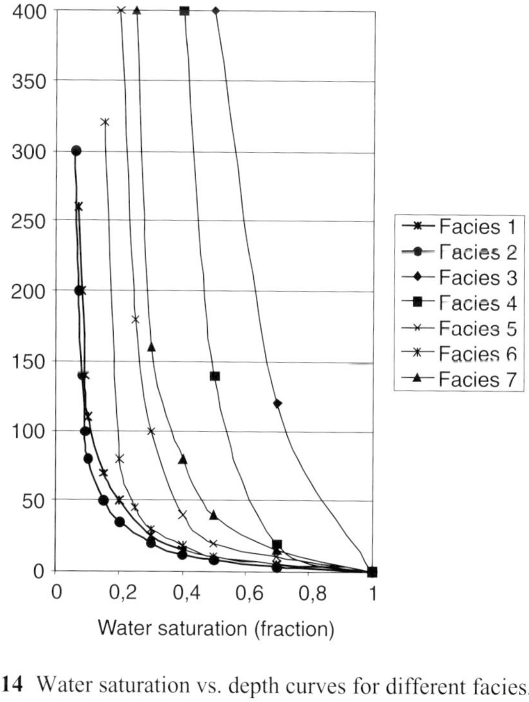

Figure 3.14 Water saturation vs. depth curves for different facies.

3.3.2.3 The Concept of Facies

As discussed in Chapter 1, most of the reservoir data available to the geoscientist belong to what has been defined as **macroscale** (Fig. 1.2). Within this scale domain, data actually refer to different or very different support volumes (e.g., core and log porosity measurements), however, from a practical point of view, these scale problems are often ignored. With few exceptions, the macroscale is therefore the smallest practical domain for reservoir description and characterization. The whole importance of the concept of facies is that it provides a means to integrate all the reservoir macroscale data in a simple, flexible and comprehensive classification system. In this context, the facies can therefore be considered the **practical elementary reservoir volume** and represents the basic building block for 3D geological modelling.

The concept of facies is particularly suited for integrated reservoir studies. Once a classification scheme has been defined and the facies have been characterised by integrating core, log and possibly seismic data, the facies can be utilised in a number of phases of a typical study. Some of the possible applications are listed briefly below.

- **3D modelling**. The facies can be applied to 3D reservoir description, through the utilisation of geocellular or stochastic modelling. As already mentioned, this is the most typical and significant application of the concept of facies.
- **Log quantitative interpretation**. A particular interpretative model, in terms of mineralogic (grain density) and/or saturation parameters (m, n) can be defined for each facies or group of facies. These models can then be applied to the quantitative interpretation of all the wells, thus providing a detailed and consistent evaluation of the petrophysical properties of the reservoir.
- **Upscaling**. The facies classification provides a robust starting framework for upscaling operations. In the geological to simulation grid upscaling process, the concept of facies may actually help in minimising the *smoothing* effect of the procedure, since the geometry of the simulation grid can be based on the small scale 3D facies distribution. The more homogeneous the facies distribution within a single simulation layer, the less *destructive* will be the upscaling process applied to the petrophysical properties. This also ensure that geology and petrophysics will be preserved at best at higher scale. Tools like Vertical Proportion Curves can be used in such a process, as is shown in Fig. 3.15, relative to a porosity upscaling procedure.
- **Rock types definition**. Even though no direct upscaling can be performed on the facies (which are *discrete* variables), the facies distribution can be used as a template in the simulation model to define the main zones for the attribution of particular sets of saturation functions (capillary pressure and relative permeability). This phase is commonly referred to as *rock types* definition.

What should emerge from these points is that the facies classification scheme, when properly defined and characterised, is an essential tool for integrated studies.

The possibility of utilising the facies through different phases of the study, makes it particularly attractive to the geoscientist. The reservoir rock can be defined and fully characterised at small scale and this classification system can be utilised in different contexts of the project. From this viewpoint, the facies can be considered as a tool which can be used to

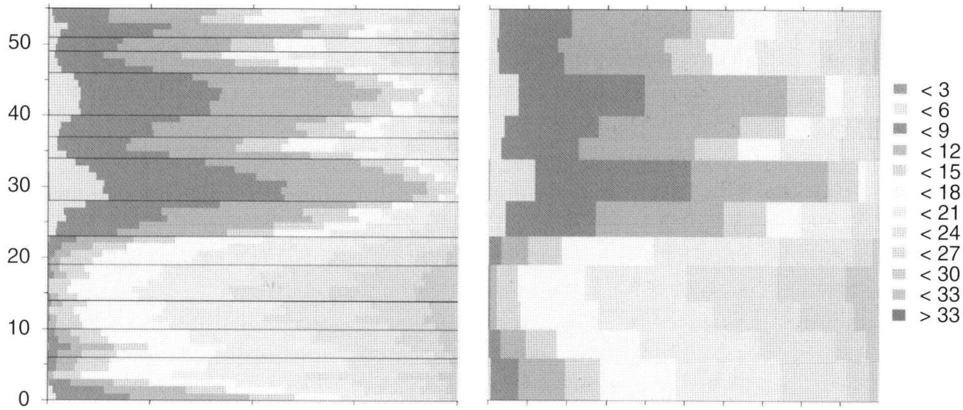

Figure 3.15 Use of Vertical Proportion Curves for upscaling.

transfer the geological information through different stages of the study, up to the simulation model. While preserving the geological and petrophysical characterization, the facies will ensure the overall consistence of the workflow.

From this point of view, the process of facies classification and characterization is a critical phase of the project, which calls for the tight integration of the different disciplines involved. The project manager has to pay careful attention during this phase, which may well represent one of the cornerstones of the reservoir study.

3.3.3 Facies Distribution

Once the facies classification scheme has been defined and the wells have been described with a vertical profile of facies, the subsequent step is to generate a 3D distribution of facies for the whole reservoir. The geometric framework for such distribution is the stratigraphic grid, as has been defined in paragraph 3.2.3.

Three dimensional facies distributions are typically obtained through stochastic modelling. These methods are relatively recent and are intimately related to the growing computing and visualisation capabilities of the geoscience applications.

Many different stochastic techniques have been developed in the last 15 years and some of them are presently available in most commercial software. The rapidity with which vendors have been implementing such applications in their geological packages testifies to the interest in these methods and now most geologists working in reservoir studies have become familiar with stochastic modelling.

A complete review of the available techniques, their advantages and their limitations would take in itself a whole book. The interested reader should make reference to summary papers on the argument [14, 15] and in general to the huge existing bibliography.

Here, after a short general introduction, the discussion will be limited to the more popular types of stochastic modelling approaches, i.e., pixel-based and object-based.

3.3.3.1 The Stochastic Approach

Stochastic modelling, in current geoscience literature, refers to the generation of synthetic geologic architectures and/or petrophysical property distributions, which are conditioned to the available quantitative (hard) and qualitative (soft) reservoir information.

These models produce non-unique, equiprobable realisations that share common statistical properties and that represent possible images of the geological complexity of the reservoir. There is no *a priori* method to choose one among the theoretically infinite realisations of the stochastic model and, to some people, this represents the disturbing element of the technique. On the other hand, the study of the statistical variability of the different reservoir images, performed on a significant number of realisations, will provide the geoscientist with a measure of the uncertainty related to the geological description (given the stochastic model used and its parameters).

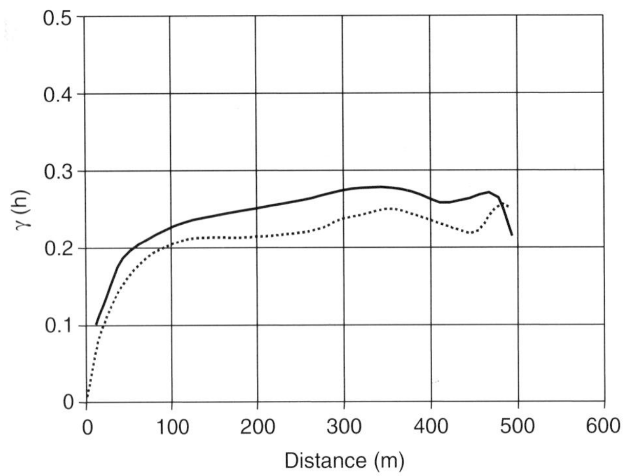

Figure 3.16 Experimental variograms for two facies.

In all cases, stochastic modelling represents an approach to reservoir description that is particularly suited for integrated reservoir studies, since it provides a means to integrate most of the information that is usually available to the geoscientist. The following points can be noted:

- **Geological knowledge**. The spatial distribution of the geological units within the reservoir is synthesised, in the stochastic model, by means of the distribution functions of the various facies (e.g., vertical proportion curves and variograms, Fig. 3.16). These functions define the average vertical and lateral extensions of individual facies, as well as their interrelationships, and they are built using the well conditioning (hard) data.

The depositional model of the reservoir provides an additional means to infer the facies correlation length (e.g., the variogram range) or the average units dimensions, when hard data are scarce or insufficient. Educated guessing, based on geological analogs or outcrop studies (soft data), can also be input to the building of such correlation functions.

- **Structural model**. While megascale structural features, e.g., major faults, are defined deterministically on seismic data, macroscale features like minor faults and fractures can be simulated through stochastic modelling. Characteristic parameters of these features like density and orientation are defined on core and log data and then extrapolated to the whole reservoir using various stochastic approaches [16]. Fig. 3.17 shows an example relative to a reservoir description obtained by combining megascale deterministic faults and small scale stochastic heterogeneities (stylolites).

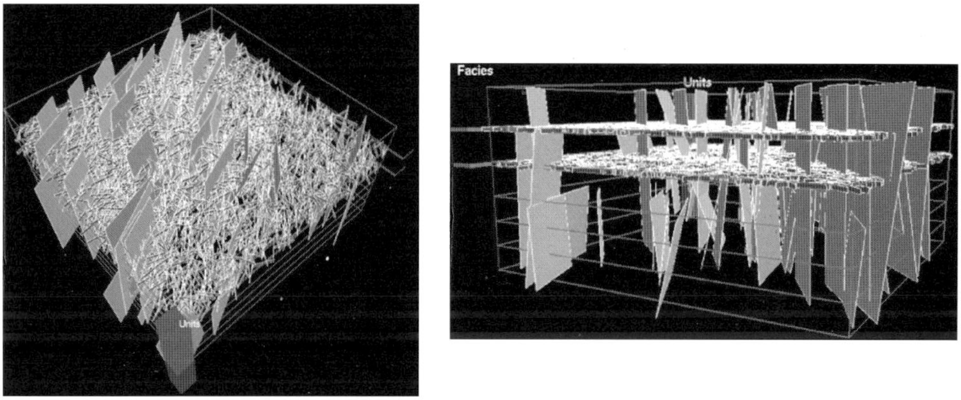

Figure 3.17 Megascale deterministic faults and small scale stochastic stylolites (Courtesy of Beicip-Franlab).

- **Petrophysical model**. The petrophysical model of the reservoir, defined in the log quantitative interpretation phase, can be extended to the whole reservoir by means of the stochastic modelling approach. This can be done by attributing average petrophysical values or probability distribution functions to each facies within the reservoir. Alternatively, when lithology is relatively homogeneous throughout the field, the facies distribution phase can be avoided and stochastic modelling can be directly applied to the petrophysical properties.
- **Seismic data**. With respect to any other reservoir data, geophysics offers the invaluable advantage of providing an insight over the lateral distribution of the geological bodies. For this reason, attempts to integrate this type of data in stochastic modelling have attracted the interest of researchers for many years. The integration of seismic data can be performed in a variety of ways, from the simple computation of the facies spatial distribution functions on the time, amplitude or impedance images, to the direct integration of these data in the actual simulation algorithm [17].

- **Dynamic data**. The integration of dynamic data (well tests and production) represents the ultimate challenge of stochastic techniques, since they provide large scale, flow-related information that are essential in the construction of a reliable reservoir model. Presently, most research concentrates on this issue. While a solution to this problem is not yet available off the shelf, a number of promising approaches are being proposed [18].

The possibility of integrating all the different available reservoir data makes the stochastic modelling approach a powerful tool for reservoir characterization. The continuous development of these techniques guarantees that in the future, even more than today, such methodologies will represent the reference approach to geologic reservoir modelling.

3.3.3.2 Pixel-Based vs. Object-Based Modelling

Currently, pixel-based (or continuous) and object-based (or boolean) algorithms represent the most widely used stochastic models for reservoir characterization.

In the pixel based model, the variable to be simulated is assumed to be the realisation of a continuous random function, whose distribution (often Gaussian) is characterised with fixed thresholds, which identify different facies or different petrophysical ranges. The most popular of these algorithms are probably the Truncated Gaussian Random Functions [19] and Indicator Kriging [20].

The method works best in the presence of facies associations that vary smoothly across the field, as it is often the case in deltaic or shallow marine reservoirs. No assumption is made about the shape of the sedimentary bodies. Often, this approach is preferred to the object-based one when the overall Net/Gross ratio is high.

Figure 3.18 shows an example, derived for a fluvio-deltaic reservoir, obtained through the Truncated Gaussian Random Functions algorithm. These kinds of models show a high degree of geological consistency, especially when a large number of conditioning wells are available and when reliable distribution functions can be established.

Figure 3.18 Truncated Gaussian simulation of a fluvio-deltaic reservoir (Courtesy of Beicip-Franlab).

The object-based algorithms generate spatial distributions of sedimentary bodies, which are obtained through the superposition of simplified geometries like sheets, discs or sinusoids, typically simulated within a shaly background facies. The parameters of these objects (orientation, sinuosity, length, width …) can be estimated on the basis of the assumed sedimentological model, seismic data, outcrop analogs or well test interpretations.

In some depositional environments, especially in a fluvial meandering setting, where sand channels are the main reservoir target, these models may provide very realistic images of the reservoir facies architecture. Generally, the method works best in the presence of low Net/Gross ratios.

Figure 3.19 shows a boolean simulation, again performed on a fluvio-deltaic reservoir. The sharper character of the simulated sedimentary bodies is evident, compared to the more noisy appearance of the pixel-based model.

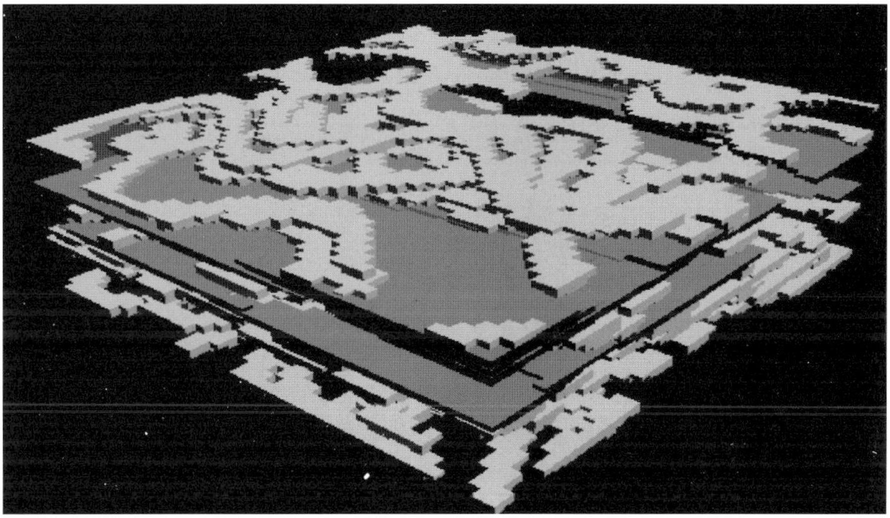

Figure 3.19 Boolean simulation of a fluvio deltaic reservoir (Courtesy of Beicip-Franlab).

Despite the debates that take place between the supporter of either methods, there is no *a priori* way to prefer one approach to the other. The choice is eventually in the hands of the geoscientist, who has the responsibility to decide which algorithm best fits his vision of the reservoir facies architecture. It is obvious, of course, that a degree of subjectivity is present in this process, which, as we will see in the next section, contribute to the overall uncertainty of the reservoir description.

3.3.3.3 Geological Uncertainty Assessment

Stochastic models, as previously mentioned, offer the possibility to quantify the uncertainty related to the geological description. Infinite possible realisations of the random function can be obtained just varying the generator seed and the comparison of a sufficiently large number of geological images will provide a measure of the uncertainty which is associated with the assumed geological model.

One of the most interesting applications of uncertainty quantification concerns the computation of the oil in place. The combination of several realisations of the various geological parameters provide a useful insight into the uncertainty existing in the oil in place figure. Fig. 3.20 shows an example of some cumulative distribution functions of the oil in place obtained using different geological realisations for a given reservoir unit. Note that the significant spread around the mean value is in this particular case related to the limited number of available wells.

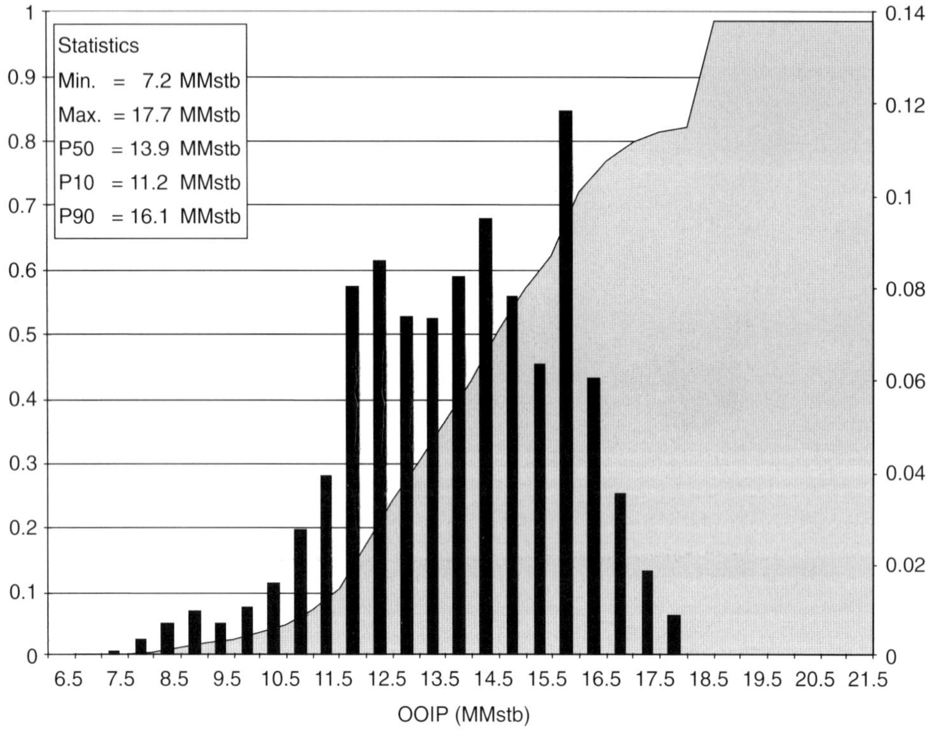

Figure 3.20 Uncertainty assessment of the OOIP value.

One important point should be highlighted, which refers to the actual **space of uncertainty** that is being sampled. This concept is often overlooked in such evaluations, and geoscientists tend to identify the uncertainty resulting from simple modifications of the random seed with the actual, or total uncertainty.

The global uncertainty space related to the geological model of the reservoir is actually much larger than that explored by the statistical variability of the random function. A recently published paper [21] demonstrates how the application of stochastic modelling may not capture the critical uncertainty of the geological model when the dynamic behaviour of the field is considered.

At least four major sources of uncertainty can be defined in a typical geological model:

1. **Uncertainty related to data quality and interpretation**. We know that basic reservoir data carry an inherent degree of error, but we hardly bother to define what the impact is on the final results. As far as uncertainty assessment is concerned, these data are assumed as error-free. The same holds for the interpretation stage.
2. **Uncertainty related to the structural and stratigraphic models**. While it is known that the structural interpretation of the field represent a major source of uncertainty (paragraph 3.1.3), this phase is virtually always carried out through a deterministic approach, which does not allow for any uncertainty estimation. As far as the stratigraphic framework is concerned, the uncertainty is basically related to the reliability of the deterministic correlative surfaces that are drawn through the wells, and its impact is therefore related to the number of wells and the particular depositional environment of the reservoir.
3. **Uncertainty related to the stochastic model and its parameters**. The same geological unit can be represented using different stochastic models, e.g., simulated annealing or indicator simulation. In general, each stochastic model will give different results and, as such, each model will explore a different part of the uncertainty space (even though the space sampled by different algorithms should significantly overlap). Note also that the choice of the model to use depends on the views of the geoscientist performing the study, since no specific rules exist to choose one model or another. Another major source of uncertainty is related to the choice of the parameters of the stochastic model, e.g., the type and range of the covariance function or the geometrical parameters of the sedimentary bodies in a object-based model. The little published information reveal that the uncertainty related to the these parameters may be quite important [22].
4. **Uncertainty related to equiprobable realisations**. This is the uncertainty that is usually explored in stochastic modelling studies and that is represented in Fig. 3.20. Overall, the uncertainty related to different realisations of a given random function is probably minor compared to the previously discussed factors.

A comprehensive evaluation of the uncertainty related to a given geological model should therefore take into account a considerable number of parameters. While this is probably not practical to do, it is important to understand, at least from a qualitative viewpoint, that what we usually refer to as uncertainty assessment is really just a limited view of the problem.

3.4 RESERVOIR HETEROGENEITY

In the previous chapters, we have discussed how to build a sound geological model of the reservoir, focussing on the integration of the available static information. To the geologist, such a comprehensive description, based on the definition of a structural, a stratigraphic and a lithological model, provides a sufficiently detailed characterization of the overall geological complexity of the reservoir.

However, such characterization may still prove to be unsatisfactory when the dynamic performance of the field is being considered. If the geological description does not capture the main features related to fluid flow we may not have properly taken into account one of the most relevant factors affecting the dynamic behaviour of the field: **reservoir heterogeneity**.

Reservoir heterogeneities are small to large scale geological features, that may or may not be significant from a strictly **static** reservoir characterization point of view, but do have a significant impact on fluid flow. Therefore, reservoir heterogeneity is not, or at least not only, a truly static issue.

An interesting and possibly surprising implication of such an observation is that the impact of reservoir heterogeneity is related to non-geological parameters, like mobility ratio, PVT properties, aquifer strength and development strategy. In other words, the same degree of geological reservoir heterogeneity may be important, for example, when the reservoir fluid is oil but may not be relevant in the case of a gas reservoir.

Likewise, for the same reservoir, heterogeneity description may be critical in some phases of the study, i.e., before the implementation of a secondary recovery project, but may prove to be unnecessary in the simulation of the primary depletion phase.

From this point of view, the relationship between reservoir heterogeneity and dynamic field parameters is one of the key issues of an integrated study, since it determines the degree of detail and accuracy to be attained in the geological description.

In the framework of a reservoir study, where the final objective is a dynamic-oriented reservoir characterization, the study of the types of reservoir heterogeneity and their impact on fluid flow is a mandatory work, which should always be addressed. Being a complex, interdisciplinary issue, the study and characterization of reservoir heterogeneity calls for the co-operation of the different professionals assigned to the study, from the geophysicist to the reservoir engineer.

This section is organised in 2 basic parts. In the first part, a classification of the types of reservoir heterogeneities will be provided, according to their geological scale and the impact on fluid flow. In the second part, some of the static and dynamic techniques that can be used to identify and characterise these features will be discussed.

3.4.1 Classification of Reservoir Heterogeneities

A reservoir is intrinsically heterogeneous. Differences in lithology, texture and sorting, as well as the presence of fractures, faults, baffles and diagenetic effects of different nature are the principal factors responsible for what we call, with a general term, reservoir heterogeneity.

The existence of these features affects the fluid flow at different scales, from the micro to the megascale. In particular, they have a considerable impact on the effectiveness of the displacement process and, consequently, on the value of the residual oil saturation and the final recovery factor. A correct evaluation of reservoir heterogeneity is therefore an essential issue in field development and exploitation, and must be explicitly taken into account in the construction of the reservoir simulation model.

A classification of the types of reservoir heterogeneities can be based on the scale, the genetic origin and the influence on fluid flow. Such a description has been provided by

Reservoir heterogeneity type	
Sealing fault Semi-sealing fault Non-sealing fault	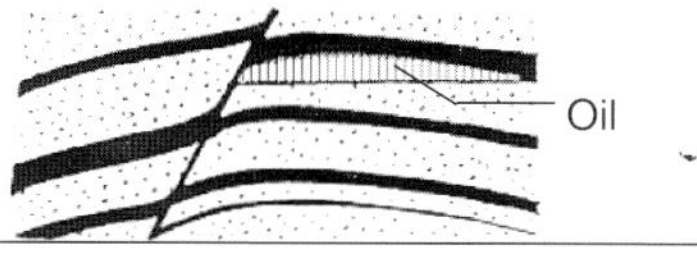
Boundaries genetic units	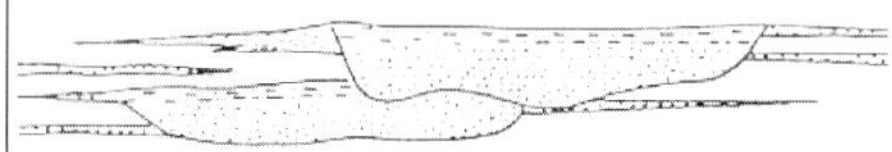
Permeability zonation within genetic units	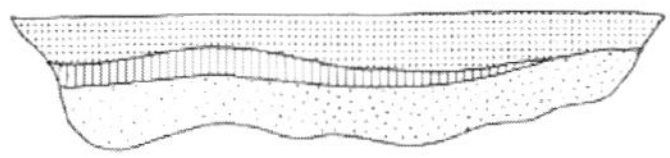
Baffles within genetic units	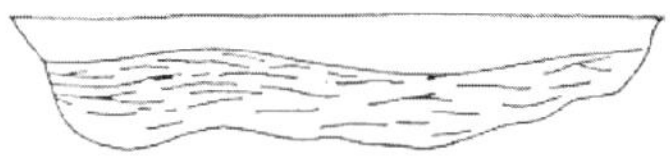
Lamination, cross-bedding	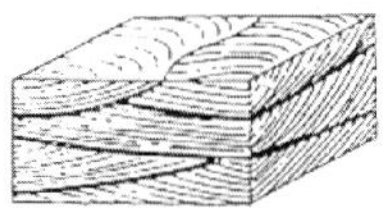
Microscopic heterogeneity, textural type, mineralogy	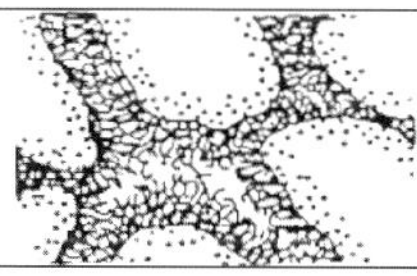
Fracturing tight, fracturing open	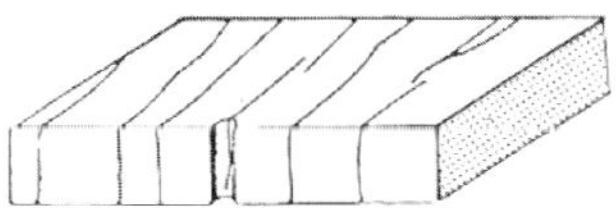

Figure 3.21 Classification of reservoir heterogeneity types [23].

Weber [23] and it is summarised in Fig. 3.21. Seven basic types of heterogeneities can be identified, referring to different scale of magnitude. They can be both of stratigraphic and structural origin.

In the next paragraphs, the main types of heterogeneities at small and large scale will be discussed, with particular emphasis on the implications for the reservoir model building.

3.4.1.1 Small Scale Heterogeneities

In actual depositional systems, small scale heterogeneities can virtually always be recognised on the available core material.

At the scale of the pore (**microscale**), heterogeneities are basically related to the occurrence of a mixture of pore types. This is clearly observed in carbonate systems, where different types of primary and secondary porosity are often associated.

At the scale of the cores (**macroscale**), heterogeneities are often related to lamination and cross-bedding. In fact, from a sedimentary point of view, the only depositional unit that can be considered intrinsically homogeneous is the lamina (see Table 3.1). Being the product of a single, geologically instantaneous depositional event, the lamina is internally free of significant heterogeneities. Laminae, however, are very thin (few millimetres to 1–2 centimetres in thickness). Laminasets or beds, which are higher order genetic units in sequence stratigraphy terms, may present a significant degree of heterogeneity.

In actual core samples, it is not uncommon to measure permeability differences of an order of magnitude or more between individual laminae [24]. Fig. 3.22 shows an example of small scale heterogeneities in a core, related to differences in texture of the sediment and to the presence of carbonate streaks. The differences in minipermeameter readings are also illustrated. Obviously, these small-scale features have a significant impact on microscopic flow efficiency, hence in the recovery factor, due to capillary trapping phenomena.

Inevitably, these observations raise a question: how do we deal with such heterogeneities and how do we represent them in the much larger numerical simulation gridblocks?

Several papers of Heriot-Watt University researchers have addressed the problem of small scale heterogeneity and its impact on fluid flow [25]. Ideally, small scale heterogeneities should be explicitly taken into account, and proper upscaling procedures should be applied to preserve at higher scale the impact of such heterogeneities on fluid flow. This phase, however, can be very time consuming, since it requires the use of numerical modelling to correctly describe the process and derive adequate pseudofunctions.

In practice, it is very seldom performed and the facies are characterised at the macroscale with average petrophysical values that are computed without much concern about small scale heterogeneities. The implicit assumption is therefore that the rock can be considered homogeneous at smaller scale.

While this assumption is in most cases the only practical approach, the presence and the impact of small scale heterogeneities should not be neglected *a priori*. In particular, parameters like permeability anisotropy (*Kh*/*Kv*) and residual oil saturation (*Sor*) should be investigated, since they may prove to be relevant in the global dynamic behaviour of the reservoir.

3.4.1.2 Large Scale Heterogeneities

Large scale (megascale) heterogeneities are the most important types of internal reservoir discontinuity. They can represent barriers to fluid flow and be responsible for what we can refer to as the **compartmentalisation** of the reservoir. Alternatively, they may represent preferential flow paths with respect to a homogeneous, lower permeability background rock. In either case, their impact in the reservoir dynamics may be so strong to dominate field performance, therefore their assessment is a mandatory task in all reservoir studies.

Referring to Fig. 3.21, the main types of large scale heterogeneities are faults, either sealing or not, boundaries of genetic units, high or low permeability streaks and shale baffles. Fractures, either open or sealed, represent another important type of reservoir heterogeneity.

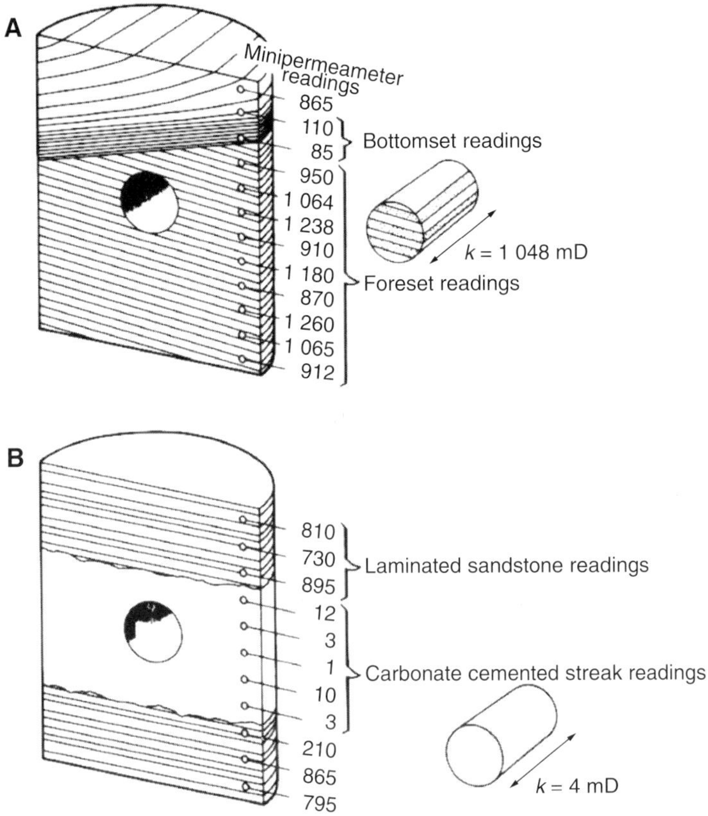

Figure 3.22 Small scale heterogeneities in a core sample [24].
A. Festoon cross – bedded sandstone sample. **B.** Barrier bar sandstone sample with carbonate cemented streak.

The main characteristics of these features and their impact on fluid flow will be briefly discussed in the following sections.

A. Faults

Faults are typical structural discontinuities. A fault can be sealing, partially sealing or conductive and therefore it may represent a barrier, an impediment or a conduit to fluid flow. Hence, when faults are concerned, the geoscientists is faced not only with the problem of identifying them, but also with the assessment of their seal potential.

From a theoretical point of view, four main mechanisms of fault seal have been recognised [26]:

1. **Juxtaposition**. Reservoir units are juxtaposed against low permeability units, i.e., shales.

2. **Clay smear**. Entrainment of clay into the fault plane, generating a low permeability surface (Fig. 3.23).
3. **Cataclasis**. Crushing of sand grains to produce a fine grained material surface, with a high capillary entry pressure.
4. **Diagenesis**. Preferential cementation that creates an hydraulic seal by completely or partially removing the original porosity.

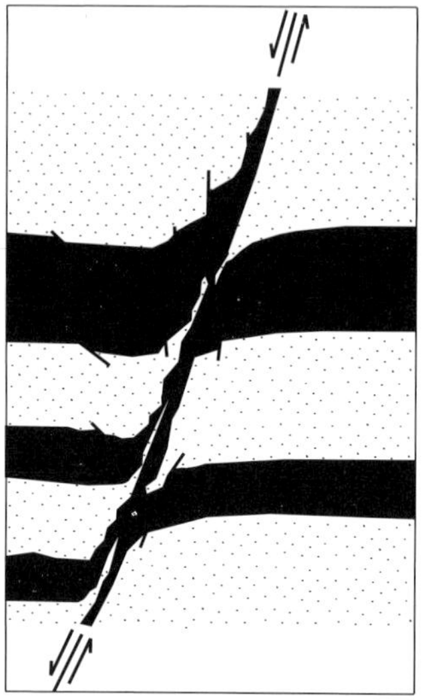

Figure 3.23 Clay smear mechanism.

Note that, with the exception of juxtaposition, all the other mechanisms generate a seal potential at a sand-sand interface.

When faults are recognised, several theoretical methods have been developed to assess their seal potential. Juxtaposition for example, can be studied by means of graphs that allow for the reconstruction of the fault throw and predict the occurrence of given lithologies along the fault plane [27].

The likelihood of clay smear to form a barrier that impedes fluid flow can also be evaluated, by means of a parameter called Clay Smear Potential [28]. An interesting application of these theoretical models in an offshore Luisiana Field has been recently described [29].

In general, however, these methods are difficult to apply, because in most cases the geometry of the fault plane is too complex and cannot be defined with sufficient precision. Furthermore, other occurring processes like cataclasis or diagenesis are spatially variable, frequently interrelated and difficult to quantify.

Therefore, from a practical viewpoint, the interest of such methods is greater in evaluating exploration prospects, when few information is available. Within the framework of an integrated reservoir study, the assessment of the fault seal potential is normally done by means of other methodologies, as it will be discussed later in this chapter.

B. Genetic Unit Boundaries

Genetic unit boundaries are virtually always encountered in real depositional systems. They represent stratigraphic discontinuities whose sealing potential is variable and generally related to a number of factors (contrast in lithology or petrophysical properties between the units, existence of an erosional surface or hard grounds ...).

Isolated bars or channels, sand or conglomeratic lenses within shaly units are typical examples of genetic unit boundaries that generate reservoir discontinuities in alluvial deposits with low-sinuosity channel fills, coastal distributary channels or in deep marine slump deposits. Deterministic log correlation in these environments can be difficult, so that stochastic models can sometimes be utilised to reconstruct the 3D distribution of these bodies. The degree of interconnection of these sand bodies determines the stratigraphic compartmentalisation of the reservoir.

Even in the absence of strong lithology contrast, however, in most cases genetic boundaries are boundary surfaces for fluid flow. As it was discussed in paragraph 3.2.1, this also represents one of the main reasons why sequence stratigraphy is used to define the main correlation surfaces of the reservoir geometric framework.

C. Shale Baffles and Permeability Streaks

In clastic reservoirs, the presence of shale or silty baffles is often observed within the main genetic units. These features can be detected on cores or well logs and their lateral extension depends on the sedimentary environment. In a marine setting, for example, the lateral continuity of these baffles can be considerable and span the typical well spacing. In other depositional environments, however, typically in continental settings, the continuity of these shale bodies can be very limited (Fig. 3.24). In the reservoir literature, these baffles have been frequently referred to as *stochastic shales*, since their occurrence within the reservoir cannot be defined with any certainty.

Carbonate streaks represent another frequently encountered discontinuity within genetic units. They can be found either in carbonate systems, usually at the interface of cycles of deposition, or in clastic reservoirs, where they are related to diagenetic processes. In either cases their extension is variable, even though it is not uncommon to observe fairly large lateral continuity.

Both shale baffles and carbonate streaks mainly impact the vertical communication within the reservoir. In the presence of such heterogeneities, especially when their lateral extension is large, vertical permeability within the genetic unit can be virtually reduced to zero. Therefore when their presence is not detected and explicitly modelled in the simulator this will result in an incorrect representation of the displacement process.

D. Natural Fractures

Natural fractures represent a particular type of reservoir heterogeneity. They can be found at all reservoir scales, from the small scale domain (microfractures, stylolites), up to the reservoir megascale (joints, regional fracture trends).

Even though natural fractures can be sealed and represent a barrier to fluid flow, in most fractured reservoirs they form an interconnected network which enhances reservoir perform-

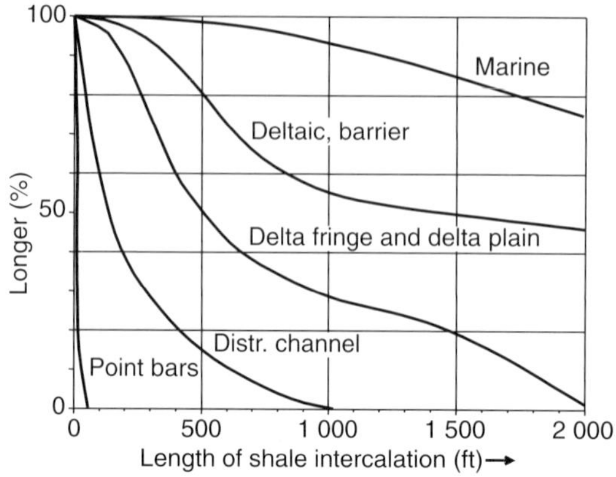

Figure 3.24 Continuity of shales in different depositional environments [23].

ance. In many cases, e.g., in low porosity carbonate reservoirs, the dynamic behaviour of the field may be completely related to the fracture system, while the matrix only plays a minor role.

In such fields, fractures can no longer be considered as reservoir heterogeneities, since the characterization of the fracture network, in terms of fracture distribution, density, aperture, porosity and permeability, actually becomes the main focus of the whole geological description.

In siliciclastic and porous carbonate reservoirs, on the other hand, fractures may occur in specific areas, e.g., in isolated swarms that locally increase reservoir properties. In such fractured areas, the hydraulic behaviour of the fracture system can be dominant compared to the background matrix and may cause uneven sweeping of the reservoir. In particular, when water or gas injection projects are active, the injected fluids may channel through the fracture system and lead to early breakthrough in the producing wells.

In these cases, natural fractures can be properly treated as heterogeneities and their impact in the overall reservoir behaviour must be explicitly quantified.

3.4.1.3 Heterogeneity Impact in Oil Recovery

As a conclusion of this section dedicated to the description and classification of reservoir heterogeneity, it is interesting to see what is the expected impact of the various reservoir heterogeneities in the global field performance.

Table 3.2 illustrates the relative importance of a number of heterogeneities with respect to some important dynamic properties of the reservoir. While inspecting this table, it is important to remember that the actual impact of these features is also related to other reservoir parameters, like fluid type and recovery process, which are not considered here.

3.4.2 Reservoir Heterogeneity Identification

This chapter deals with a number of techniques that can be used to identify the presence of reservoir heterogeneity and, more generally, to evaluate the internal geometry of the reservoir, in terms of continuity and communication of the formation genetic units.

Table 3.2 Impact of reservoir heterogeneity type for oil recovery [24, mod.].

Reservoir heterogeneity type	Reservoir continuity	Sweep efficiency		
		Horizontal	**Vertical**	**Microscopic**
Large scale				
Sealing fault	×	×		
Partially sealing fault	o	×	×	
Non-sealing fault	o	×	×	
Boundaries of genetic units	×	×	×	
Permeability zonations		o	×	o
Baffles and streaks		o	×	o
Open fractures		×	×	
Tight fractures		×	×	
Small scale				
Laminations and crossbedding		o	o	×
Mineralogy and texture			×	
Open microfractures		×	×	×
Tight microfractures		×	×	×

o = moderate effect × = strong effect

From a general viewpoint, several techniques can be utilised, involving both static and dynamic data. They can be grouped in four broad categories which are listed below, in the order they become available during a typical exploitation project:

1. **Geophysics**. 2D and 3D seismic surveys are the most important source of information as far as internal reservoir description is concerned. They are more relevant early in the field life, when information coming from other disciplines is scarce. Other and more sophisticated geophysical techniques, like Vertical Seismic Profiles and Crosswell Seismic can be collected later in the field life, with the specific objective of clarifying particular areas of the reservoir.
2. **Fluid data**. Differences in fluid contact depths and spatial variations in oil and formation water compositions can normally be detected early in the field life, through the analysis of the collected samples. These differences may be the expression of a reservoir compartmentalisation.
3. **Well testing**. Pressure transient tests are performed throughout the field life, with the objective of assessing well deliverability and damage. When good quality data are available, these tests can also provide useful information about the internal reservoir geometry. In some cases, unconventional tests like extended well testing are performed early in the field life, with the specific objective of evaluating reservoir continuity and reducing the risk related to unfavourable reservoir connectivity.

4. **Production data**. Well production performance is often the ultimate and most reliable source of information about reservoir compartmentalisation. Unfortunately, these data become available when the development phase is in its final stage or has already been finalised. For this reason, they are useful especially in later reservoir studies, e.g., when the implementation of a secondary recovery project is under consideration.

In the next sections, we will show how these methods can be practically used in evaluating the impact of reservoir heterogeneity and assessing the degree of internal compartmentalisation.

3.4.2.1 Geophysics

Geophysics is an invaluable source of information in the assessment of reservoir heterogeneity. Seismic derived images can actually be considered to be the most important tool to locate megascale reservoir heterogeneities, both of structural and stratigraphic nature. The technique is especially relevant during the first phases of a field development, when other types of information, e.g., production data, are not yet available.

The application of geophysics to reservoir characterization is relatively recent, since these techniques have been traditionally applied to the delineation of the external geometry of the field (see paragraph 3.1). However, since the beginning of the decade of 1990, the availability of less expensive and better quality data, as well as the development of innovative techniques, has generated a new interest about this subject. As early as 1992, the first monograph entirely dedicated to reservoir geophysics was published [30], contributing to the spreading of this new culture.

Nowadays, a number of geophysical techniques are routinely used in reservoir characterization studies, and there is little doubt that in the future, they will become more and more the key technologies for internal reservoir description.

The remaining of this chapter is divided into 3 main sections, each devoted to a different type of seismic acquisition (surface, borehole and crosswell surveys). We will see how they can be typically used in defining the internal geometry of the reservoir.

A. Surface Geophysics

Surface geophysics can be used to investigate the degree of internal heterogeneity of reservoirs, provided that good quality seismic data are available. In particular, as already commented in paragraph 3.1.2.1, seismic data can be used to identify and locate structural and stratigraphic features which could generate internal compartmentalisation, even though they do not provide information about the sealing capability of such heterogeneities.

The main issue with surface geophysics is resolution, since it ultimately defines what can be identified with seismic data. Seismic resolution is the ability to separate effects related to different subsurface objects, and it depends upon the velocity and the frequency content of the wavetrains: it is higher when velocity is low and frequency is high. Therefore, shallow reservoirs, normally characterised by lower rock velocity and higher frequency content, can be resolved much better than deeper reservoirs.

Reservoir heterogeneities can be identified by means of the traditional time interpretation (cross-sections and time slices), but analysis of the seismic attributes is also becoming more and more widespread.

The most commonly utilised attribute is possibly seismic amplitude, since reservoir discontinuities normally have a strong effect on the amplitude of a given reflector. Other attributes, like dip azimuth and dip magnitude, provide a pseudo 3D view of an interpreted surface, and allow for the detection of small scale features, that are hardly visible in traditional 2D time sections.

Another seismic attribute that has been recently introduced is the coherency cube, which identifies trace to trace discontinuities within the seismic cube. As in the case of amplitude analysis, the idea is that the coherency of the seismic signal is disrupted when faults or other discontinuities are present.

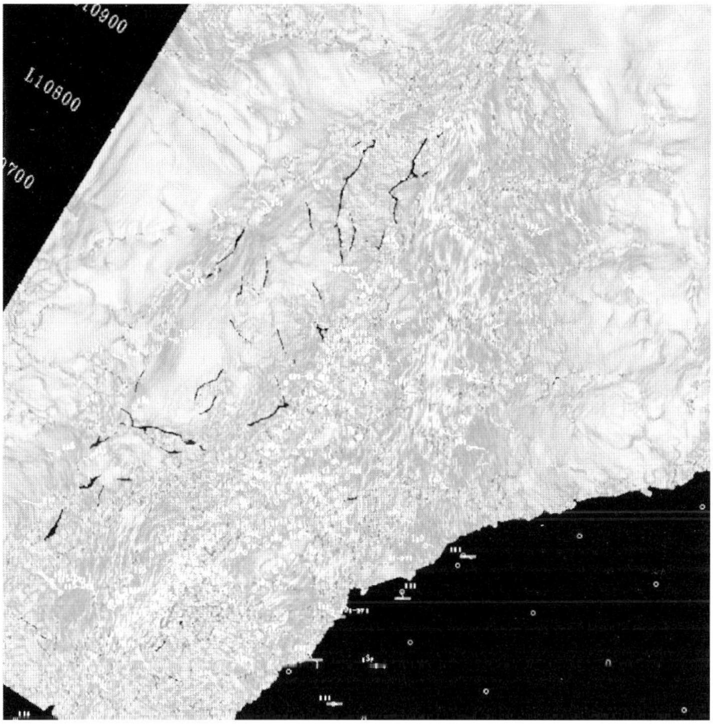

Figure 3.25 Edge map (Courtesy of PDVSA E&P).

Figure 3.25 shows an example of attribute map that can displayed and analysed in a workstation during the interpretation phase. The attribute shown in this example is the so-called *edge*, a parameter which reflect somehow the *rugosity* of the surface and that can be used to detect faults and fractures. Of course, one major issue in this kind of studies is the actual meaning of the interpreted patterns, since in many cases spurious signals can easily be interpreted as reservoir features.

Other recently introduced technologies allow for a deeper insight into reservoir internal geometry. Some of them are already available to the geoscientists, others are still in the development phase. Among the former are 3D-3C (3 components) seismic surveys, where the 3 spatial components of elastic waves are recorded, allowing for the availability of a volume of shear waves (P_S), in addition to the conventional compressional waves. Other processing techniques that are being increasingly applied are Amplitude vs. Offset (AVO), which in some cases allows for an improved description of the reservoir.

One example of the application of such methodologies is given in Ref. [31]. In this case, a 3D-3C multicomponent acquisition survey was interpreted and subject to AVO analysis. This analysis using both techniques allowed for the distinction to be made between clay-filled and sand-filled channels in a cretaceous incised valley system.

B. Borehole Geophysics

Until approximately 20 years ago, borehole seismic was a limited branch of geophysical acquisition. Velocity surveys, also called *checkshot surveys*, were the only measurements collected in the borehole environment, with the aim of tying the time-related surface seismic data to the depth-related well information.

During the eighties, it became apparent that when the full elastic wave field was collected, the traditional velocity survey could evolve to a structural imaging tool. Shooting a surface source at the well location and at some distance from it, while recording the wave-trains in the borehole, provided 2D images of the reservoir that could be compared and integrated to the surface seismic.

These types of acquisitions are called Offset Vertical Seismic Profiles (OVSP). An alternative geometry uses variable offset distances, shooting continuously away from the well. This is achieved in practice by the movement of the source boat along a pre-determined direction, and the resulting survey is hence called Walkaway Vertical Seismic Profile (WVSP).

Offset and Walkaway VSP's have evolved considerably in recent years, especially as far as the configuration of the borehole geophones is concerned. Widely used throughout the world, they provide the most valuable large scale information that can be collected in the borehole environment. A further evolution of these traditional configurations is represented by 3D VSP [32].

The main advantage of borehole seismic lies in the ability to record cleaner seismic information, since rays travel only once through the overburden formations. In addition to that, the recorded signal has a higher frequency content and hence better resolution than the surface seismic.

An interesting application of VSP to reservoir characterization is presented in Fig. 3.26 (from [33]). This is the imaging of the Pueblo Viejo Fault, one of the main structural lines in the Maracaibo Lake, Venezuela. The figure shows the results of the offset VSP, compared to the correspondent crossline of the surface seismic volume. The improvement in the definition of the image is noteworthy: while in the surface data at best a fault *zone* could be interpreted, in the VSP the fault plane is clearly imaged.

Such applications prove to be invaluable in the study and development of a field. For this reason, while still rather expensive in comparison with traditional formation evaluation techniques, borehole seismic is an important source of information for reservoir characterization. The evolution of current techniques towards more sophisticated source-receiver configurations will allow for more and more information to be available, while optimising execution time and cost.

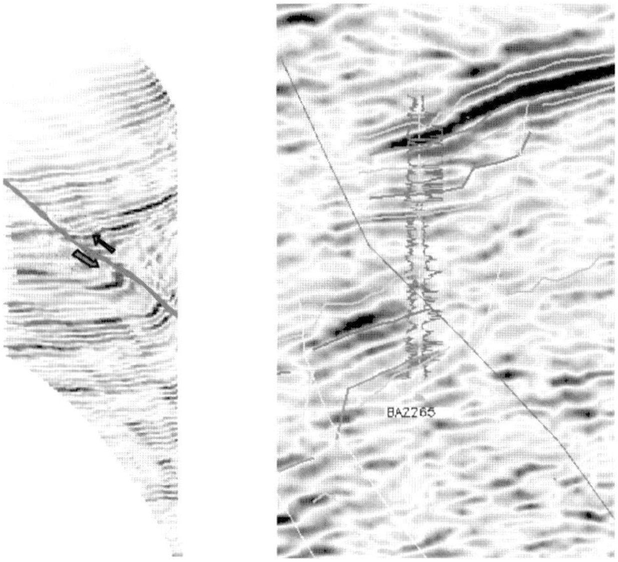

Figure 3.26 VSP (left) and surface seismic images of a fault plane [33].

C. Crosswell Geophysics

Crosswell geophysics consists of imaging the reservoir section between 2 or more wells by inducing a seismic wavetrain in one well (source well) and recording the arrivals in an offset well (receiver well).

The source is normally piezoelectric and operates at a frequency range of 200-2 000 Hz, that is considerably higher than any surface seismic. This allows for a much greater resolution, although over shorter distances. The full wavetrains can be recorded at the observation well, and the relevant processing allows for the possibility to produce a velocity image (also called *velocity tomogram*) as well as reflection images of P (compressional) and S (shear) waves.

The advantages of this type of acquisition are obvious, in terms of definition and resolution of the seismic images. However, there are still a number of drawbacks, starting from the cost of this kind of operations, which is still rather high. Also, the processing of the raw data, as well as the subsequent interpretation, is often not straightforward.

For these reasons, so far crosswell seismic has been applied mostly in the framework of shared research projects, where other independent reservoir data acquisition campaigns are normally performed. The results of one of these industry projects, which also included sequence stratigraphy, core analysis, facies classification and geostatistical modelling, are summarised in a recently published paper [34].

Despite the relative immaturity of crosswell seismic, considerable interest of the oil industry and service companies is clearly perceivable, which testifies to the growing potential of this technique. The key factor about crosswell seismic is that it potentially offers a way to solve one of the historical challenges of reservoir engineers, which is reservoir con-

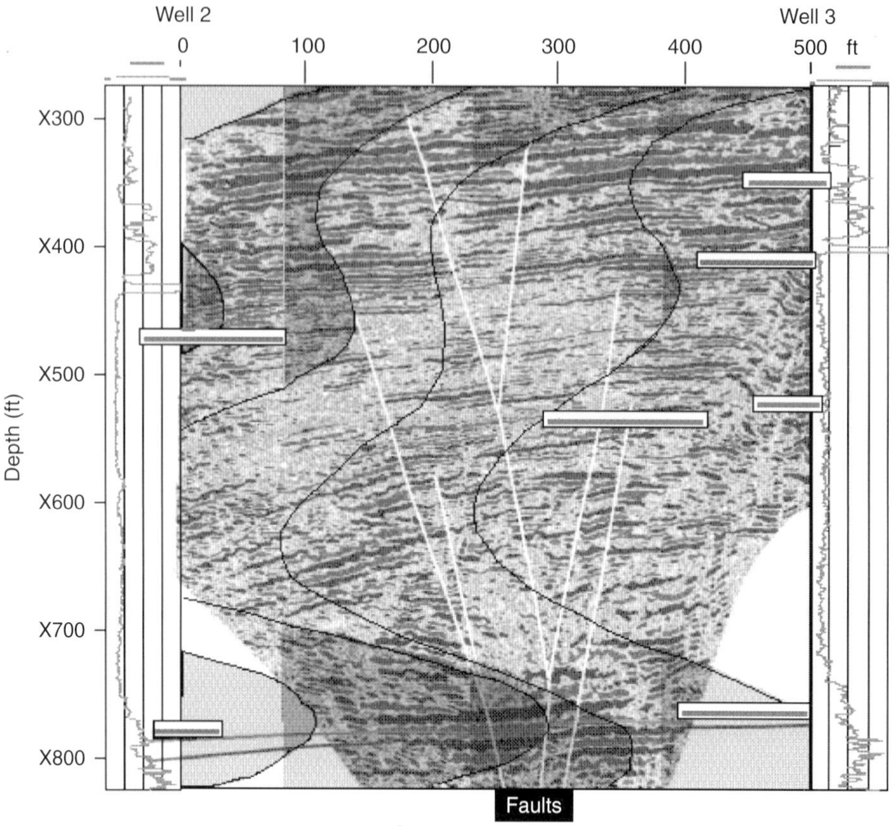

Figure 3.27 Crosswell seismic reflection section [35].

nectivity. As a matter of fact, when properly processed, the seismic information gathered in a crosswell seismic acquisition may give valuable information about intra-reservoir connectivity. Some vendors already offer special processing techniques, that allow for so-called *connectivity mapping*. The process makes use of the amplitude and frequency content of P-wave and S-wave travelling through the reservoir layers and the energy that is lost in the path. The idea in this case is that seismic connectivity relates somehow to fluid flow connectivity, even though this is not always obvious.

In the last few years, the technical literature about crosswell seismic and its practical application has been considerably increasing. An interesting example of crosswell seismic acquisition and interpretation has recently been published, about an offshore carbonate field in the Arabian Gulf [35]. Results are summarised in Fig. 3.27, which illustrates the seismic section recorded between 2 wells. In this configuration, the piezoelectric source was placed in well 3 while the receiver string was located in well 2. The interpretation of the crosswell survey is shown together with the original surface seismic data. It can be noted that the vertical resolution of the crosswell seismic is about 2 feet, at least one order of magnitude better than the surface survey and comparable to log resolution. In the figure thinning and pinchout of reflectors, as well as small offset faults can be observed.

3.4.2.2 Fluid Data

Fluid data, either hydrocarbons or formation waters, provide reliable and often overlooked information about reservoir heterogeneity. Many techniques are available to exploit such data and the integration of the results may provide the key to a better understanding of the reservoir compartmentalisation.

The point of interest concerning reservoir fluids, hydrocarbons and formation waters, is the spatial variation of some parameters, like composition or PVT properties. This variation, if it exists, may be the expression of the existence of fluid barriers within the reservoir.

Spatial variations in the reservoir fluid distributions are the product of processes that happen over a very long, geological time scale. In this respect, fluid data differentiate from any other reservoir information and can be considered as **quasi-dynamic data**, in contrast with truly static data, such as wireline log or seismic data, and truly dynamic data, like production parameters.

The focus of this section is to analyse how reservoir heterogeneity can be inferred from differences in the areal and vertical distribution of fluids. The common case of different gas/oil/water contacts will be reviewed first, then it will be shown how PVT properties, chemical compositions and more sophisticated geochemical techniques can be used, in different stages of a field life, to assess the existence of ancient and present barriers to fluid flow in the reservoir.

Reference [36] is a summary paper that offers a comprehensive description of how these techniques can be used and combined to improve, early in field life, the existing reservoir characterization.

A. Fluid Contacts

At the time of discovery, reservoir fluids are in hydraulic equilibrium, and they are vertically distributed according to their density at reservoir pressure and temperature. The interface between these fluids is horizontal and therefore, if the reservoir is hydraulically connected, all the wells will encounter these fluid contacts at the same depth. As a consequence, if different wells drilled in the same reservoir encounter fluid contacts at different depths, the reservoir is likely to be compartmentalised.

This simple rule has some notable exceptions, though, the most common being the existence of a tilted contact. The presence of an active hydrodynamism, for example, or lateral variations in the petrophysical properties of the reservoir rock, may generate true or apparent tilts in the oil-water interface, which are not necessarily related to reservoir heterogeneities.

On the other hand, it should be noted that the existence of a common fluid contact in all the wells drilled during the appraisal phase does not guarantee in itself reservoir continuity. In some cases, barriers to fluid flow may have been generated only after the hydrocarbon migration phase, as a consequence of diagenetic effects related to circulation of fluids in the reservoir. In this case, reservoir barriers are normally detected only after the beginning of the exploitation, observing for example different rises of the fluid contacts in different blocks, as a result of reservoir fluids withdrawal.

In the majority of cases, however, the general rule holds and differences in the contacts depth can be interpreted as evidences of a degree of reservoir compartmentalisation.

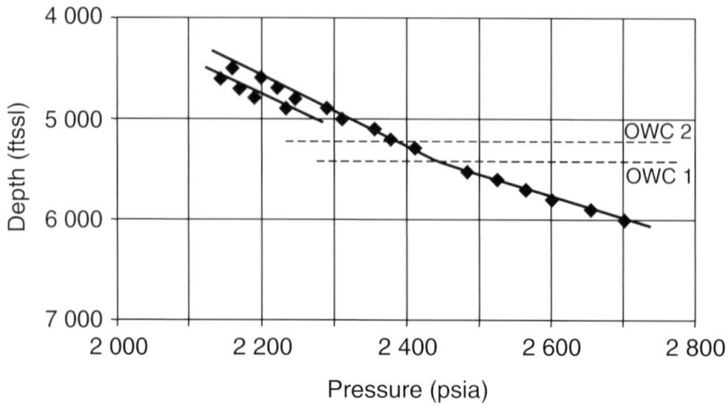

Figure 3.28 WFT measurements and position of the OWC.

Several types of data can be used to locate fluid contacts, from wireline logs, to routine core analysis, to pressure measurements. Without going into further detail on this basic issue, it should be appreciated that WFT (Wireline Formation Tester) pressure measurements are one of the most effective way to identify fluid contacts, at any stage of field life.

An example of such measurements is shown in Fig. 3.28, which shows pressure data collected in 2 different wells at the time of discovery. In this field, the presence of an intra-reservoir fault had been suggested on the basis of the seismic interpretation. One of the 2 wells encountered an Oil Water Contact (OWC1), which is clearly visible at 5 500 ft depth in the WFT graph. The second well did not reach the contact, however data show that it is hydraulically separated from the first well. Note that if the two wells are connected to a common aquifer, the position of the contact in the second well can be inferred from the extrapolation of the pressure data to the aquifer gradient (OWC2).

Another interesting point that can be observed in the graph is that the 2 wells have the same oil gradient, which can suggest that the 2 reservoir blocks are filled with the same oil. In addition to that, it can be noted that at any depth within the oil column, the 2 wells exhibit a constant pressure difference of about 60 psi. This means that, whatever the seal mechanism of the intra-reservoir fault, its sealing potential is higher than 60 psi. On the other hand, this does not necessarily guarantee that, under the viscous pressure differentials induced by the production (usually far greater than 60 psi), this fault will still behave as a seal. When uncertainty exists over the sealing potential of a fault, the only information that can wipe out any doubt comes from the production performance of the field.

B. Variation in Hydrocarbon Composition

Variations in hydrocarbon composition within a reservoir can be related to a number of mechanisms. Some of them act during the first phases of the petroleum generation, while others occur after the filling of the reservoir.

The most common process responsible for field-wide variations in hydrocarbon composition is possibly related to the maturation of the source rock, which tends to generate hydrocarbons with gradually changing composition, and to expel progressively lighter and more mature oil. These mature oils fill the part of the reservoir which is closest to the kitchen area, normally the lowest flanks, while the heavier oils generated first will accumulate towards the top of the structure.

Other mechanisms that can generate variations in hydrocarbon composition are biodegradation and leakage of gas. They both tend to remove the lighter fraction from the oils and generally occur in a non uniform way across the reservoir.

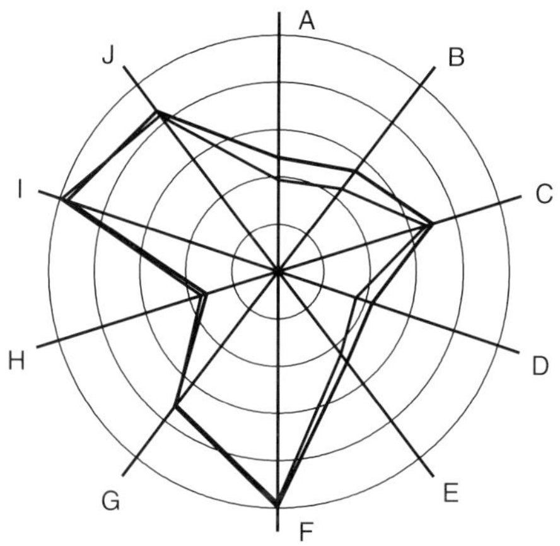

Figure 3.29 Starplot of chromatography peaks height ratios for oil fingerprinting.

Under normal conditions, these variations tend to homogenise and disappear with geologic time, as a consequence of mixing processes related to diffusion and convection of the fluids. Diffusion and convection are slow processes that act continuously from the moment of the filling of the reservoir. Analytical and numerical models have shown that convection is by far the most important mechanism to re-equilibrate the spatial inhomogeneities in fluid composition [37]. This mechanism, also called **gravitational segregation**, or density overturn, acts to re-establish the gravitational equilibrium of the reservoir, which at the end of the migration phase is filled with heavier hydrocarbon in its structural highest part.

In reservoirs with average permeability, filled with medium API oil, these mixing processes would equilibrate most density differences within a million years over a distance of 1-2 kilometres. Consequently, the presence of important differences in fluid densities within a reservoir is a strong indicator that mixing has not happened and that barriers to fluid flow are likely to exist.

The analysis of lateral variations of fluid composition can be achieved through a variety of methods. The most obvious is the direct comparison of chemical composition of oils sampled at different locations within the reservoir. With this aim, a number of geochemical techniques are available to highlight differences and similarities among the oil samples [38].

One of the most widely used of these techniques is gas chromatography, which provides what is often referred to as oil fingerprinting. The method is fast and relatively inexpensive compared to other reservoir characterization techniques. The composition of oils collected in different parts of the fields can be compared by means of starplots of the type shown in Fig. 3.29. In this figure, each axis represents a typical and significant ratio of chromatographic peaks and the shape of the resulting figure is the typical fingerprint of any oil (2 oils are compared in this example). The use of this kind of plot allows for the identification of possible compartments within the reservoir. An interesting application of the technique is given in Ref. [39].

When complete analyses are not available or are not sufficient, other parameters can be used, that also fingerprint oil composition. Among them, we can note common PVT data like bubble point pressure, volume factors, density and solubility ratio at reservoir conditions. The advantage of PVT data is that they are collected early in the field life, therefore they are available at a stage when strategic decisions have to be made about the development of the field. The drawback is that only few samples are normally collected, so that the information available over spatial variability is scarce. Furthermore, PVT samples have often problem of reliability, because, both in the cases of down-hole sampling and surface recombination, there is uncertainty over the representativeness of the obtained sample compared to the actual reservoir fluid (see paragraph 6.2.3).

An example of the use of PVT data in assessing reservoir compartmentalisation is illustrated in Fig. 3.30. The 2 clouds of points of saturation pressure represent samples collected in two different units of a field, vertically separated by an extensive layer of marine shale. These distributions show a degree of internal consistency, which is probably related to gravitational segregation of the oil, but they are markedly different from each other, thus confirming the lack of communication between the 2 reservoirs.

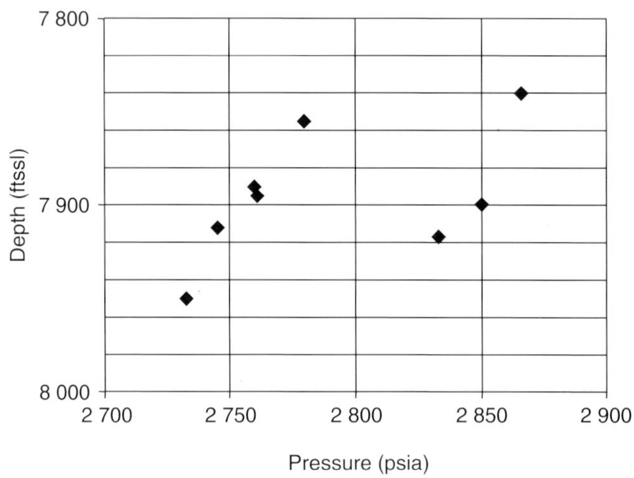

Figure 3.30 Saturation pressure vs. depth, compartmentalised reservoir.

Later in the field life, more information about produced hydrocarbons become available, which can also be used to confirm the existence of barriers to flow within the reservoir. Stock Tank API gravity is the most commonly used parameter, since it is routinely measured in all the producing wells.

Figure 3.31 shows an example of API gravity map, where two distinct regions are clearly visible, with consistently different average values of API gravity. This field has been on stream for over 40 years, and these data have been available for a long time. In fact, the existence of a NE-SW fault separating the two areas had been postulated long before the first 3D seismic survey, shot in recent years, finally confirmed its presence and exact location.

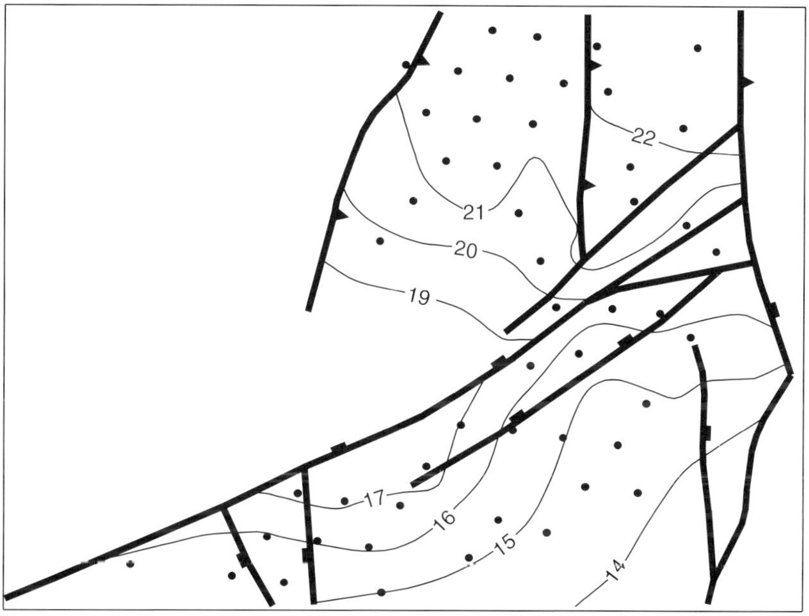

Figure 3.31 API gravity map.

C. *Variation in Formation Water Composition*

Most reservoir rocks record a complicated history of post-depositional processes, like diagenesis, cementation and precipitation of authigenic minerals, which are related to the interaction between the rock and circulating waters. As a consequence of these processes, the composition of formation waters is also constantly changing. As a matter of fact, gradual as well as abrupt changes in reservoir water composition have been frequently reported in the literature [40]. In some cases these spatial variations are so strong as to have a significant impact over the calculated OOIP.

As in the case of petroleum, diffusion and convection are the main mechanisms responsible for homogenising these variations in the composition of formation waters. When present, the existence of an active dynamism in the aquifer can effectively help the mixing process. Therefore, also in the case of formation waters, the evidence of variations in the composition may testify to the lack of hydraulic communication.

Early in the field life, the information on formation water composition is normally collected during DST or other types of well testing. Later, when production has started, more information can be gathered from separator samples for routine resistivity measurements and/or chemical analysis. Another source of data for formation waters comes from core samples, especially when cores have been recovered using low invasion techniques, which guarantee minimal drilling mud contamination.

One interesting technique that has been introduced recently is Residual Salt Analysis (RSA), whereby salts precipitated by formation waters in the pores are re-dissolved and analysed to determine the Strontium isotopic ratio, $^{87}Sr/^{86}Sr$ [41].

This parameter can be used to identify variations in formation water compositions, both in the water and in the oil leg. Fig. 3.32 illustrates an example of application of RSA analysis to a North Sea oilfield aquifer. A discontinuity is clearly visible in the $^{87}Sr/^{86}Sr$ curve in correspondence to a shale, that also relates to a pressure difference of 340 kPa. Since diffusivity would have homogenised this ratio difference in few thousand years, the interpretation is that the separating shale, despite having a thickness comparable to other shale layers, has greater lateral extent and represents a barrier to fluid flow.

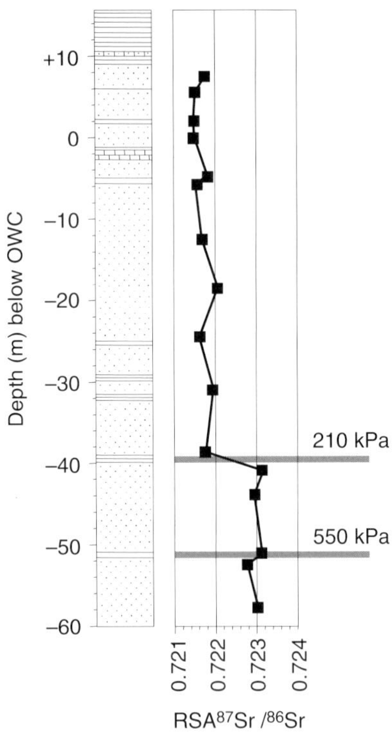

Figure 3.32 RSA analysis for a North Sea aquifer [37].

When the drilling fluids contamination is not an issue, this kind of analysis has several advantages over other geochemical techniques. It proves to be statistically robust, it can be applied to unpreserved core samples and, not least, it is not expensive.

3.4.2.3 Well Testing

Well testing has changed considerably in the last 20 years. Traditionally, wells were tested to determine produced fluids, borehole damage, deliverability and some basic reservoir parameters like pressures and permeability. In recent years, the advent of high accuracy electronic gauges, together with the availability of PC-based interactive type-curve matching software, has transformed well testing into a powerful discipline for reservoir characterization.

The evolution of pressure gauges is particularly important in this respect. Special tests like pulse and interference tests, which rely on the recognition of very small pressure disturbances, are now viable characterization techniques in most fields. Also, the installation of stable permanent down-hole gauges, already common in many mature areas, now allows for the acquisition of continuous pressure profiles, which in some cases can also be used to improve reservoir description in real time [42].

As far as integrated reservoir studies are concerned, the importance of well testing can be significant. When high quality pressure data are available, modern interpretative techniques allow the engineer to identify megascale reservoir heterogeneities and to infer the underlying geological model, thus providing invaluable input and/or feedback to other reservoir characterization techniques. One interesting application in this respect concerns stochastic modelling, since a well test interpretation may provide an estimation of some of the input parameters, e.g., channel width in a object-based model.

In the next sections, we will see how the analysis of traditional transient tests like drawdown and build-up can be used in identifying reservoir heterogeneities and assessing the internal reservoir geometry. We will also review some particular applications like Extended Well Testing, Multiwell Testing (Interference and Pulse Tests) and Tracer Tests. A general treatment of the use of well testing for reservoir management purposes can be found in Ref. [43].

A. Build-up and Drawdown Transient Tests

Transient tests are performed by introducing a change in the surface production rate of a well and recording the associated changes in bottomhole pressure. These pressure disturbances extend into the formation and are affected in various ways by reservoir features. For example, a pressure disturbance will have difficulty entering a tight reservoir zone but will pass unaltered through a high permeability zone and it may diminish or even disappear when entering a gas cap. Therefore, a record of wellbore pressure response produces a curve whose shape is dependent upon reservoir characteristics.

In many cases, these variations in pressure response are small and hardly recognisable or even interpreted as noise. This is particularly true for old well tests, where data were collected by means of poor resolution mechanical devices like the *amerada*. However, in recent years, the availability of high performance gauges allowed for the acquisition of stable, precise and accurate downhole pressure measurements. Furthermore, the introduction of the pressure

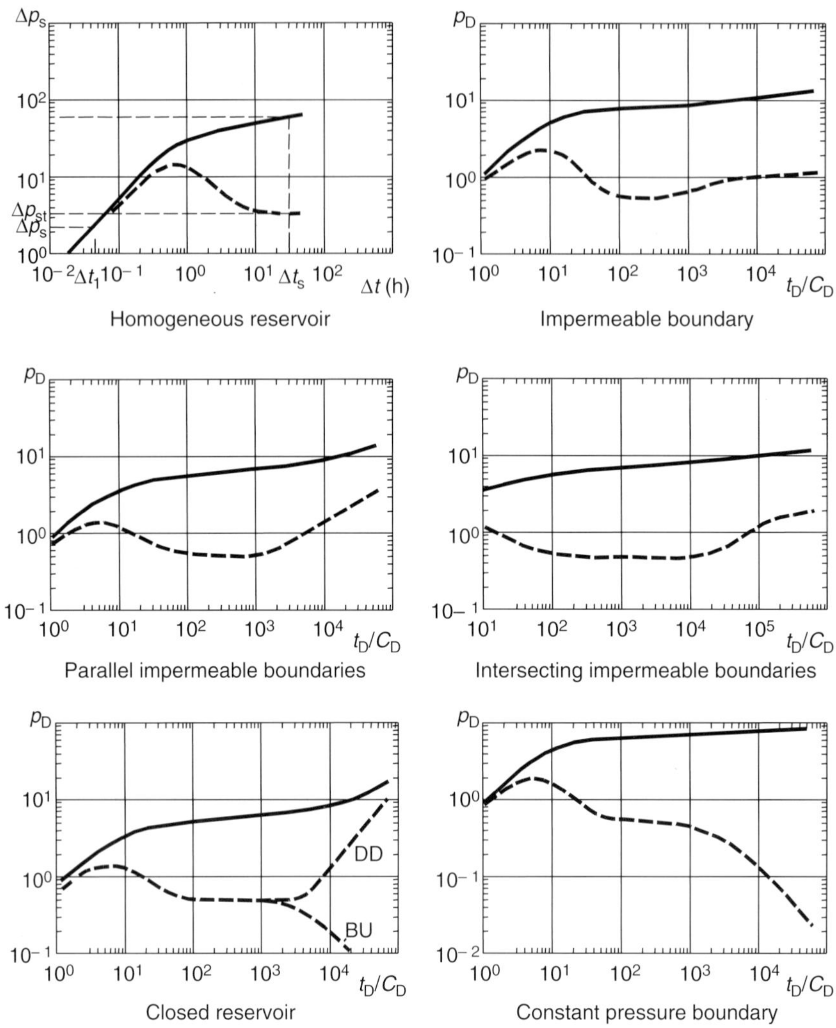

Figure 3.33 Reservoir models and pressure transient behaviour.

derivative curve in the log-log diagnostic tools, now permits the recognition of subtle variations in the pressure trend, that are related to particular well and reservoir configurations.

Currently, several libraries of type curves are available, which allow for the recognition of a large number of theoretical reservoir models. Fig. 3.33 provides the typical shapes of the pressure curves in a log-log diagnostic for a number of simple reservoir configurations. We refer the reader to reference textbooks for a more comprehensive treatment of the subject [44].

In all cases, it should be appreciated that the application of these models to real well test interpretation has to be done with caution. Sometimes the quality of the data does not allow for a clear model selection, while in other cases the inherent complexity of the reservoir generates a pressure distribution that is difficult to interpret. In addition, it should be noticed that the use of these analytical curves generally leads to a simplified description of the geological heterogeneities around the well.

In fact, transient testing is an inverse problem (i.e., find the model that satisfies a *given* solution), and as such it is subject to the problem of non-uniqueness. In other words, many possible input scenarios can provide the same, measured pressure response.

For these reasons, the interpretation has to be done with a clear understanding of the structural and stratigraphic setting of the reservoir. Actually, the existing software for well test interpretation are flexible tools, where the user has control over a number of tuning parameters and the danger is to fit the data perfectly to an unrealistic reservoir model.

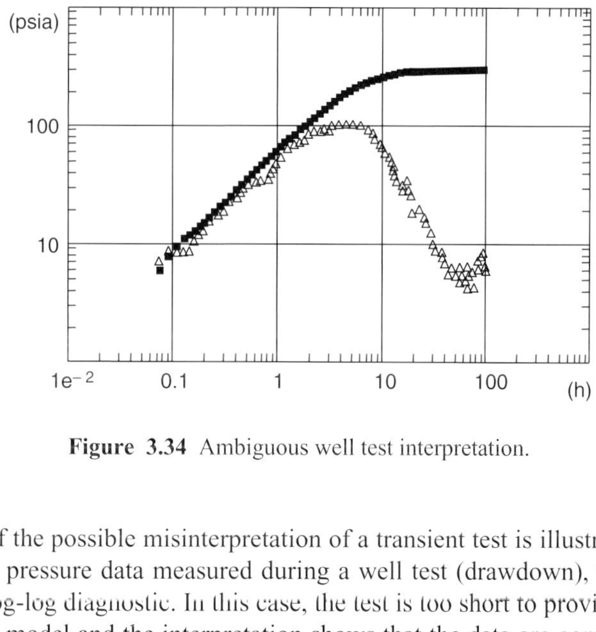

Figure 3.34 Ambiguous well test interpretation.

An example of the possible misinterpretation of a transient test is illustrated in Fig. 3.34, which shows the pressure data measured during a well test (drawdown), together with the derivative, in a log-log diagnostic. In this case, the test is too short to provide adequate indications about the model and the interpretation shows that the data are compatible with several possible models: sealing fault, dual porosity or closed reservoir.

The choice of the correct model cannot be made just on the basis of the best fit of the measured pressures. The knowledge of the reservoir structural and stratigraphic setting is necessary to indicate to the engineer the most likely model and to rule out unrealistic solutions. In this case a fault had already been interpreted by the geophysicist close to the well, while the dual porosity and closed models appeared to be unlikely, on the basis of the geological knowledge. Therefore, the first model was selected as the most probable for this reservoir.

In the framework of an integrated reservoir study, the geological input to well test analysis safeguards the coherency of the interpretation. A sound procedure for transient test analysis should then include the following steps:

1. **Data pre-processing**. Raw data should be quality checked and validated, especially in the case of old tests performed with mechanical gauges. The pressure effects not

related to the reservoir should be identified and possibly removed. This includes microseismic, geotidal, gauge drift, changing liquid levels and so on. These effects can be easily be misinterpreted as reservoir characteristics. An exhaustive treatment of data pre-processing procedures can be found in Ref. [45].

2. **Model recognition**. This is the most critical phase of the interpretation. It refers to the recognition of peculiar patterns in the log-log diagnostic, that are relevant to specific flow regimes. In most instances, the analysis of the derivative trend and stabilisation allows for the recognition of the underlying reservoir model. In other cases, this recognition can be difficult and/or ambiguous. In all cases, the model should be confirmed by consulting the geoscientists working with independent data. This phase, as shown in the previous example, permits to rule out some of the possible alternative models.
3. **Parameters estimation**. In this phase, the initial parameters relevant to the selected reservoir model are interactively refined to match the measured data, normally by means of non-linear regression techniques. Results of this phase (e.g., distance from a fault) should be cross checked for consistency with independent sources.
4. **Validation**. While several drawdown and build-up phases are often carried out during a well test, the analysis is normally performed on a part of these transients. The most straightforward validation of the interpretation is to carry out a forward simulation of the whole well test using the interpreted model and parameters and observe the match. Frequently, for example, one transient can be interpreted by means of an infinite acting reservoir, while later transients would show some boundary effects. In these cases, we should reconsider the assumed model and apply the necessary corrections. Another validation technique is to verify that other well tests, carried out in the same well but in different times, are consistent with the assumed reservoir model, since the latter does not change with time.

In conclusion, well test analysis may represent an important source of information about the internal reservoir architecture, when a sound interpretative procedure is applied. Mostly, we need to make sure that the interpretation is not performed just as a matching game, and that coherency is sought between transient analysis and the actual geological model of the reservoir.

An interesting application of well testing to reservoir characterization is described in Ref. [46], where more than 400 transient tests have been interpreted in the framework of a reservoir study. The example shows how geology, geochemistry and conventional pressure measurements distribution can be integrated in the analysis, to achieve a reliable description of the reservoir.

B. Extended Well Tests

The Extended Well Test (EWT) is a particular type of test, which is often carried out with the specific aim of assessing reservoir continuity. It has a duration of several days or even several months, while a number of pressure transients are normally recorded by means of bottom hole gauges.

The usual objectives of an EWT are to test the reservoir limits and to quantify the oil in place which is connected to the tested well. For this reason, EWT are often carried out in newly discovered fields, to gain a better understanding of the internal reservoir geometry and hence of the required development plan. This is particularly true in the case of deep sea developments, where maximum information need to be gained prior to project sanction, with a limited amount of exploration and appraisal wells.

Several interesting applications of extended well testing have recently been reported in the technical literature. One of them is illustrated in Fig. 3.35 (from [47]). In this case, it was decided to carry out an EWT with the objectives of clarifying the connectivity of the reservoir and determining the number and location of the producer and injector wells. The figure shows the bottom hole pressure recorded during the test as a function of time. The test was carried out in a horizontal well and it included 2 main flow periods and 3 build-ups, with a total duration of 57 days. This test allowed some critical issues in terms of reservoir characterization to be clarified and eventually opened the way to the development of the field.

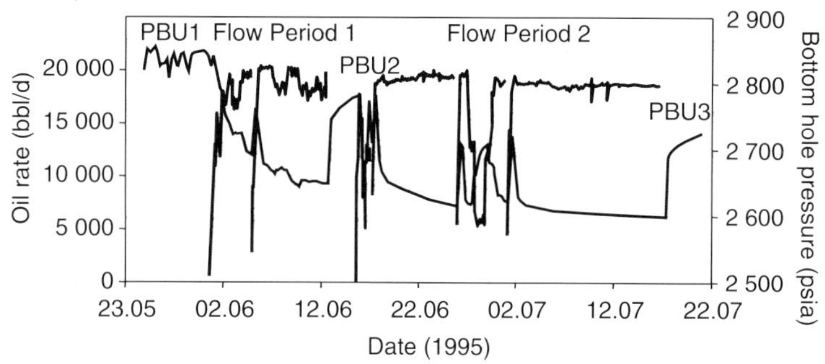

Figure 3.35 EWT pressure transients [47, mod.].

C. Multiwell Tests

In a traditional build-up, drawdown or fall-off test, pressures are measured in the same well where production or injection rates are varied. In multiwell tests, the flow rate is changed in one well (*the active well*), while the pressure is measured in one or more offset wells (*the observation wells*).

The usual objective of multiwell testing is to verify the hydraulic connectivity between wells and hence reservoir continuity. For this reason, they are frequently run prior to the implementation of secondary recovery projects like waterflooding. In addition to that, when monophasic fluid conditions exist in the tested area, they can be used to characterise the reservoir with average interwell properties.

There are some features that are characteristic to multiwell tests:

- **Weak pressure disturbance**. The pressure disturbance measured at the observation wells depends on reservoir properties and the distance from the active well, however it is always weak, often smaller then 1 psi. For this reason, the observation wells must be kept closed. This also means that accurate pressure gauges must be used and that the measured data must be filtered before the interpretation, to remove all the effects (production from other wells, drift, tides ...) that are not related to the change of flow rate in the active well.

- **Test duration**. Unlike single well tests, multiwell tests have a duration which can span several days or even several weeks. During this period it is important to accurately control the production from the surrounding wells, not directly involved in the test. Another important issue is the stability of the pressure gauges utilised, which must guarantee a minimum drift for long time periods. The duration of the test is also the main reason why these tests are not run as frequently as we would like, since there can be a significant loss of production related to the shut-in of the observation wells.

Several types of multiwell tests exist, the main ones being interference and pulse tests. A slightly different type of test that will also be briefly described is the tracer test.

- **Interference tests**. They are usually run by inducing a single, long duration production (or injection) rate change in the active well, while recording the pressure in the observation wells.
- **Pulse tests**. They are run by inducing repeated, short duration changes in the rates of the active well. These are normally a series of drawdown and shut-in periods, where the formers are also normally longer than the latters.

The choice of performing an interference or a pulse test depends upon the reservoir and fluid characteristics, as well as the objectives of the test itself. When dealing with reservoir with good mobility (k/μ) and diffusivity ($k/\phi\mu c_t$), like a permeable gas reservoir, the pressure disturbance is small but travels relatively fast through the rock, and a pulse test can be performed in a realistic time frame. On the contrary, in the case of a low permeable, heavy oil reservoir, the pressure disturbance induced by the active well will be greater, but will travel slowly through the reservoir. In this case, an interference test is preferable, since a prohibitively long pulse test would be necessary to record repeated transients.

Another point to take into consideration is the distance between active and observation wells and the sensitivity of the utilised gauges. In fact, the pressure variations in the case of the pulse test are less than in a interference test, often smaller than 0.1 psi, and could not be detectable when the distance between the wells is large or when the gauges are not extremely sensitive.

One example of application of multiwell testing to reservoir characterization can be found in Ref. [48], referring to the Fortescue field, Australia. In this case, the results of a pulse test have been used to clarify the internal geometry and the boundaries of the reservoir. To this aim, a pulse test was designed and executed, which included two 36-hours (18-hours flow and-18 hours shut-in) pulses.

Figure 3.36 shows the results of the test. The response to the pulses, filtered for the background, is very clear in the observation wells 2 and 3, and happened only a few minutes after the shut-in of well 1, which indicated a very high transmissibility of the system.

Multiwell testing is a powerful tool for internal reservoir description. As a matter of fact, even though the interpretation does not tell us much about the nature of the intra-reservoir connectivity, these tests can be used to support or reject alternative geological models derived by other techniques. For example, if the presence of a fault is suspected between two wells, the existence of pressure interference will prove that either the fault does not exist, or that it is not sealing.

When multiple tests are available, their interpretation may also allow for the building of transmissivity or diffusivity maps, that can be integrated in the reservoir description and even-

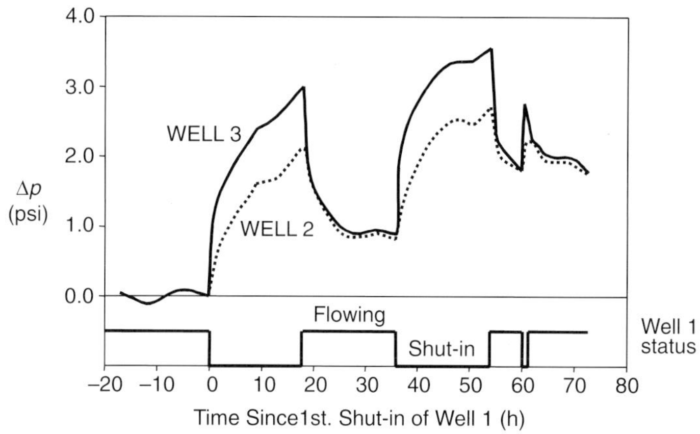

Figure 3.36 Pressure response in a pulse test [48, mod.].

tually input to the dynamic simulator. One such application is given in Ref. [49]. Another useful application reported in the literature concerns the definition of reservoir limits [50].

Tracer tests represents a different kind of multiwell testing, in that no pressure measurements are involved. In a typical tracer test, a non-reacting and non-adsorbing tracer, normally chemical or radioactive, is injected in the active well, while its breakthrough and concentration is monitored in one or more observation wells. Therefore, the results of a tracer test can provide useful information about intra-reservoir connectivity [51].

The usefulness of tracer tests is emphasised prior to EOR projects, where the analysis of the breakthrough of the tracer and the evolution of its concentration may provide useful information about the expected ultimate recovery. In this context, the relatively high expenses involved in such tests may find easier justification.

3.4.2.4 Production Data

Production data are often the definitive source of information about reservoir heterogeneity and internal compartmentalisation. Even when other methods have not shown evidence of intra-reservoir discontinuities, when exploitation starts anomalies are often observed in the production parameters, that show an unexpected, and often undesired, reservoir behaviour. At this stage of field exploitation, these anomalies are the most important source of information to the geoscientist about reservoir heterogeneity.

Production data refer to those parameters that are routinely measured during the exploitation of the field, i.e., oil rate, water cut (F_w), gas-oil ratio (GOR) and static pressures. In the next paragraph, we will comment on these parameters and we will show how they can be used to assess reservoir continuity.

A. Pressure Data

The most obvious evidence of reservoir compartmentalisation is possibly the observation of different pressure decline trends in different wells, or groups of wells. Fig. 3.37 shows an example, relative to some pressure measurements collected in an oil field during its producing life.

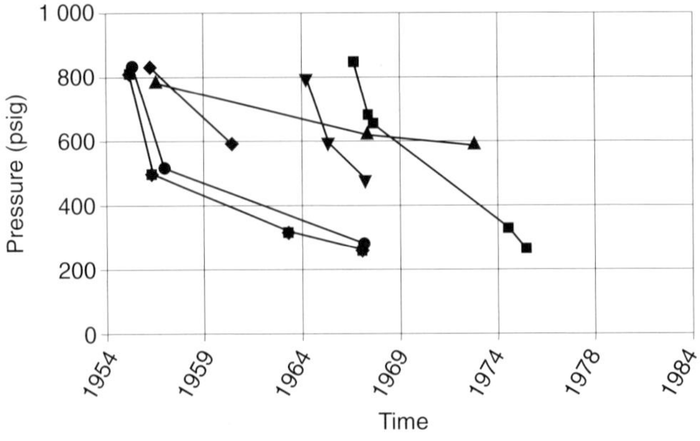

Figure 3.37 Pressure measurements in a faulted reservoir.

It can be noticed that most of the wells follow a different pressure decline trend with respect to the others. The geophysical interpretation of the area showed a complex system of intersecting faults, with 3 major visible trends, NW and NE and ENE (Fig. 3.38). All of the wells are located in different fault blocks, thus proving that the separating faults are sealing. The exception of the two wells that follow the same decline trend should also be noted, which indicates that in fact not all the faults are sealing.

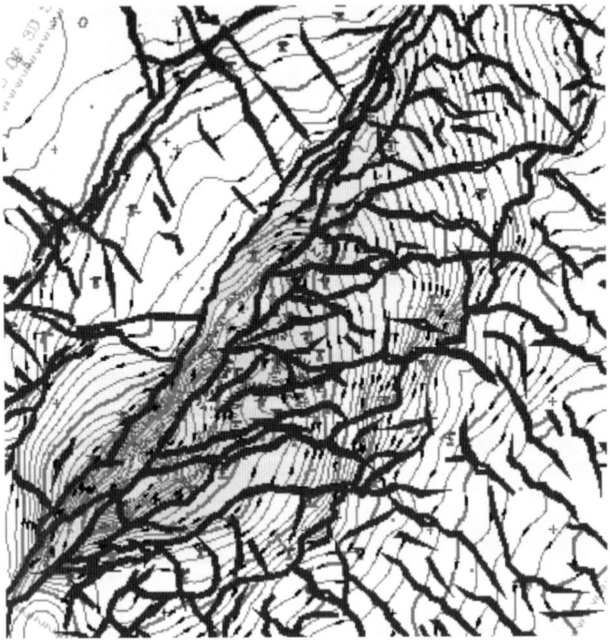

Figure 3.38 Structural map of a compartmentalised reservoir (Courtesy of PDVSA E&P).

One more feature that can be noted is that the pressure difference between the various fault blocks can be significant, which proves that the sealing mechanism is extremely strong. In other contexts, the sealing potential of the faults could be overcome by the pressure differentials imposed by the viscous forces related to production.

Another example of reservoir compartmentalisation is illustrated in Fig. 3.39, which shows the pressure data collected during the over 40 years of production history of an oilfield. Here, after a few years of rapid pressure decline, data start grouping into two distinct families, which follow different trends. This pressure behaviour was unexpected and the subsequent revision of the stratigraphy of the area, highlighted that a zone of thinning of the reservoir sand was present in the region between the two main areas of the field. It was concluded that, even though no wells had tapped a completely shaly section, a stratigraphic boundary had to exist in order to justify the observed pressure behaviour.

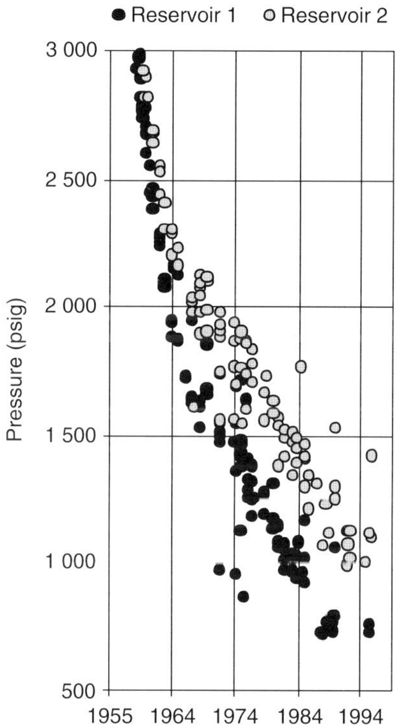

Figure 3.39 Pressure decline trends in the presence of a stratigraphic boundary.

B. Oil Rate, Water Cut and GOR

All the commonly recorded production parameters are related in some way to the prevailing average pressure and saturation conditions in the reservoir. Therefore, even in the absence of direct pressure measurements, parameters like oil and gas rates, water cut, gas or condensate oil ratio can be used to infer information about reservoir compartmentalisation.

The use of this kind of data is in some way obvious, however attention must always be paid to accurately distinguish the behaviour of the reservoir from anomalies related to well completion problems. The following typical field performance may flag the existence of reservoir discontinuities:

- **Hydrocarbon rates**. Declining rates, observed in one or more clustered wells may be the consequence of a declining average pressure level, related to an isolated compartment.
- **Gas-oil ratio**. Rising GOR's, not in line with the general field trend, may be due to the depletion of isolated portions of the reservoir, where the pressure has fallen below the bubble point. Likewise, early gas breakthrough in the producing wells may be related to the existence of high permeability paths.
- **Water cut.** Water breakthrough may happen early or, conversely, with significant delay in some parts of the field, with respect to the general advance of the water front (aquifer or injected water). As in the case of gas, these anomalies are frequently related to the existence of reservoir heterogeneities.

Finally, it should be noted that such anomalies in the field production performance allow the geoscientist to ascertain the existence and the impact of reservoir heterogeneity, but does not permit the identification and exact location of such discontinuities. From this point of view, the information coming from production data is complementary to the one derived from static data like geophysics, which provide the type and location of the heterogeneity but not its effectiveness with respect to fluid flow.

Therefore, production information should always be cross-checked and integrated with independent source information. This integration is the best approach to an effective reservoir characterization.

References

1 Abrahamsen P, Omre H, Lia O, Stochastic models for seismic depth conversion of geological horizons. SPE paper 23138.

2 Vincent G, Corre B, Thore P (1999) Managing structural uncertainty in a mature field for optimal well placement. SPE-REE, August.

3 Bu T, Damsleth E (1996) Errors and uncertainties in reservoir performance predictions. SPE-FE, Sept.

4 Dubrule O, Basire C, Bombarde S, Samson P, Segonds D, Wonham J, Reservoir geology using 3D modelling tools. SPE paper 38659.

5 Payton CE, Seismic stratigraphy application to hydrocarbon exploration. AAPG Memoir 26.

6 Van Wagoner et al. (1990) Siliciclastic sequence stratigraphy in well logs, cores and outcrops: concepts for high resolution correlation of time and facies. AAPG Methods in Exploration Series 7.

7 Vail PR, Mitchum RM, Thompson S, Seismic stratigraphy and global changes of sea level, part 3: relative changes of sea level from coastal onlap. *In:* Payton CE, Seismic stratigraphy application to hydrocarbon exploration. AAPG Memoir 26, 63-97.

8 Mulholland JW (1998) Sequence stratigraphy: basic elements, concepts and terminology. The Leading Edge, Jan.

9 Galloway WE (1998) Clastic depositional systems and sequences: application to reservoir prediction, delineation and characterization. The Leading Edge, Febr.

10 Eschard R, Lemouzy P, Bacchiana C, Désaubliaux G, Parpant J, Smart B (1998) Combining sequence stratigraphy, geostatistical simulations and production data for modeling a fluvial reservoir in the Chaunoy field (Triassic, France). AAPG Bulletin, April.

11 Martin AJ, Solomon ST, Hartmann DJ (1997) Characterization of petrophysical flow units in carbonate reservoir. AAPG Bulletin, May.

12 Davies DK, Vessel RK, Auman JB (1999) Improved prediction of reservoir behaviour through integration of quantitative geological and petrophysical data. SPE-REE, April.

13 Agarwal B, Allen RL, Farrel HE (1997) Ekofisk field reservoir characterization: mapping permeability through facies and fracture intensity. SPE-FE, Dec.

14 Haldorsen HH, Damsleth E (1990) Stochastic Modelling. JPT, April.

15 Dubrule O (1989) A review of stochastic models for petroleum reservoirs. Geostatistcs 2 (M. Armstrong Ed).

16 Sabathier JC, Bourbiaux BJ, Cacas MC, Sarda S, A new approach of fractured reservoirs. SPE paper 39825.

17 Doyen PM, den Boer LD, Pillet WR, Seismic porosity mapping in the Ekofisk field using a new form of collocated cokriging. SPE paper 36498.

18 Hu LY, Le Ravalec M, Blanc G, Roggero F, Noetinger B, Haas A, Corre B, Reducing uncertainties in production forecast by constraining geological modelling to dynamic data. SPE paper 56703.

19 Matheron G, Beucher H, de Fouquet C, Galli A, Guerillot D, Ravenne C, Conditional simulation of the geometry of fluvio-deltaic reservoirs. SPE paper 16753.

20 Journel AG, Alabert FG, Focusing on spatial connectivity of extreme-valued attributes: stochastic indicator models of reservoir heterogeneities. SPE paper 18324.

21 Fanchi JR, Meng HZ, Stoltz RP, Owen MW (1996) Nash reservoir management study with stochastic images – a case study. SPE-FE, Sept.

22 Massonnat GJ, Sampling space of uncertainty through stochastic modelling of geological facies. SPE paper 38746.

23 Weber KJ (1986) How heterogeneity affects oil recovery. *In:* Reservoir characterization (Lake LW, Carrol HB, Eds) Academic Press.

24 Weber KJ, van Geuns LC (1990) Framework for constructing clastic reservoir simulation models. JPT, Oct.

25 Pickup GE, Ringrose PS, Corbett PWM, Jensen JL, Sorbie KS, Geology, geometry and effective flow. SPE paper 28374.

26 Yelding G, Freeman B, Needham DT (1997) Quantitative fault seal prediction. AAPG Bulletin, June.

27 Knipe R (1997) Juxtaposition and seal diagrams to help analyse fault seals in hydrocarbon reservoirs. AAPG Bulletin, Febr.

28 Bouvier JD et al., Three dimensional seismic interpretation and fault sealing investigations, Nun River Field, Nigeria. AAPG Bulletin 73.

29 Alexander LL, Handschy JW (1998) Fluid flow in a faulted reservoir system: fault trap analysis for the Block 330 Field in Eugene Island, South Addition, Offshore Luisiana. AAPG Bulletin, March.

30 Sheriff RE Ed, Reservoir geophysics. SEG, Investigations in Geophysics 7.
31 Margrave GF (1998) Interpreting channel sands with 3D-3C seismic data. The Leading Edge, April.
32 Dodds K, Farmer P, Fryer A (1998) 3D vertical seismic profiles: a users' guide. JPT, Jan.
33 Lopez D, Munoz R, Gonzalez JE, Gou Y, Pascual JC, Cosentino L, Large scale integrated reservoir studies: The Bachaquero Intercampos experience. SPE paper 53996.
34 Tucker KE, Harris PM, Nolen-Hoeksema RC (1998) Geologic investigation of cross-well seismic response in a carbonate reservoir, McElroy Field, West Texas. AAPG Bulletin, August.
35 Sheline HE (1998) Crosswell seismic interpretation and reservoir characterization: an offshore case history. The Leading Edge, July.
36 Smalley PC, Hale NA (1996) Early identification of reservoir compartmentalisation by combining a range of conventional and novel data types. SPE-FE, Sept.
37 Smalley PC, England WA, Reservoir compartimentalization assessed with fluid compositional data. SPE paper 25005.
38 Larter SR, Aplin AC, Corbett PWM, Ementon N, Chen M, Taylor PN (1997) Reservoir geochemistry: a link between reservoir geology and engineering? SPE-RE, Febr.
39 Kaufman RL, Dashti H, Kabir CS, Pederson JM, Moon MS, Quttainah R, Al-Wael H, Characterizing the Great Burgan Field: use of geochemistry and oil fingerprinting. SPE paper 37803.
40 McCoy DD, Warner Jr. HR, Fisher TE (1997) Water salinity variations in the Ivishak and Sag River reservoirs at Prudhoe Bay. SPE-RE, Febr.
41 Smalley PC, Lonoy A, Raheim A (1992) Spatial $^{87}Sr/^{86}Sr$ variations in formation water and calcite from the Ekofisk Chalk Oilfield: implications for reservoir connectivity and fluid composition. Applied Geochemistry 7, 341.
42 Unneland T, Manin Y, Kuchuk F (1998) Permanent gauge pressure and rate measurements for reservoir description and well monitoring: field cases. SPE-REE, June.
43 Kamal MM, Freyder DG, Murray MA (1995) Use of transient testing in reservoir management. JPT, Nov.
44 Bourdarot G (1998) Well Testing: Interpretation Methods. Ed. Technip, Paris.
45 Mattar L (1996) Critical evaluation and processing of data before pressure transient analysis. SPE-FE June.
46 Kabir CS et al., Characterizing the greater Burgan Field: integration of well-test, geologic and other data. SPE paper 37749.
47 Richardson SM, Herbert N, Leach HM, How well connected is the Schiehallion reservoir? SPE paper 38560.
48 Braisted DM, Spengler RM, Youie RA (1993) Reservoir description through pulse testing in a mature field. SPE-FE, June.
49 Najurieta H et al. (1995) Transmissivity and diffusivity mapping from interference test data: a field example. SPE-FE, Sept.
50 Mahmoud ML, Torre AJ, Ayan C, Pulse test interpretation for Badri Field. SPE paper 25632.
51 Brigham WE, Abbaszadeh-Dehghani M (1987) Tracer testing for reservoir description. JPT, May.

CHAPTER 4

Rock Properties

Fluid flow takes place in an interconnected network of pore spaces. The characteristics of this network determine the volume of hydrocarbons contained in the reservoir, how it is distributed in relation to other reservoir fluids and how easily and in which proportion all these fluids flow towards the producing wells.

The characteristics and the properties of this porous network are related to the original grain size distribution of the reservoir rock, which in turn depends on the primary depositional processes that acted during the sedimentation. In addition, modifications of the original porous network often happen after the burial, which are related to secondary, or post-depositional processes like diagenesis, cementation, dissolution, fracturing and so on.

Understanding the close relationship existing between porous network, rock properties and fluid flow forms a cornerstone of the whole reservoir study. A correct description of the petrological and petrophysical properties of the reservoir rock is therefore a fundamental requisite in order to correctly represent the dynamic behaviour of the field in the numerical simulation study.

The chapter is organised in two main sections. The first section is dedicated to the petrophysical evaluation, i.e., to the estimation of the reservoir properties of interest at the well locations.

The discussion will concentrate on the basic reservoir parameters, i.e. those parameters that can be measured through routine core analysis, while multiphase flow rock properties (wettability, capillary pressure and relative permeability) will be discussed in the Reservoir Engineering part (Chapter 7).

The focus of this section is to review the commonly available sources of data and to highlight the advantages as well as the problems of consistency that may arise when more than one technique is used to determine the various reservoir parameters.

The second part of this chapter deals with the problem of describing the spatial variability of these rock properties. In this case too, a number of techniques are presented, which can be integrated in order to derive a consistent description at the scale of the reservoir. The issues related to 2D and 3D distributions will be reviewed, with particular emphasis on the application of geostatistical techniques as well as the integration of seismic data, with the specific aim of improving the interwell reservoir description.

4.1 PETROPHYSICAL EVALUATION

In this section, the main petrological and petrophysical features of reservoir rocks will be discussed. Several direct and indirect techniques will be presented, that allow for the determination of these properties at some fixed point in the reservoir space, i.e., the well locations. The overall objective in this case is to obtain, whenever possible, a vertical profile of each property for the greatest number of wells.

Generating vertical profiles of rock properties is the main focus of what is normally called petrophysical evaluation. Ideally, these profiles should be generated by direct measurements on core samples, since these data are possibly the most accurate. However, very few reservoirs worldwide have enough cored wells, the giant Hassi-Messaud field in Algeria being a notable example in this respect.

In fact, in the vast majority of cases, the only information available in every well is some suite of wireline logs and for this reason well logs typically form the backbone of any petrophysical model. All the other information should be used to calibrate and improve the quality of the interpretation of the well log profiles.

The definition of a consistent petrophysical model is not an easy task, especially when data are sparse and/or of poor quality. In all cases, correct integration of all the available sources allows for the best use of data. In the following sections, particular emphasis will be given to these integration issues.

4.1.1 Microscopic Rock Properties

Most of the data used in a typical reservoir study, as already noted in Chapter 1, belong to the macroscopic domain. Some investigation techniques, however, allow for the study of the reservoir rock at the microscale, i.e., the scale of individual pores.

The study of the pore system characteristics is often of primary importance in understanding the flow mechanics, the displacement process and ultimately the observed reservoir performance. Even though in the majority of cases the information gained at the scale of the pore cannot be used from a quantitative viewpoint and can hardly be extrapolated to the whole reservoir, a qualitative understanding of the microscopic rock and pore systems characteristics provide a sound basis for the subsequent characterization work.

Two microscopic rock properties will be discussed in this context: pore system organisation and mineralogic composition.

4.1.1.1 Pore System Characteristics

The pore system properties, in terms of pore types, geometry and interrelationships, govern the fluid distribution and interaction at the microscale. Their description and characterization are important to understand the fluid flow behaviour of reservoirs.

Until a few years ago, studies on the pore systems were performed only through thin section analysis and the application to operational studies was generally limited. In recent years, improvements in the investigation techniques and in particular the availability of powerful machines like the Scanning Electron Microscope (SEM) have allowed for a deeper insight into the structure of the pore system.

Observations based on thin sections and SEM permit the identification of several parameters which can be used to characterise the pore system:

- **Pore body size**. This parameter defines the average dimension of the pores, expressed in micrometers.
- **Pore body shape**. This qualitatively describes the prevalent shape of the pores, e.g., triangular, polyhedral, irregular ...
- **Pore throat size**. This is the controlling factor in terms of global transmissibility of the pore network. It is expressed in microns.
- **Aspect ratio**. The ratio of the pore body to pore throat size. It is an adimensional number.
- **Coordination number**. This parameter defines the average number of pore throats that intersect the pores.

Figure 4.1 shows an idealised pore surrounded by 4 rock fragments and illustrates the parameters defined above.

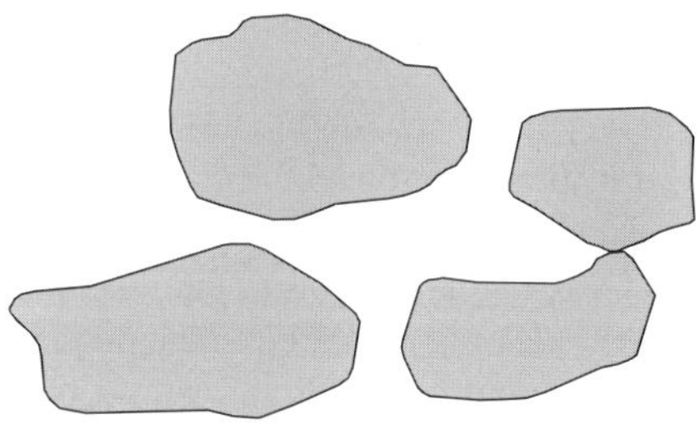

Figure 4.1 Pore system characteristics.

In particular studies, the availability of a sufficiently large number of analyses may allow for a classification of the reservoir rock as a function of the pore system characteristics. One example of such application has recently been presented [12].

While in the majority of studies the number of analyses will not permit such a detailed characterization process, the screening and the evaluation of the available data (when they exist) may often provide useful indications about the pore system geometry, which could then be useful in the definition of the petrophysical model of the reservoir.

4.1.1.2 Mineralogy

Among the thousands of minerals that have been identified in nature, only a small portion is actually commonly found in reservoir rocks, due to the selection operated by various sedimentary and post-depositional processes. However, the identification of the main mineral phases and the comprehension of their interrelationships is of great interest in the work of characterization of the reservoir rock.

From a genetic viewpoint, the main distinction that can be made while describing the mineralogical association of a rock is between allogenic and authigenic minerals. The former are the grains that constitute the main framework of the rock, as deposited by some sedimentary (primary) process. The latter have been generated within the rock itself after the deposition, as a consequence of diagenetic processes related to changes in the physical or chemical conditions.

The study of the typical mineralogical associations of a reservoir rock is very important in the framework of a reservoir study. At least two main points should be noted here:

- **Influence on well logs readings.** The mineralogical association of a reservoir rock has an impact on the interpretation of well logs, since the interpretation of most log curves should be performed with a prior knowledge of the lithology of the reservoir. Therefore, errors in the assumed lithology result in severe bias in the interpretation. Attention must also be paid to the presence of heavy minerals, like siderite and pyrite. The latter has not only an influence on the density, but also on the resistivity logs, since it is highly conductive. Its presence results in lower than expected resistivity reading and, when not taken into account, too high water saturation values. Finally, the presence and nature of clay minerals affects practically all of the wireline logging tools, and a prior knowledge of the types and distribution of clay minerals greatly help in the quantitative interpretation phase.
- **Influence on petrophysical characteristics**. The distribution of the mineralogic phases within the rock framework and their relationship with the pore space has a direct influence on the petrophysical properties of the rock. The typical example in this respect is the presence of dispersed authigenic clays within the reservoir rock. Even when the volumetric percentage of these clays is negligible, their impact on the producing characteristics of the rock can be significant. Fig. 4.2 shows typical distributions of dispersed clays within the pore system of a rock and their relative impact in the petrophysical properties [1]. Other examples are related to diagenetic process that affected one or more of the original mineralogic phases. At Prudhoe Bay, the dissolution of primary chert has generated a widespread microporosity system superimposed on the original intergranular system. This microporosity average 40% but, due to the small pore throat radii, is filled with hydrocarbons only at a significant height above the free level. This generates a dual-porosity system that, in turn, has an impact on the production performance of the reservoir [2]. Finally, in carbonate systems, even more than in siliciclastic reservoirs, the study of the diagenetic history that affected the reservoir rock often provides the key to understanding the production behaviour of a field.

The mineralogic information gathered in core samples or cuttings, should also be used to calibrate the wireline logs interpretation, in terms of volumetric analysis. Most of the interpretative packages nowadays allow for a multimineral computation and the resulting lithologic column may provide a reliable representation of the distribution of the main mineralogical phases along the wellbore profile. The correct interpretation of the volumetric composition of the rock is the best guarantee of a reliable petrophysical interpretation and may also be of great help to the geoscientists in the geological correlation phase.

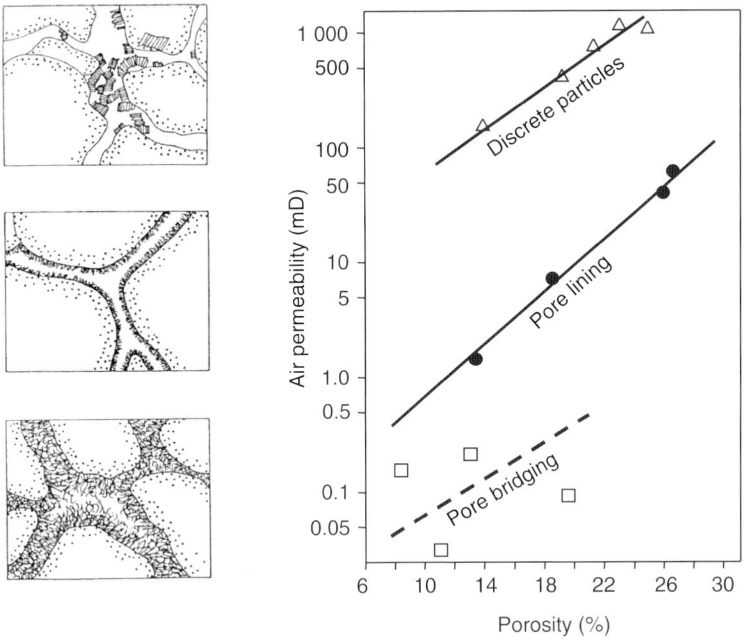

Figure 4.2 Authigenic clays and their impact on petrophysical properties [1].

4.1.1.3 Investigation Techniques

Many laboratory techniques are available to study the pore system characteristics and the type, abundance and location of the mineralogical phases of a reservoir rock.

- **Thin section petrography**. This still represents the most commonly applied technique. The analysis can be performed on samples from cuttings, conventional and sidewall cores and allows for a valuable insight into the mineralogy, as well as the pore system of the rock. Thin section petrography is the main technique used to study petrological features like fabric, texture and sorting and provides a means to deduce the diagenetic history of the sediments. The presence and the relative importance of different types of pore systems can also be evaluated, together with their interrelationships with authigenic minerals or cement phases.
- **Scanning Electron Microscope (SEM)**. As mentioned, this relatively recent technique allows for detailed investigations into the pore system and related mineralogy. It provides accurate information on the amount, size and morphology of the pores present in the reservoir rock. Together with thin section petrography, this technique allows for a comprehensive analysis of the depositional and diagenetic processes that affected the reservoir rock.
- **X-Ray Diffraction (XRD)**. This is the most widespread technique for the determination of bulk mineralogical composition of core samples. The technique is fast and accurate and also allows for a precise determination of the clay fraction.

Other recent laboratory techniques that have found useful application in such domain are Cross Tomography (CT) scanning and Nuclear Magnetic Resonance (NMR).

4.1.2 Grain Size and Sorting

Textural analysis, though often neglected, is an essential piece of information in the understanding of the pore space network. The dimension of the grains that constitute the reservoir rock and their size distribution are among the most influencing factors when reservoir parameters like storage capacity and transmissibility are concerned. In fact, even though the petrophysical characteristics of a rock are obviously dependent upon the size and distribution of *pore spaces*, it has been shown that they are also related to their solid counterpart, i.e., grain size and distribution.

Grain size and sorting are normally measured in the laboratory by means of sieve analyses, which provide the dimension and the distribution of the grains that make up the reservoir rock. Results are usually displayed in the form of cumulative distribution functions (CDF) of the type shown in Fig. 4.3. Note that the shape of the CDF allows for a quick estimation of the average value of the grain size distribution, as well as the evaluation of the spread, which in turn is a measure of the sorting of the sediment. Other measurement techniques includes laser-light scattering on disaggregated samples.

Recently, attempts have been made to derive textural parameters from borehole measurements, using the Nuclear Magnetic Resonance tool. Empirical relationships between the relaxation time T_2 and grain size have been derived, which in some cases reasonably honour well core measurements.

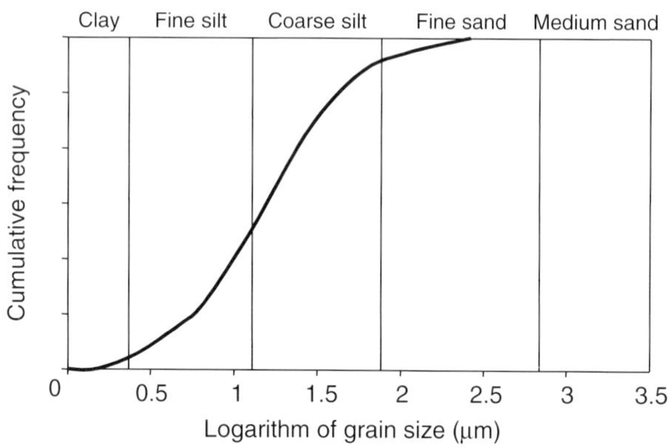

Figure 4.3 Typical sieve analysis graph.

The main interest in studying the textural parameters of the rock is that they can often be related to the petrophysical properties of the reservoir and, in some circumstances, can provide reliable quantitative assessment of some of the most interesting properties.

The relationship between petrophysical properties and texture of the rock has been the object of both theoretical and experimental investigations for many years.

The classical relationship that relates porosity, permeability and texture of a rock is the Kozeny-Carman equation [3]. This relationship assumes that flow through a porous medium can be represented by a laminar flow through a bundle of capillary tubes of different radii and it allows for the determination of the permeability k of a given sample, as a function of porosity ϕ and other textural parameters of the rock:

$$k = \phi^3 / K_0 \, \tau \, s^2 \tag{4.1}$$

where s is the pore surface area per unit pore volume, τ is the tortuosity of the sample, equal to $(L_a/L)^2$, the ratio between the actual length of the flow path and the length of the sample, and K_0 is a shape factor which varies between 2 and 3, with a typical value of 2.5.

Note that the Kozeny-Carman equation implies that permeability is directly proportional to porosity and inversely proportional to the surface to volume ratio of the rock. This makes sense intuitively, since fluid flow characteristics are dominated by large pores, where the ratio s is low.

An alternative form of Eq. 4.1 can be obtained if the parameter s is defined instead as pore surface area to grain volume:

$$k = \phi^3 / K_0 \, \tau \, s^2 \, (1 - \phi^2) \tag{4.2}$$

A further rearrangement of this equation, based on simple geometrical transformations, allows for the definition of an equivalent spherical diameter, d:

$$k = \phi^3 \, d^2 / K_0 \, \tau \, (1 - \phi^2) \, 6^2 \tag{4.3}$$

In this form, the Kozeny-Carman equation can be used to compute permeability by means of granulometric data, and has some merit in the cases of well sorted sediments.

It should be emphasised that the Kozeny-Carman equation is basically a pore-size model. Other theoretical models have been developed, which link the grain size and sorting to permeability. These models can be used to estimate permeability when porosity and grain size measurements are available. Among the most interesting are the models of Berg and Van Baaren [4, 5]. Generally, these models make use of porosity and grain size parameters, like the median and the spread of the distribution, as predictors to derive permeability. Note that the spread of the grain size distribution (the sorting parameter) accounts for the reduction in permeability which is to be expected when small grains plug the pore spaces, thus increasing the surface to volume ratio, as predicted from the Kozeny-Carman equation.

Among the experimental analyses, one of the most important works that relates texture and petrophysical parameters has been carried out by Beard and Weyl [6]. They sieved a number of sand samples and found a general increase in permeability and porosity as the degree of sorting progresses from very poor to well sorted. From a physical point of view, this can be explained with the plugging of progressively smaller pores as grain size decrease, reducing both porosity and permeability. Furthermore, they observed a regular increase in permeability as grain size increases from very fine to coarse, while porosity remained constant. This reflects the well known concept that, for well sorted samples, porosity is independent of grain size, while permeability is not.

As far as Carbonates are concerned, investigations conducted on non-vuggy rocks showed results similar to those seen above for siliciclastic rocks [7]. Data collected from selected, non-vuggy carbonates are plotted in Fig. 4.4, whereby the fundamental dependence

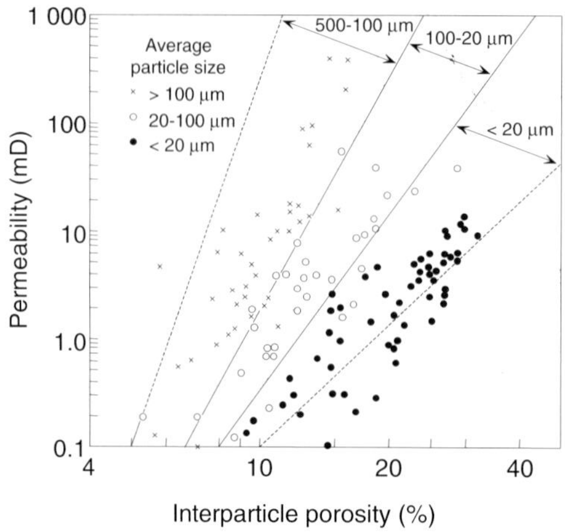

Figure 4.4 *K*-ϕ relationships from selected, non-vuggy carbonates [7].

of permeability on grain size dimension can easily be appreciated. The presence of vugs would increase total porosity, arguably without influencing permeability too much.

In conclusion, the texture of the sediments is one of most important factors in determining the petrophysical properties of the reservoir rock, provided that the pore space network is not altered by diagenetic effects.

Nevertheless, the application of such relationships in a reservoir study is not often straightforward, since grain size measurements are normally available on a limited number of cored wells. Therefore, a link with other, more readily available data is necessary.

Fig. 4.5 shows an example relative to an oilfield where sufficient grain size measurements were available to develop a facies classification scheme [8]. In the figure, facies are ordered with increasing grain size. As can be observed, there is a clear, direct relationship between grain size and horizontal permeability, the only exceptions being the shales and the coarser lithofacies [7, 8], where the representativeness of the permeability measurements is questionable. In this study, the relationship between grain size and lithology provided the critical link to extend the classification scheme to the whole field.

4.1.3 Porosity

Porosity is defined as the ratio of the pore space volume to the bulk volume of reservoir rock. It is an adimensional parameter and can be expressed in fraction or percent.

However, in spite of such a simple definition, porosity can be a difficult parameter to quantify, since the porous volume of a reservoir rock is often a complex network of spaces of different shape, dimension and origin. As a consequence of this complexity, several classification systems can be considered.

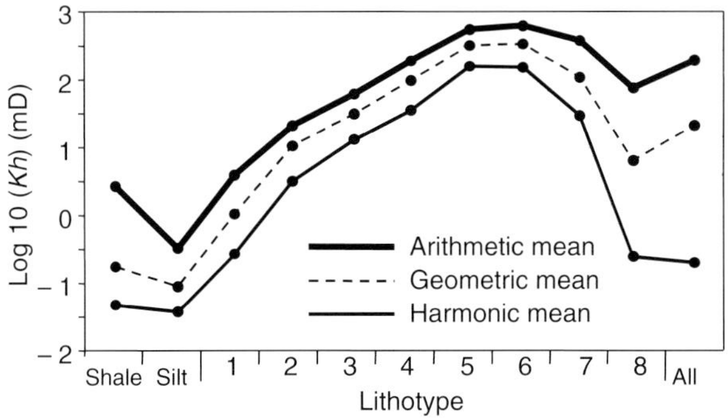

Figure 4.5 Core-plug average horizontal permeability by grain size class [8, mod.].

A general and simple classification of the pore system can be based on the genetic process responsible for the formation of porosity. From this point of view, we can distinguish 2 fundamental types of porosity, primary and secondary.

Primary porosity is the original porosity preserved in the sediments after deposition and initial compaction. It is strongly dependent on the textural characteristics of the sediments (grain size, shape and sorting) and tends to decrease with time and depth of burial. Sometimes it is preserved by early migrations of hydrocarbons.

Secondary porosity is related to tectonic stresses that affected the sediments after the burial and/or to the circulation of underground waters. The former processes tend to generate fractures, stylolytes and joints, while the latter are responsible for dissolution, deposition, recrystallisation, leaching and dolomitisation processes that may affect the reservoir rock after the deposition. Secondary porosity is normally more important in carbonate rocks than in siliciclastic sediments, due to the fragility of these minerals and their relatively high solubility.

Another popular and simple way to classify the porosity of reservoir rocks makes use of the image of connected and isolated pores, which leads to the distinction between total and interconnected, or effective, porosity.

The situation is shown in Fig. 4.6, which illustrates the idealised pore system of a clean sandstone that underwent a simple diagenetic history. Reservoir rocks often show some degree of isolated porosity, due to the presence of cementing materials that seal off some portions of the pore space. The main issue concerning these isolated pores is that they do not contribute to fluid flow.

Where carbonate rocks are concerned, the description of the pore system can be even more complicated. In the majority of cases, several types of porosity coexist and can be identified in the same carbonate rock. The most useful classification scheme for reservoir description purposes is probably the one proposed by Choquette and Pray [10]. This classification distinguishes 7 types of porosity, on the basis of their origin and dimension: interparticle, intraparticle, intercrystal, moldic, fenestral, fractures and vugs.

The description and the quantification of porosity is a very important stage in the reservoir characterization process. In the next sections, the main characteristics and the limitations of the available techniques will be reviewed.

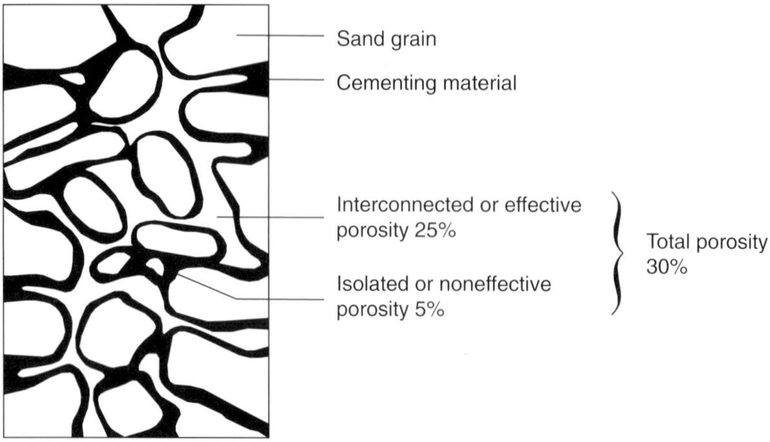

Figure 4.6 Effective, isolated and total porosity of a clean sandstone [9].

4.1.3.1 Core Porosity

Porosity is routinely measured in the laboratory using small plugs of reservoir rock, sampled in a more or less systematic way from a core.

Historically, several methods have been used to measure porosity, based on the determination of two of the three basic volumetric parameters: bulk volume, grain volume and pore volume. Ref. [11] provides an exhaustive treatment of these methods.

Currently, most of the techniques that are used in the laboratories are based on the measurement of pore volume, which yields an estimation of effective, or interconnected, porosity. These techniques are based on either the extraction of a fluid from the rock sample or the introduction of a fluid in the pore volume of the rock. Other types of analyses, based on computing the volume of a sample crushed to grain size, can provide measurements of total porosity.

Extraction techniques make use of devices based on Boyle's law. The most commonly used fluids are gases that have negligible adsorption on rock surfaces, like nitrogen or helium. The samples must be cleaned and dried before the measurements but can normally be utilised for further analyses. These techniques have been in use for over 40 years now, and they are still probably the most widespread. They usually provide very reliable results and can be applied in measurement at reservoir conditions.

The methods based on the introduction of a fluid into the pore space, also called saturation methods, consist in saturating a clean and dried sample with a fluid of known density and determining the pore volume from the measured change in weight of the sample. For its simplicity and accuracy, this is still a frequently utilised method. Limitations arise when complete resaturation cannot be achieved and in the case of very high permeability samples, when the saturated weight determination may be difficult to perform. Furthermore, typically only ambient condition measurements are made.

Accuracy and Representativeness of Core Porosity

Core data is usually taken as the reference measurement for porosity. While this is undoubtedly correct in the majority of cases, a number of factors need to be taken into account, which can severely affect laboratory measurements. In particular, secondary porosity may require special attention, since the processes responsible for its development may act at a much larger scale with respect to the core plug dimension.

The following points will detail some of the critical issues that should be considered when evaluating the representativeness and the accuracy of the available core data.

- **Accuracy**. Measurements performed in different laboratories and/or using different techniques may give different results. In a comparative study performed by 5 major laboratories, porosity has been measured in a number of samples of different lithology and petrology [12]. As expected, results showed a spread in the measured porosity, which was primarily related to the measurement technique used. The average difference between the various methods was 0.8 p.u., but in some individual samples the difference was as high as 2 p.u., which may represent a non negligible error in porosity estimation. The results of more recent investigations on the precision of laboratory measurements are described in [13].
- **Support volume**. Laboratory techniques usually provide reliable porosity determinations for siliciclastic rocks, where primary, intergranular porosity is by far the predominant pore system. In this case, even though the measurements are performed in a very small volume of the reservoir rock, the results represent the average of thousands of pores and have therefore statistical significance. This holds also in the case of carbonates, when the predominant porosity is primary (intercrystalline or oolitic). However, when secondary porosity is present, care must to be taken in evaluating the laboratory results. Secondary porosity is often related to processes that are spatially variable to a much larger scale compared to the core plug dimension and in such cases the results may not be considered representative. Vugs and solution cavities, for example, are normally much bigger than the original intercrystalline pore spaces, and may be irregularly distributed within the reservoir (Fig. 4. 7). In some cases, if measurements are performed on full-size cores, results are more representative since the support volume is bigger. This is true for vuggy carbonates but also for clastic rocks like conglomerates, where the constituting clasts may sometimes be bigger than the typical size of a core plug. The Brae formation in the Jurassic North Sea fields is one example of such reservoir conglomerates.
- **Fractures**. Another critical point concerning core measurements is the presence of fractures. To date, there is no satisfactory method that provides a reliable estimate of the fraction of secondary porosity that is related to fractures. The problem in this case is two-fold. On one hand, even more than in the case of dissolution, fracturing is a process that has to be characterised at a much larger scale than the core (plug or full-diameter) sample. On the other hand, recovering and handling the core before the arrival to the laboratory inevitably alters the fracture volume, especially open fractures (the most important). For these reasons, fracture porosity is normally estimated by means of correlations and graphs based on Poiseuille's law [14].

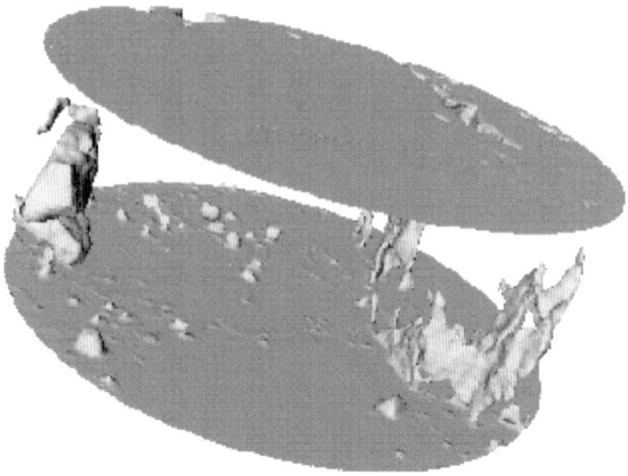

Figure 4.7 3D image of porosity distribution in a vuggy carbonate (Courtesy of Institut Français du Pétrole).

- **Measurement conditions**. In most cases, porosity measurements are performed at laboratory (ambient) conditions. Therefore, a correction needs to be applied to take into account the increase in porosity due to the decompression that occurred when the core was taken at the surface. In the case of consolidated rocks, this porosity correction may or may not be important, depending on the rigidity of the grain framework and the pressure differential. However, when dealing with unconsolidated or partially consolidated formations, this correction is always significant and has to be applied. Compaction corrections can be applied using theoretical models [15] or can be established by measuring the effective porosity at reservoir pressure and temperature on some samples and deriving an empirical correction law.

The above points highlight that errors or lack of representativeness may affect core derived porosity data. Consequently, attention must be paid in the evaluation of the available information and, whenever deemed necessary, corrections must be applied.

Once every possible source of error has been ruled out and the convenient corrections have been applied, core data become the reference information as far as porosity is concerned. Every other data, typically log interpretation, should be compared and calibrated against it.

4.1.3.2 Log Porosity

The most common procedure to evaluate the porosity of a reservoir is through log interpretation, since the data required are normally available in the majority of the wells.

Nowadays, a number of logging tools are available that can provide an (indirect) estimate of porosity, through the measurement of some physical properties of the reservoir rock. They normally work in an open hole environment, but some tools can also be run in cased holes.

As a consequence, especially in the case of old fields, the geoscientist is often faced with the problem of integrating porosity profiles that are derived from different tools, or different generations of the same tool, possibly run in different environmental conditions. Even before any comparison with core data, it is therefore important to understand the kind of log information available, and make sure that every effort is undertaken in order to calibrate the data and hence to arrive at a sound and reliable log porosity interpretation.

The description of the various logging tools, their advantages and their limitations is obviously beyond the scope of this work. Ref. [16] and [17] provide exhaustive treatments of the logging tools and the interpretation techniques that are used and have been used in the past. In the following sections, the discussion will be limited to some of the most popular logging tools and their characteristics. For the sake of clarity, the treatment is divided in two parts, dealing with traditional and recent tools respectively.

A. *Traditional Tools*

The traditional porosity logging tools are the sonic, the neutron and the density logs. Different generations of these tools have been utilised in field operations for more than 30 years and they still represent the most frequently available measurements.

The sonic tool measures the velocity of acoustic wavetrains in rock formations. Velocity of compressional waves is a function of the rigidity and density of the material: the more rigid the medium and the lower its density, the higher the velocity. Porosity tends to lower the overall rigidity of the rock and it is therefore inversely proportional to velocity. The empirical relationship that allows the calculation of porosity is the well-known Willye relation:

$$\phi = \frac{(t - t_{ma})}{(t_f - t_{ma})}$$

where t, t_f and t_{ma} are the measured, fluid and matrix travel times, respectively. Therefore, the computation of porosity requires an *a priori* knowledge of the lithology of the reservoir rock and the saturating fluid.

In the case of well consolidated rocks, particularly carbonates, the Willye relation holds reasonably well and the sonic log provides good estimates of porosity. However, in the case of shallow, unconsolidated sandstones and in general in all cases when the rock has not reached its full degree of compaction, the sonic log can severely overestimate formation porosity. The same happens in the presence of gas or light hydrocarbons, since they significantly lower the acoustic wave velocity.

One interesting feature of the sonic log is that sometimes, especially in carbonate rocks, it can provide an estimate of secondary porosity. In fact, the tool tends to ignore isolated vugs or fractures, since acoustic waves will follow the shortest interconnected path through the rock fabric. When an independent density-neutron porosity measurement is available, the difference between the 2 types of porosity may provide a qualitative indication of the degree of secondary porosity developed in the reservoir.

The density tool provides an estimate of the bulk density of the rock by measuring the attenuation of gamma rays between a source and a receiver. Gamma rays are scattered and absorbed in the formation as a function of the electron density of the formation, which is closely related to bulk density. In turn, density is related to porosity by the simple equation:

$$\phi = \frac{(\rho_{ma} - \rho_b)}{(\rho_{ma} - \rho_f)}$$

where ρ_{ma}, ρ_b and ρ_f are the matrix, bulk (measured) and fluid densities, respectively. Note that also in this case, as for the sonic log, the determination of porosity from density measurements requires the prior knowledge of lithology and fluid type.

Also, if the formation is saturated with gas or light hydrocarbons, the measured bulk density will be affected and the computed porosity will result far in excess of reality.

Contrary to the sonic, the density tool allows for the determination of total porosity, since the measured bulk density is an average of all types of pore spaces present in the formation rock.

The neutron tool measures the formation absorption rate of fast neutrons continuously emitted into the formation by a radioactive source. These neutrons travel in the formation and are slowed by collisions with nuclei, until they reach a very low (thermal) energy level and are finally adsorbed. The most effective element that contributes in slowing these neutrons is hydrogen, since its mass is similar to that of a neutron. Therefore, the neutron population in the formation is inversely proportional to porosity, since, in clean formations, all of the hydrogen is contained in the pore fluids.

As the sonic and the density tools, the neutron is sensitive to the lithology of the formation, because the matrix contributes to the slowing and capture of the neutrons. In addition to that, the tool is particularly sensitive to environmental conditions and to the presence of gas, which lowers the hydrogen density of the pore space. As the density tool, the neutron measures the total porosity of the formation.

With the exception of clean sandstones saturated with water or oil, the neutron log has to be used with caution in the determination of porosity. However, the combination of the density and neutron log allows for a much more reliable calculation, since porosity can be determined without a prior knowledge of the lithology. Moreover, the combined use of the 2 tools allows for corrections for clay and gas effects.

For these reasons, the density/neutron combination still represents the most frequently utilised logging technique for porosity determination.

B. Recent Tools

In the last years, a new generation of tools has been introduced in the market that provide useful data for porosity evaluation: pulsed neutron and magnetic resonance tools.

Pulsed neutron tools are the evolution of the Thermal Decay Time (TDT) and Carbon Oxygen (CO) logging tools, which have been used since the 1970's for saturation determination in cased holes. As with the neutron log, these tools are based on the introduction of fast neutrons into the formation and the subsequent determination of thermal neutron densities or gamma rays spectra. Therefore, a porosity curve having the same properties of the classical open hole neutron porosity can be obtained. This porosity curve has the same limitations already discussed for the neutron log and should be used with cautions in gas saturated or shaly formations.

The main advantage of pulsed neutron logging is the possibility of cased-hole logging. No other tool offers this possibility. It may therefore be run in old wells, where no porosity curves were recorded, providing a saturation and a porosity profile at the same time.

Nuclear magnetic resonance tools are probably the most interesting and most promising borehole logging tools of the last decade. Even though magnetic resonance principles have been known for years, the evolution of the hardware has been slow and only in recent years its application has become widespread.

Principle and physics of NMR (Nuclear Magnetic Resonance) logging are discussed in detail in Ref. [18]. Basically, the tool generates a strong pulsed magnetic field into the formation, which determines an alignment of the hydrogen protons. A second magnetic field, perpendicular to the previous, is then induced, that rotate the protons 90° with respect to the orientation induced by the permanent magnet. The tool measures the time needed by the protons to dephase from the second magnetic field (T_2 time), as well as the time required to return to their original position (T_1 time), once the magnetic fields are released. The T_2 time, also called **relaxation time**, is a strong function of the surface/volume ratio in the pore space, and hence of porosity.

Fig. 4.8 shows the fluid distribution in the pore space and the relevant NMR interpretation in the T_2 curve of a typical sandstone. Two main components can be distinguished: the Free Fluid Index (FFI), which is the percentage of pore space occupied by moveable fluids (water + hydrocarbon) and the Bulk Volume Irreducible (BVI), which is the total pore space containing irreducible water. The latter is composed of the Clay Bound Water (CBW), which is the water bounded to the clay minerals, plus the Bulk Volume Capillary (BVC), which represents the water trapped by capillary forces.

Curent NMR tools are able to discriminate the hydrogen in the interconnected pores (free fluids) as well as the capillary bound water, thus measuring a sort of effective porosity. Microporosity and clay-bound water have a much higher surface/volume ratio, which yields a very low T_2 value. Research is ongoing, and the next generation of NMR tools will be able to determine all the components of the total rock porosity [19].

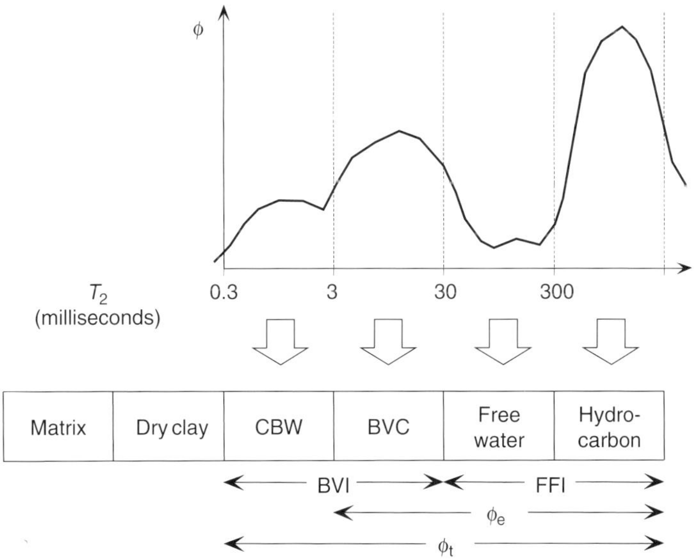

Figure 4.8 T_2 curve and fluid distribution in the pore space [20].

As far as carbonates are concerned, the NMR measurements may allow for the distinction between intercrystalline and vuggy porosity [21]. Actually, due to their dimensions, vugs are often characterised by a very low surface/volume ratio, and therefore cause a typical T_2 response larger than that of the smaller, intercrystalline pore spaces.

Recent application of NMR logging has shown that this tool can provide reliable porosity estimation in a number of different lithological environments. In fact, one of the main advantages of this tool is that, in contrast to all other porosity devices, measurements are independent of lithology.

A recent paper [22] summarises the experience gained in the San Jorge Basin (Argentina), where more than 4 000 NMR logs were run.

4.1.3.3 Integrating Core and Log Porosity

Cores and logs are practically the only sources for porosity estimation and properly integrating these data is essential in deriving a reliable interpretation.

In principle, this integration can be difficult to achieve. The complexity of such a task can be appreciated when considering that each source of data provides a piece of information that is not necessarily comparable to other sources. Fig. 4.9 illustrates a standard rock porosity model on a generic shaly sandstone. It can be observed that different tools and measurements techniques are sensitive to different portions of the pore system. In the upper part of the figure, the sampled fraction of the pore space is shown for a number of downhole logging tools and laboratory techniques. The understanding of the characteristics and the limitations of each technique is obviously an essential requisite for a correct calibration of the data and hence for the integration of the available information.

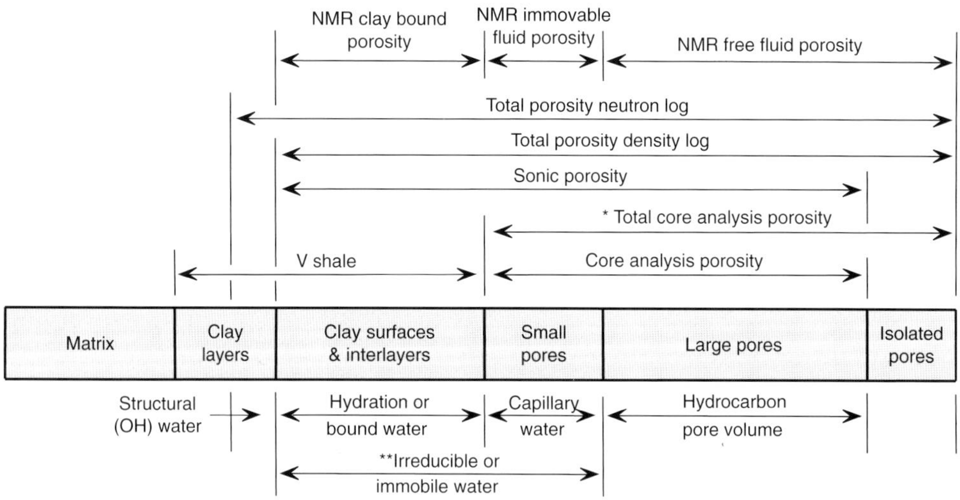

* If sample is completely disaggregated during measurement.

** Varies as a function of height above the free water level.

Figure 4.9 Standard rock porosity model for a shaly sandstone.

A general integration procedure for porosity estimation is hard to define, since it depends upon the available data and the reservoir under study. Nevertheless, from a general point of view, the following points should be included:

- Review all the available core data, paying attention to the measurement techniques that have been used. Eliminate those data that are deemed to be unreliable for any reason. Check the importance of overburden effects and, if necessary, apply the relevant correction.
- Collect and review all the porosity logs run in the field under study, including cased hole curves if existing. Check the quality of the data and eliminate those curves that do not satisfy the minimum requirements. Verify the response of the tools in selected lithological markers (e.g., shales) and, if necessary, apply the necessary calibration and/or normalisation to the curves. Compute porosity profiles for all the wells that have reliable logs.
- Perform an accurate core-log depth matching.
- Compare the core and log porosity results. This can be done in several ways, the most common being a simple cross plot of the two measurements. These cross plots normally allow the identification of systematic errors in the log interpretation that can then be easily compensated. Another familiar technique is to overlay the two porosity curves as a function of depth. In this case, interpretative problems related to particular lithological zones can be easily detected and corrected. Corrections should be applied keeping the cores as the reference data.

It should be emphasised that, when good quality data are available, porosity interpretation is not a critical issue in a reservoir study, especially where primary porosity is concerned.

There are exceptions, though. Sometimes, for example, porosity logs are missing and/or only suites of old logs are available: in such cases the calibration and relevant interpretation phase can prove to be difficult.

Other examples include complex lithology reservoirs where core data are missing or scarce. Since the most common porosity logs require a prior knowledge of lithology, the lack of core data may lead to errors in porosity interpretation that may have a non negligible impact in the evaluation of the oil in place.

Porosity interpretation can be challenging also in the case of carbonate reservoirs, when secondary porosity represent a significant portion of the total pore space. In particular, the interpretation may become critical in the case of fractured reservoirs, when matrix porosity is low and fracture porosity represents a significant percentage of total porosity.

In all cases, properly integrating the available information helps to reduce the impact of such uncertainty factors.

4.1.4 Water Saturation

The pore space of reservoir rocks is filled with fluids, normally water and hydrocarbons. The relative spatial distribution of these fluids depends on a number of factors that are related to the physical properties of both the rock and the fluids, as well as to rock-fluid interactions.

The determination of the saturation conditions of the reservoir formation is one of the most important tasks in a reservoir study. In fact, not only do these conditions affect the calculation of the hydrocarbon in place, but also the fluid mechanics and hence the expected producing performance of a field. Unfortunately, fluid saturation is more difficult to determine than porosity, as in most cases its evaluation is subject to different sources of uncertainty.

A number of techniques are available to ascertain the saturation conditions of a reservoir rock. Some of them are based on direct measurements of the quantity of fluids present in the pore spaces, others are based upon indirect measurements performed either on core samples or in the borehole environment.

In the next sections, some of the most commonly utilised techniques will be reviewed, which allow for the determination of vertical saturation profiles at the well locations. In the second part of this chapter, it will be shown how 2D and 3D distributions of saturations can be generated, starting from these typical vertical profiles.

4.1.4.1 Core Saturations

Fluid saturations can be determined on core data either by measuring the quantity of fluids extracted from a core sample, or by means of capillary pressure measurements.

A. Extraction of Fluids

Extraction of fluids from core samples is based on the determination of water and hydrocarbon quantities existing in a native state core sample. The most accurate method of fluid extraction is probably the Dean-Stark method, whereby the reservoir water contained in a native-state core plug is boiled at just above 100 degrees for a period of many hours and collected in a separate graduated tube. Later, porosity is measured on the clean sample and water saturation is calculated as the fraction (or percentage) of pore space filled with water.

The quality of the saturation data that can be derived from extraction techniques is extremely variable. From a general viewpoint, the fluid content of a core plug at the moment of such measurements has been affected by two major processes:

1. **Core invasion**. During drilling operations the pressure differential between the mud column and the formation causes the mud filtrate to invade the core and to displace some of the original fluids.
2. **Fluid expansion**. During the recovery of the core from bottom hole to the surface, the confining pressure is constantly decreasing, thus allowing the expansion of the entrapped water, oil and gas. The latter, having the greater coefficient of expansion, will tend to displace the other fluids outside the core.

The effects of these two processes are illustrated in Fig. 4.10 (from [11]) for cores cut with water (above) and oil-based mud (below).

When the drilling fluid is water, the invasion process results in a significant increase in the core water saturation, which is difficult to quantify. For this reason, extraction data for a well drilled with a water-based mud are not useful for precise saturation determination.

However, when the drilling fluid is an oil-based mud, the invasion process has normally little or no effect on water saturation. It is also considered that gas expansion has a minor impact on original water saturation.

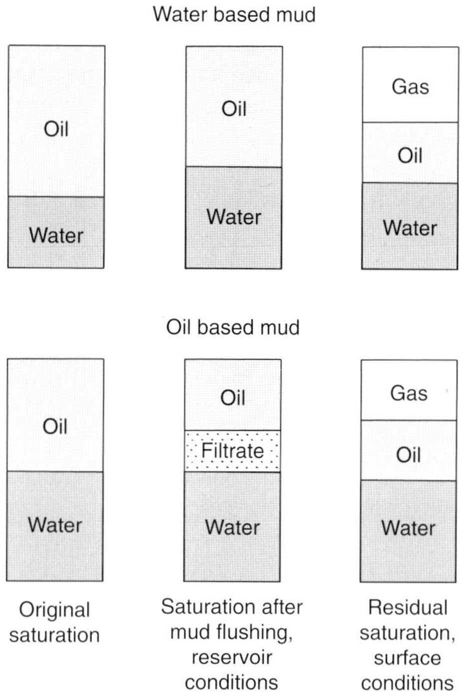

Figure 4.10 Typical saturation changes in a core before and after recovery (Courtesy of McGraw-Hill Companies) [11, mod.].

A comprehensive work on the reliability of oil-based core saturations, based on more than 8000 samples from Prudhoe Bay Fields, confirmed these assumptions [23]. Companion plugs were cut at the edge and at the centre of a number of cores, to investigate the effect of mud-filtrate invasion. Significant differences were not observed. Gas expansion was measured on pressure-retained cores and again results showed a negligible influence. The impacts of other factors, like rate of penetration and handling conditions, were also investigated.

Results of the investigations carried out at Prudhoe Bay showed that, under certain circumstances, extraction data are probably the most reliable source of information as far as fluid saturations are concerned. The limitation of this kind of information is that it can be considered reliable only in the irreducible water saturation zone, i.e., outside the transition zone. In fact, when mobile water is present, this can be displaced by the mud-filtrate and the extracted quantity will be too low with respect to the in-situ water saturation.

B. Capillary Pressure Measurements

Saturation data from cores can also be obtained in the laboratory through capillary pressure measurements.

Capillary pressure occurs whenever two fluids coexist in the pore space of a reservoir rock and it is defined as the difference in the pressure measurable in the two phases.

There is an inherent relationship between capillary pressure and water saturation, because water is retained in the pore spaces by capillary forces. Therefore, a vertical distribution function for water saturation may be obtained through the prior knowledge of the capillary pressure distribution within the reservoir.

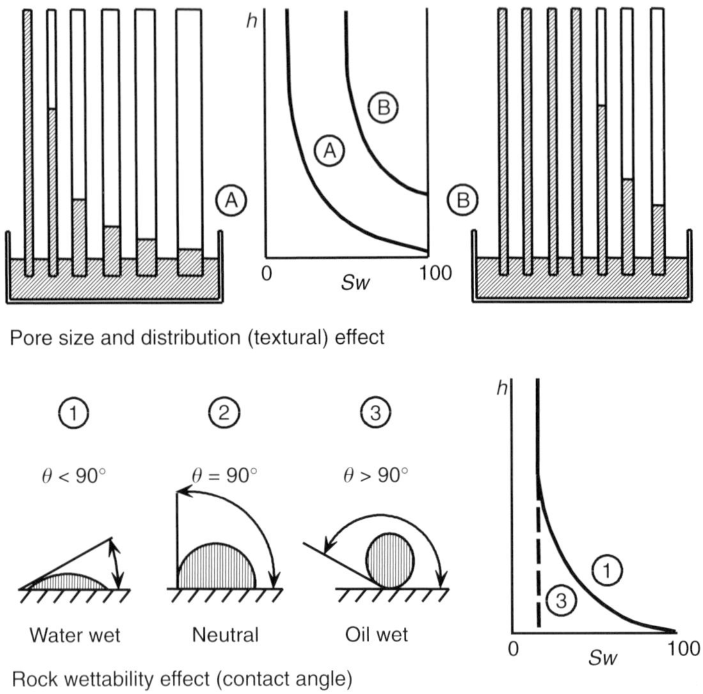

Figure 4.11 Controlling factors on capillary pressure.

From a theoretical viewpoint, capillary pressure is expressed in the following terms:

$$Pc = 2\sigma \cos \theta / r \tag{4.4}$$

where σ is the fluids interfacial tension, cos θ is the contact angle and r is the capillary radius. Therefore, capillary pressure is a function of fluid properties (interfacial tension term), rock properties (the capillary radius) and rock-fluid properties (the contact angle, i.e., wettability). Any variation of these factors throughout the reservoir causes a variation of capillary pressure and hence water saturation. The dependence of capillary pressure on so many parameters makes water saturation modelling such a difficult challenge.

In actual reservoirs, the capillary pressure increases indefinitely upward from the free water level, where it is null by definition. The rate of increase mainly depends on the density differences between the two fluids. Therefore, in field units, capillary pressure can also be stated as:

$$Pc = 0.0069\, h\, \Delta\rho \tag{4.5}$$

where h is the distance in feet from the free water level and $\Delta\rho$ is the density difference in pounds per cubic feet between the two fluids. This is the relationship that allows for the comparison between laboratory measurements and field data.

The influence of the various rock and fluid parameters on capillary pressure can be appreciated in Fig. 4.11 and Fig. 4.12. Pore size and distribution (i.e., texture), can be considered the equivalent of the capillary radius in real reservoir rocks. Texture actually represents one of the most important controlling factors: small pores will tend to generate higher capillary pressure for a given water saturation, while the shape of the curve is influenced by the sorting of the sediment. Another very important parameter is wettability: oil-wet rocks tend to show reduced transition zones as well as lower irreducible water saturations. Fluid density differences also play an important role on capillary pressure functions: the smaller the difference, the bigger the transition zone. The inverse relationship holds for interfacial tension, the smaller the tension, the smaller the transition zone. Finally, capillary pressures depend on the direction of flow (drainage or imbibition, referring to decreasing and increasing wetting phase saturations respectively).

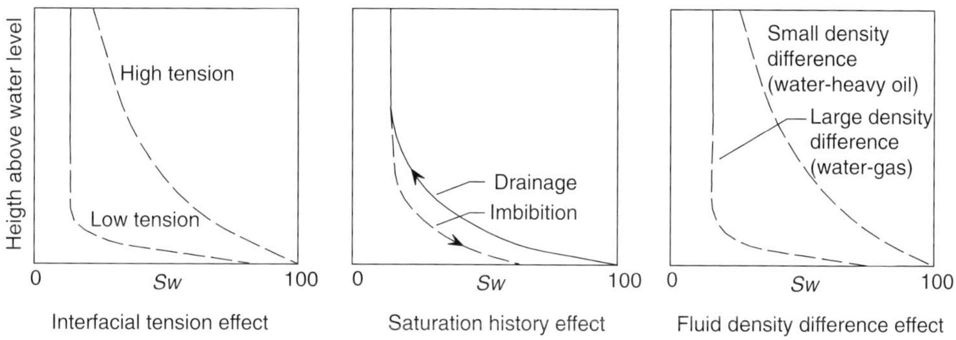

Figure 4.12 Controlling factors on capillary pressure.

There are different methods for determining capillary pressure in the laboratory, the most commonly used being the Restored State Cell, the Mercury Injection and the Centrifuge methods.

- **Restored State Cell**. A clean and 100% water saturated sample is placed in a cell where increasing pressures of a non-wetting fluid are imposed. Displaced water is evacuated from a semi-permeable membrane of uniform pore size distribution located at the base of the sample. Enough time is allowed for each pressure step in order to reach a static equilibrium and saturations are measured gravimetrically. This method normally provides very accurate results and furthermore any pair of fluids can be used (air-water, air-oil, oil-water), even though for practical purposes usually only the air-water system is run. The drawback is that typically several days are required to reach equilibrium at each pressure point. Therefore, only 6 to 8 points of saturation are taken, as these tests are rather expensive.

- **Centrifuge**. In this case, cleaned and 100% water saturated samples are placed in a centrifuge, which rotates at various constant speeds. Samples are effectively subject to increased gravitational forces and thus water is expelled from the end of the core. The quantity of water that at any step is forced out of the samples is read with a stroboscope and water saturation is calculated volumetrically. This method is possibly less accurate but much faster than the Restored State and also higher effective pressures can be obtained. In this case too, all the fluid systems can be tested.
- **Mercury Injection**. This is the simplest and the quickest way of obtaining capillary pressure data in the laboratory. Mercury is injected into a clean and dry rock sample at different pressures and at each step the quantity forced into the sample is measured. Since Mercury is always a non-wetting phase, water saturation can be indirectly computed. This method allows for a fast determination of capillary pressure and furthermore it provides typically 20 to 30 measurements for each sample. The main drawback is that the mercury-air data have to be converted to the air-brine system (wettability correction), which introduces a degree of uncertainty. Moreover, the rock samples cannot be used for further testing.

The utilisation of capillary pressures data measured in the laboratory is not as straightforward as in the case of porosity measurements. A number of steps have to be performed in order to use these data in engineering calculations.

1. **Reservoir condition conversion**. Laboratory tests are performed using fluids that do not have the same physical properties as the reservoir fluids, therefore a conversion must be applied. The most frequently used formula for this conversion is:

$$Pc_R = Pc_L \ (\sigma \cos \theta)_R / (\sigma \cos \theta)_L \tag{4.6}$$

   where the subscripts R and L refer to reservoir and laboratory conditions, respectively. Note that this equation is derived by equating the capillary pressure expressions and recognising that the capillary radius does not change. This corrected capillary pressure can then be converted in height above the free water level by means of Eq. 4.5, thus giving:

$$h = Pc_R/\Delta\rho = Pc_L/\Delta\rho \cdot (\sigma \cos \theta)_R / (\sigma \cos \theta)_L \tag{4.7}$$

   Therefore, the conversion of a laboratory capillary pressure curve into a water saturation vs. height curve require the knowledge of the reservoir fluid gradients ($\Delta\rho$), as well as the wettability conditions of the rock $(\sigma \cos \theta)_R/(\sigma \cos \theta)_L$.
2. **Data normalisation**. More frequently than not, the available capillary pressure measurements for a given field have been performed in different times and by means of different techniques. Therefore, these data must be normalised. When available, restored state measurements are usually considered as the reference information. In other cases, in oil-based mud wells, the normalisation procedure can be performed against irreducible water saturation values as determined from Dean-Stark extraction data.
3. **Data averaging**. The measured capillary pressure curves are strongly dependent upon the petrophysical properties of the rock sample, i.e., porosity and permeability. Some engineering applications like numerical simulations require the availability of an average capillary pressure curve for each rock type. The most commonly used equation for capillary pressure averaging have been proposed by Leverett [24], and is called ***J*-Function**:

$$J(Sw) = Pc/\sigma \cos \theta \cdot (k/\phi)^{1/2} \tag{4.8}$$

Other averaging techniques include empirical correlations derived for the particular field under study by means of multivariate regressions. When dealing with water saturation distribution (paragraph 4.2), the practical application of some of these averaging techniques will be discussed.

In all cases, care must be exercised in using directly these data in reservoir engineering calculations. As it appears clearly from Eq. 4.7, the derivation of a water saturation function with depth from capillary pressure data require the knowledge of a number of parameters and this may generate significant uncertainty.

From this point of view, the critical concern is often the wettability of the core samples. As a matter of facts, all these laboratory techniques test clean samples, i.e., samples where the original reservoir fluids have been completely extracted. A clean sample is often strongly water wet.

In many instances, uncertainty exists over the actual wettability conditions in the reservoir. More often than not, the core recovering and handling procedures generate alteration in the original wettability [25]. In such cases, the term $(\sigma \cos \theta)_R$ in Eqs. 4.6 and 4.7 is difficult to quantify, while the usually applied literature values can only be considered as large approximations.

The bottom line is that the utilisation of capillary pressure data to estimate reservoir saturations requires special attention. As usual, the geoscientist must be aware of the representativeness of the data and their reliability before using them in a reservoir study.

4.1.4.2 Log Saturations

Water saturation can be indirectly measured in the borehole environment by means of 2 types of logging devices, i.e., resistivity and pulsed neutron tools.

A. Resistivity Logging

The common practice of evaluating water saturation for a given field is through the interpretation of log resistivity curves. The principle behind this methodology is that the electrical conductivity of the formation is entirely due to the water present in the pores, since both the rock matrix and hydrocarbons are perfect insulators. Therefore, the interpretation consists in comparing the measured resistivity of the formation with its theoretical resistivity if it contained only water. If the former resistivity is higher than the latter, the presence of hydrocarbon is inferred.

The equation that allows for the calculation of water saturation from resistivity measurements is the well known **Archie** equation. Originally presented in 1942 [26], today this equation still forms the backbone of clean formation interpretations:

$$Sw^n = \frac{a}{\phi^m} \cdot \frac{R_w}{R_t} \tag{4.9}$$

As can be appreciated, water saturation is a function of 6 independent variables (n, a, ϕ, m, R_w, R_t). With the exception of R_t, which is measured through resistivity tools, all these parameters have to be computed or estimated independently. In the remaining part of this section, these parameters will be discussed, with particular emphasis over the uncertainty they carry.

Formation resistivity, R_t. The uninvaded formation resistivity is the main parameter of the Archie equation. It is measured in the borehole by means of resistivity logging tools. The description of various logging tools and their features is beyond the limits of this work and can be found in any logging textbook. In this context, it is useful to note some of the critical factors concerning formation resistivity measurements.

- The Archie equation requires true, or uninvaded, formation resistivity. Since drilling mud invasion always affects a significant portion of the formation around the borehole, it is important to evaluate the invasion profile and, if necessary, perform the necessary corrections.
- An old log analysis rule says that the deeper the investigation ratio, the lower the vertical resolution. This means that the determination of R_t can be a problem in the case of thin beds. It is always good practice to make sure that the resistivity devices in use have the resolution that we need.
- The different resistivity logging tools have different application fields, depending on formation porosity, bulk resistivity and borehole fluids salinity. Even though in most cases any device could give satisfactory results, in particular conditions the *wrong* tool can provide misleading results.

It should be noted, however, that under normal conditions accurate estimates of R_t can be obtained. Most of the uncertainty related to water saturation calculation often comes from other sources.

Formation water resistivity, R_w. Formation water resistivity can be estimated in a number of ways. The most common practice is to solve Eq. 4.9 for R_w in the water leg, where $Sw = 1$. This is normally good practice, since water salinity changes only slightly with depth. However, examples have been reported, where the connate water in the oil leg has a different composition than the aquifer water. In this case, water saturation estimates based on bottom aquifer readings can be in serious error.

Another way of estimating R_w that provides good results is through resistivity and/or salinity measurements on produced water samples, as well as extracted samples from cores. When none of the above information is available, R_w can still be estimated by means of crossplots based on some rearrangements of Eq. 4.9. The popular Pickett and Hingle plots are examples of such crossplots.

One further point that should be considered is that the normal assumption of using one single value of R_w for the field needs to be checked with the available data. Indeed, recent publications have shown that large variations in water salinity can be encountered, both vertically and laterally, which can cause significant errors in water saturation computation [27].

In any case, care must be exercised in estimating water resistivity. When more than one source of information is available, the relevant data should always be compared and calibrated.

Porosity, ϕ. In most cases, porosity is not a major source of uncertainty. Here, it is sufficient to note that, being part of the Archie equation, an overestimation of porosity determines not only a direct overestimation of the oil in place, but also an indirect overestimation linked to the saturation term, since in Eq. 4.9 higher porosity generates lower water saturation.

Formation factor. The formation factor is defined as the ratio of the resistivity of the rock completely saturated with brine to the resistivity of the brine:

$$F = \frac{R_o}{R_w}$$

In a 100% water saturated sample, since water is the only conductive media, the formation factor will increase as the volume available to water decreases. Therefore, it is expected that porosity as well as the degree of cementation should be the main factors in controlling the passage of current. The general relationship that expresses such concepts has been proposed by Archie in 1942:

$$F = a\phi^{-m} \quad (4.10)$$

where the terms a and m, already encountered in Eq. 4.9, are the cementation factor and the cementation exponent, respectively. The value of the cementation exponent m reflects the tortuosity of the interconnected pore space and normally ranges between 1 and 2.3. The more complex the current path, i.e. the higher the cementation, the higher the value of m.

The general relationships between porosity and formation factor are illustrated in Fig. 4.13, where the value of the cementation exponent m represents the slope of the straight line.

Historical relationships for the formation factor are the following:

$$F = \phi^{-2} \qquad \text{for limestones}$$

$$F = 0.81\ \phi^{-2} \text{ or } F = 0.62\ \phi^{-2.15} \qquad \text{for sandstones}$$

The second equation for sandstones is commonly known as **Humble** formula. The cementation factor a represents the consistency of the pore geometry. When a is different from 1, pore geometry would change with porosity or, in other words, different types of porosity are present in the reservoir rock [29]. Note that this is expected to be the common situation, since in general cementation modifies pore geometry. However, most petrophysicists prefer to eliminate one parameter by choosing a value of 1 for the cementation factor a. In this case, the straight line of Fig. 4.13 is forced through the origin.

The formation factor F of a rock can be routinely evaluated in the laboratory by measuring the electrical properties of a brine saturated sample. When several measurements are available, laboratory results may allow for the calibration of both the a and m terms, as illustrated in Fig. 4.14. Particular attention should be paid to the possibility of determining different values of these parameters for different facies, or groups of facies, as identified in the geological study.

However, care must be taken when utilising core-derived formation factor values in log interpretation. Shaly sands for example, may sometimes provide unrealistic formation factor values, since the clay minerals act as separate conductors from the brine. Also, the effect of overburden pressure on resistivity should be taken into account, since this causes a significant increase in resistivity. Nowadays, electrical measurements are often performed with confining pressure, but most of the electrical resistivity data available for relatively old fields have been determined on unconfined core samples. In such cases, a preliminary normalisation procedure should be applied.

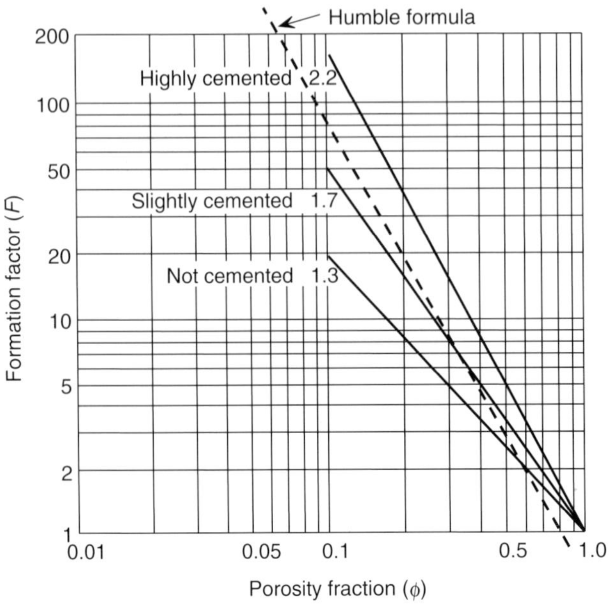

Figure 4.13 General porosity vs. formation factor relationships [28, mod.].

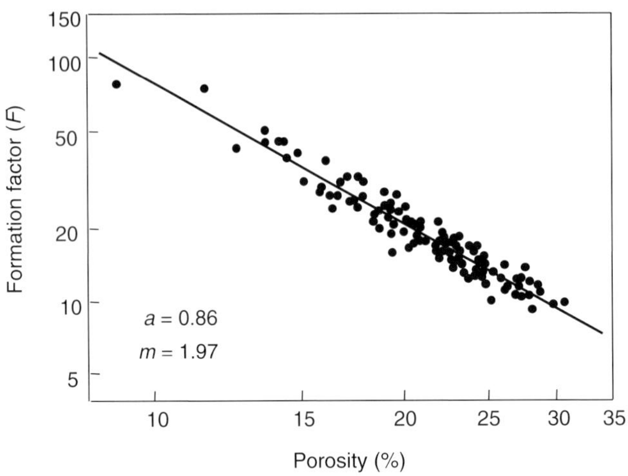

Figure 4.14 Formation factor vs. porosity.

Saturation exponent, *n*. The current flow path in the rock depends not only on the tortuosity of the pore system, but also on the distribution of fluids within the pore system. Since oil and gas are not electrical conductors, their presence and distribution will reduce the cross-sectional area available for the current and will increase the actual resistivity of the

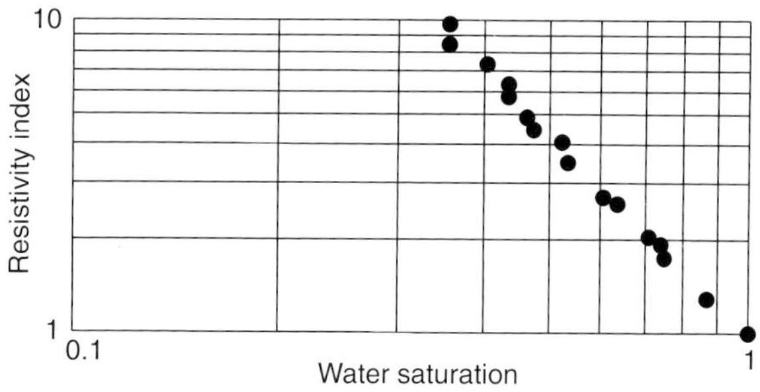

Figure 4.15 Resistivity index vs. water saturation.

system. Therefore, a resistivity index can be defined as the ratio of rock resistivity at any saturation to rock resistivity when completely saturated with water:

$$I = \frac{R_t}{R_o} = Sw^n$$

Note that, as expected, this relationship is similar to the equation of the formation factor. The parameter n is called saturation exponent, and it expresses the increment in resistivity due to the distribution of non-conductive fluids into the pore spaces. The typical value of the saturation exponent is 2. However, while this value is adopted by many petrophysicists for most formations, deviations often occur, particularly in shaly sands.

Like the formation factor, the resistivity index can be computed in the laboratory, by measuring the resistivity of a sample under different saturation conditions. Fig. 4.15 shows the relationship between resistivity index and saturation conditions of a rock sample, as obtained from a typical laboratory experiment.

Laboratory tests in general provide reliable values of the saturation exponent to input into the Archie equation. However, the computed values of n are dependent upon the wettability of the rock sample. If it is suspected that wettability has been altered during the cleaning of the samples, the computed values of n could be in serious error. Analysis conducted on preserved samples from oil-based cores, can often provide more accurate results. Ref. [23] reports an interesting example from Prudhoe Bay, whereby the values of the saturation exponent have been calibrated against the water saturations measured through Dean Stark extraction data on oil-based cores. In this case, routine laboratory measurements performed on cleaned samples proved to be unreliable, probably due to wettability modifications.

Shaly formations. The above discussion focussed on the inherent uncertainty that exists in evaluating water saturation for clean formations from log measurements. In the case of shaly formation, the usual Archie equation is normally replaced by more complex formulations like the *Indonesian, Simandoux, Poupon, Waxman-Smits, Dual Water*, not to mention the most used. Without going into unnecessary detail, it is worthwhile appreciating that these equations require further parameters to the evaluation of water saturation, like shale volume, shale resistivity and/or cation exchange capacity. Of course, in this case, the overall uncertainty is even higher. Ref. [23] provides a methodology to estimate the global uncertainty in water saturation measurements as a function of the fractional uncertainty of each individual parameter.

B. Pulsed Neutron Tools

The other family of logging devices that allow for a determination of water saturation is Pulsed Neutron Capture (PNC) logs. Two basic tools are currently used, the Thermal Decay Time and the Carbon/Oxigen tools.

The principle of TDT is similar to that of the neutron logs. Pulses of fast neutrons are emitted by a radioactive source into the formation, and these particles are slowed down by collisions with other nuclei. When the neutrons reach a stable low energy level, they are susceptible to being captured by other atoms. The probability that neutrons are captured by some particular atom is referred as capture cross section of that material. Among the most commonly found elements in sedimentary rocks, Chlorine has the greatest capture cross section, therefore thermal neutrons are more readily absorbed by chlorine than any other material. However, all the rock and fluid components contribute to neutron absorption, therefore the global capture cross section Σ_t is a volumetric sum of the type:

$$\Sigma_t = \Sigma_{ma}(1 - \phi) + \Sigma_w Sw\phi + \Sigma_{hc}(1 - Sw)\phi \tag{4.11}$$

where the subscripts t, ma, w, hc refer to total, matrix, water and hydrocarbon, respectively. Rearranging the equation, water saturation can be determined.

Note that this determination requires the knowledge of 5 independent parameters, most of which can only be inferred from literature tables. Therefore, the interpretation is normally performed by calibrating the individual Σ values in a water section and/or against the results of open hole interpretations. Another disadvantage of this technique is that in low porosity and low water salinity formations results are less reliable.

The main advantage of TDT measurements is that they can be run in cased hole and through tubing. This makes these logs the main tools for reservoir monitoring. In addition to that, they can distinguish between oil and gas (while resistivity logs do not) and are not influenced by pore geometry, so that the interpretation does not require the knowledge of parameters like *a*, *m* and *n* of the Archie equation.

When used in a time-lapse fashion, TDT logs interpretation becomes simpler. In fact, since porosity and matrix capture cross section do not change with time, Eq. 4.11 can be rewritten as:

$$\Delta Sw = \Delta\Sigma/\phi(\Sigma_w - \Sigma_{hc}) \tag{4.12}$$

This equation can be used to monitor the changes in water saturation that occur in the reservoir as a consequence of aquifer or injected water encroachment.

The Carbon/Oxygen tool, as the TDT, works with a pulsed neutron source but in this case inelastic gamma ray spectra are used to determine the relative concentration of Carbon and Oxygen in the formation. A high C/O ratio indicates oil bearing formations, while low C/O ratio indicates water bearing formations.

This tool has a distinct advantage with respect to the TDT tool, in that it can work in low salinity formations. Recent advances in the technology of this tool allow for use through-tubing and the collection of high quality data.

4.1.4.3 Integrating Core and Log Measurements

Integrating core and log measurements to evaluate water saturation is not straightforward, since there is no general rule that can be applied. As discussed in the previous sections, three main sources of information are normally available to the geoscientist, i.e., core extraction data, capillary measurements and log data interpretation. Each of these can provide accurate or biased information, depending on a number of factors. Nevertheless, keeping in mind that the objective is always to generate reliable saturation profiles for the greatest number of wells, a few general guidelines can be laid down that can applied in most cases.

- The first task is always to evaluate the reliability of all the sources, as a function of the reservoir under study and the type and amount of core and log data available. Capillary pressure data must be normalised and carefully converted to reservoir conditions, in order to be compared with other available data.
- When available, Dean-Stark extraction data from oil-mud cores should be regarded as the best approximation of actual reservoir irreducible water saturation. In these cases, any other data should be matched against extraction data above the transition zone.
- Log interpretation should always be compared with the available capillary pressure measurements. The global water saturation profile can also be split into curves of *Sw* vs. depth for different facies or porosity classes, in order to facilitate the comparison with similar groups of core measurements.
- When consistent and reliable core data are available, capillary pressure measurements should be used as reference information to calibrate the log interpretation, especially when uncertainty exists in some of the interpretative parameters of Archie's equation.
- When uncertainty exists over the reliability of capillary pressure data (wettability alteration), log interpretation should be preferred. Furthermore, the latter is representative of the actual reservoir scale, while core plugs are based over a small support volume.
- Pulsed Neutron logs should always be calibrated to open hole log interpretations.

An example of integration of independent sources for water interpretation evaluation is shown in Fig. 4.16, for a North Sea oilfield. Dean Stark core extraction data were available for this well, and they have been used to calibrate the log interpretation in the irreducible water zone. It is interesting to note that while approaching the oil-water contact, the two series of data diverge. Actually, in the transition zone, the Dean-Stark values underestimate water saturation as a result of the displacement of the movable water induced by the drilling fluid.

4.1.5 Permeability

Permeability is defined as the ability of a rock formation to conduct fluids. Beyond any doubt, it is the most important petrophysical property of a reservoir. Most of the parameters used to define the economics of a development project are, more or less directly, related to the permeability of the reservoir.

Unfortunately, permeability is also the most difficult parameter to describe in a reservoir study. Indeed, permeability in many cases represents one of the paradoxes of conventional

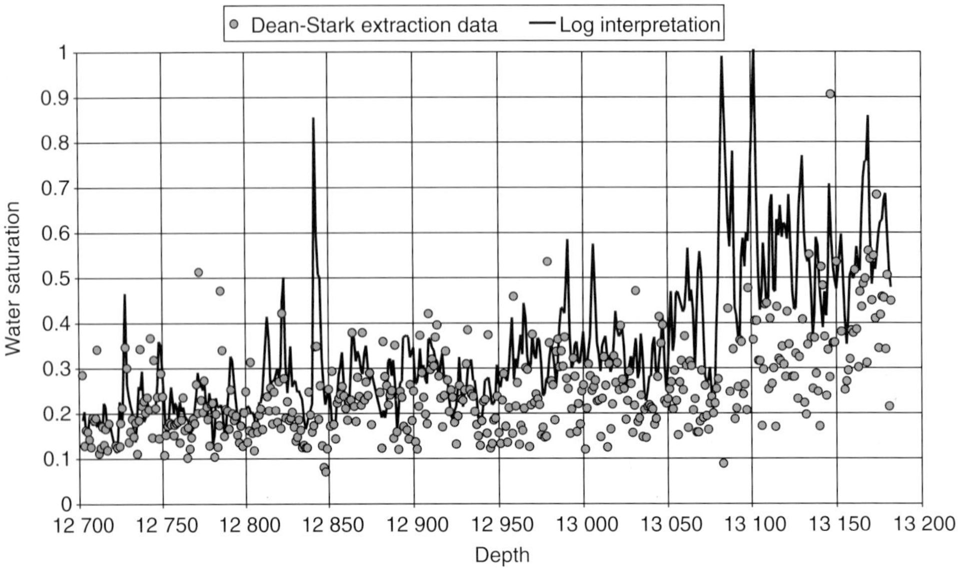

Figure 4.16 Water saturation evaluation for a North Sea oilfield.

studies. Geologists and petrophysicists often plead with management about the necessity of acquiring more information to be able to accurately describe the reservoir flow units and they usually devote a lot of technical effort to this modelling issue. On the other hand, reservoir engineers often do not hesitate to sacrifice the results of such description in the simulation phase, for the sake of history matching the reservoir performance. From this point of view, permeability estimation is one of the critical tasks in the framework of an integrated reservoir study, since in many cases a synergistic approach allows for the definition of accurate and robust permeability models.

However, dealing with permeability, the concept of integration becomes particularly difficult to apply, because a deep understanding of all the static and dynamic implications of such a parameter is required. As already mentioned in Chapter 1, many techniques exist, that provide direct or indirect information about permeability, but each of these techniques offer a particular piece of the mosaic and the final integration, in all cases, is far from being trivial.

As a matter of fact, each professional tends to have its own perception of permeability, depending on the techniques he is used to deal with. However, when integration is concerned, distinctions must be made among absolute and effective permeability, the scale and type of measurements, the environmental conditions that affect the data and so on. These differences must be clarified and understood before any synergy is sought.

In line with what has been done with porosity and water saturation, in the course of this chapter we will review most of the techniques that are routinely used to describe absolute permeability at the well locations (relative permeability will be discussed in Chapter 6).

After a short introduction, the section has been organised into 4 parts, dealing with the most frequently used sources for permeability determination, i.e., core analysis, wireline measurements, well testing and empirical correlations. A short digression will also be dedicated to the recent application of Neural Networks. A final section will discuss the issues related to the integration of the different sources.

4.1.5.1 Generalities

Back in 1856, while studying the public fountains of the town of Dijon, the French engineer Darcy derived a simple, empirical equation to calculate water flow rates:

$$Q = KA\frac{\Delta h}{L}$$

where Q represents the volume rate of flow of water downward through a cylindrical sand pack of cross-sectional area A and length L, while Δh is the hydraulic head. In his relationship, Darcy also defined a constant of proportionality K, which is a characteristic of the sand pack. Later investigators showed that such law could be extended to other fluids by substituting the constant of proportionality K with the term k/μ, where μ is the viscosity of the fluid and k is a property of the rock alone, called **permeability**. In this formulation, permeability is independent of the saturating fluid and therefore k can be better defined as **absolute permeability**.

In the more general form of the Darcy's equation, applied to a horizontal flow, the pressure differential ΔP substitutes the hydraulic head Δh, thus giving:

$$Q = \frac{K}{\mu} A \frac{\Delta P}{L} \tag{4.13}$$

Darcy's experiments were conducted with sand packs 100% saturated with water, where the dominant conditions of flow were laminar and no reaction was observed between fluid and rock. Therefore, to apply his empirical equation, the following conditions must be verified:

1. Laminar, or viscous flow. The velocity of the fluid is sufficiently low to be directly proportional to the potential gradients.
2. The fluid phase is unique and is 100% saturating the pore space. When more than one fluid is present, permeability is no longer a function of rock medium only, but also of the relative saturations of the fluid phases. The concepts of **effective** and **relative permeability** will replace absolute permeability in this case.
3. No reaction between fluid and rock.

The condition of laminar flow can be easily appreciated when the Darcy equation is expressed in terms of velocity V:

$$V = \frac{Q}{A} = \frac{K}{\mu}\frac{\Delta P}{L}$$

This equation suggests that a plot of fluid velocity V vs. the potential $\Delta P/L$ should yield a straight line through the origin, where the slope is equal to fluid mobility K/μ, as shown in Fig. 4.17.

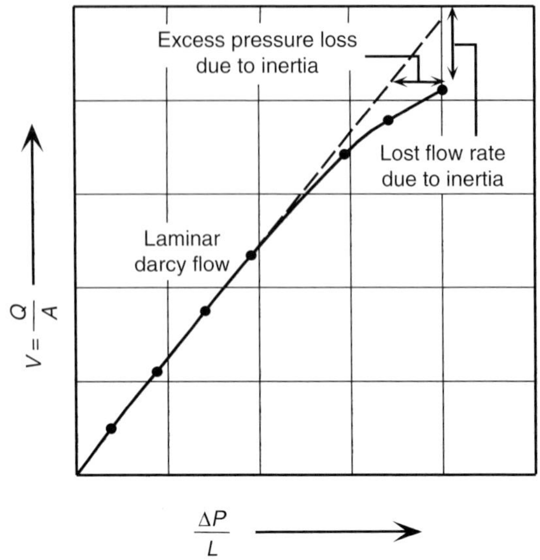

Figure 4.17 Fluid velocity vs. pressure drop.

However, when the velocity of the fluid phase increases, inertial effects appear and the linear behaviour is no longer verified. These inertial effects, observed and modelled by Forchheimer, are caused by fluid acceleration at the pore throats and deceleration at pore bodies and become increasingly important with the pressure differential. In opposition to the previous, this behaviour is also called **non-Darcy flow** and is particularly important when the fluid phase is gas.

Another peculiar characteristic of permeability is that, in contrast with other petrophysical properties, it is a directional quantity and it often shows distinct anisotropy. In fact, permeability is primarily dependent upon the textural characteristics of the rock, which in turn are the result of distinct depositional processes. The arrangement of the grains that constitute the sediment have a strong impact on the characteristics of flow. Grain orientation and alignment, the presence of shale or silty horizons, laminated features like stylolites or fractures are just the more familiar characteristics of a reservoir rock responsible for the existence of a permeability anisotropy. An example of matrix and fracture permeability anisotropy is illustrated in Fig. 4.18. The mathematical aspect of this anisotropic behaviour is that permeability in fact is to be described by a tensor.

4.1.5.2 Laboratory Measurements on Core Samples

Core analysis allows for a direct measurement of absolute permeability in the laboratory, under different experimental conditions. In general, the formation samples that can be tested may have different support volumes. The smallest volumes concern samples recovered from sidewall cores, where the typical plug dimension is < 1 inch long. Slightly bigger, 1 to 1-1/2 inches, are the traditional plugs taken from conventional rotary cores. Full-diam-

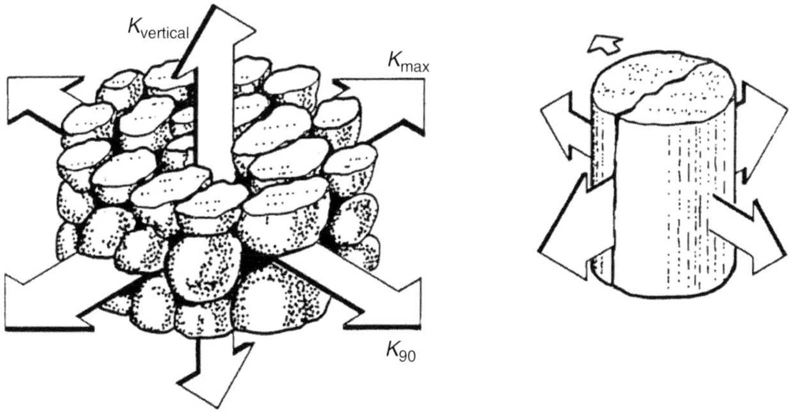

Figure 4.18 Matrix and fracture permeability anisotropy.

eter samples, up to several inches long, can also be analysed in the case of very heterogeneous formations.

Steady state and unsteady state measurements are currently performed in most laboratories. Steady state measurements have been the standard for more than 30 years and are still widespread. In this case, a gas (normally air) is flowed through the core sample at different rates, while pressures are measured. Sufficiently low fluid rates are used, in order to satisfy the viscous flow conditions (straight line of Fig. 4.17). Since the geometry of the sample is known, Darcy's law can be solved for permeability.

The accuracy and the reliability of laboratory measurements is normally good. However, a number of issues must be taken into account before utilising these data:

- **Cleaning procedures**. Measurements are performed on clean and dried samples, where all the original saturating fluids have been removed by retorting or solvents. In some cases, these cleaning procedures may not remove all of the heavy hydrocarbon fractions, thus leading to underestimated results.
- **Representative volume**. There is a scale factor related to conventional plug measurements. Plugs are usually taken every foot and they are implicitly considered to be representative of that particular volume section of the core. However, the ratio between the volume of the plug, where permeability is measured, and the volume of 1 foot of core is between 100 and 150 and therefore the representativeness of that single measured permeability value is doubtful. Should a slightly different position of the plug have been chosen, a different permeability value would have resulted for the same core volume, especially when small scale heterogeneities are present.
- **Gas slippage effect**. Measurements have to be corrected for the effect of gas slippage. This phenomenon is related to the different physics of flow in the porous network between liquids and gases. When the diameter of the pores approaches the mean free path of the gas, the gas molecules tend to have a finite velocity at the pore wall, while normally this would not happen with liquids. This phenomenon is therefore called gas slippage (Fig. 4.19). The effect of gas slippage is to increase the volumetric flow rate,

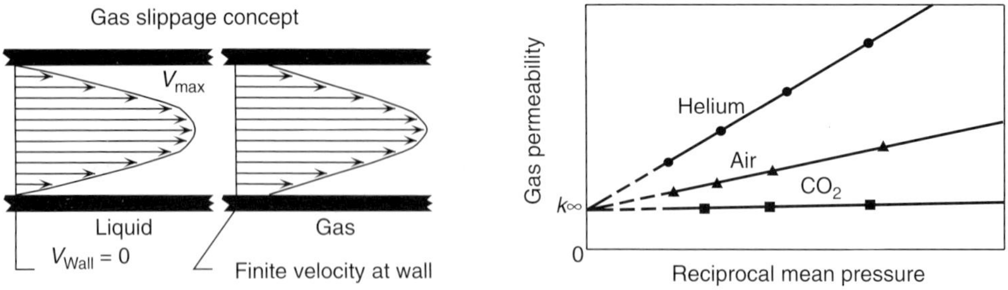

Figure 4.19 Gas slippage effect and Klinkenberg correction.

which is more pronounced in low permeability rock and low molecular weight gases. Klinkenberg [30] has reported variations in permeability when different gases are used in the flow tests. The data obtained with the lowest molecular weight gas (Helium) yield the straight line with the greatest slope, thus indicating a greater slippage effect. However, when extrapolating the different gas lines to infinite mean pressure, a common point can be identified, which has been designated k_L or equivalent liquid permeability. Routine permeability measurements in the laboratory should always be corrected for the Klinkenberg effect.

- **Overburden correction**. Measurements should be corrected for the overburden pressure. When the core is brought to the surface, all the confining forces are removed and the rock tends to expand in all directions. In turn, this expansion causes a modification of the pore geometry which may have a strong impact on rock permeability, depending on the pressure differential, the consolidation state of the rock itself and the clay content. In unconsolidated rocks, for example, where the compressibility of the pore system is very high, a consistent reduction in permeability can be expected. Also, when authigenic clays are present, permeability can suffer a drastic reduction when overburden pressure is applied, since the clay can bridge at the pore throats, thus impairing the fluid circulation. Interestingly, these reductions in permeability are often related to small porosity reductions. Since in most rocks the observed reduction in permeability values is not negligible, an overburden correction should always be applied to the routine data. When overburden measurements are available a correction function for the routine data can easily be constructed, otherwise empirical relationships can be applied.

Core analysis is the most typical source of permeability data for most reservoirs. Since the technique allows a direct measurement of permeability, it is considered by many geoscientists as the base data to which all the other different source of information should be calibrated.

However, it should be noted that core permeability is, at best, an accurate representation of a particular core sample under specific laboratory conditions and its direct use in reservoir engineering calculation can be seriously misleading. Several factors, typically the presence of large scale heterogeneities, may impair the representativeness of core data for reservoir simulation applications. Therefore, when independent information is available, it should always be compared with the core analysis data, and all discrepancies should be explained.

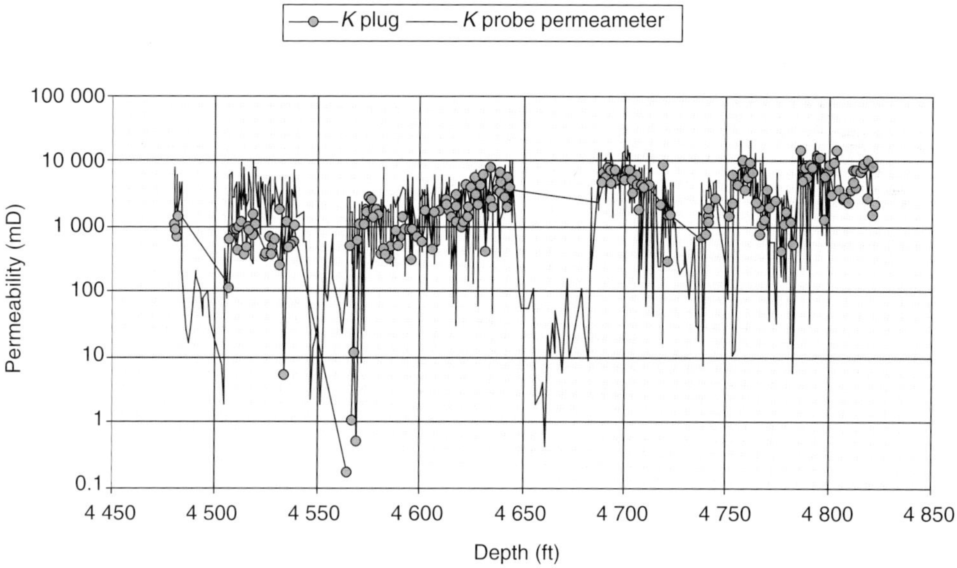

Figure 4.20 Conventional plug permeability and probe permeameter data comparison.

Another technique that is often applied in the laboratory on slabbed cores is the probe permeameter. This type of devices have been in use now for many years and are routinely utilised in the field to measure permeability variations in outcrops. More sophisticated, unsteady state versions of the tool are utilised in the laboratory to compute high density permeability profiles. These data are especially useful when the existence of small scale heterogeneities is suspected. These probe permeameters provide absolute permeability values, corrected for gas slippage effect. Measurements can only be performed at ambient conditions [31].

In most cases, the probe permeameter results are comparable to the routine permeability measurements performed on plugs. However, it is not unusual to observe significant discrepancies between the 2 types of measurements, as illustrated by the upper section of the well of Fig. 4.20, between 4 505 and 4 540 feet depth.

The reasons for these differences may be several. First of all, the sampled volume in the case of the probe permeameter is much smaller and the measurement is statistically more influenced by small scale heterogeneity. In fact, probe permeability data frequently provide extreme values which are not observed in conventional plugs. Other differences are related to the physics of the technique: probe permeameter measurements are subject to high local gas velocities and therefore to significant inertial effects even at low pressure differential. In addition, the geometry of the streamlines within the small sampled volumes are completely different compared to plug measurements, where all the injected gas flows from one end to the other end of the sample. Ref. [32] provides a discussion about observed differences between plugs and permeameter measurements.

4.1.5.3 Wireline Measurements

Permeability can be estimated in the borehole by means of wireline tools. Typical measurements are made by the Wireline Formation Tester tools (WFT) and Nuclear Magnetic Resonance tools (NMR). In both cases, data refer to indirect measurements at reservoir pressure and temperature.

A. Wireline Formation Testing

The Wireline Formation Tester is a tool that can be run in the borehole, positioned at a desired level and actuated to measure reservoir pressure and flow capacity, from which permeability can be calculated. The procedure can be repeated at various depth levels, therefore almost continuous pressure and permeability profiles can be obtained along the borehole. At the same time, reservoir fluid samples can also be obtained. In its simplest configuration, the WFT tool is shown in Fig. 4.21. More complex configurations can also be adopted that utilise multiprobe tools. These tools allow for the determination of an anisotropy ratio *Kv*/*Kh*, in addition to the horizontal permeability component.

During a WFT test, the probe is positioned on the sandface and sealed with a concentric packer by hydraulic action. A small piston is then withdrawn to disrupt the mud cake and establish communication with the formation, allowing reservoir fluids to flow into the tool. These fluids successively enter 2 small pretest chambers, while pressures are continuously monitored.

This phase, in analogy with conventional well tests, is called *drawdown*. When the pretest chambers are full, pressure builds back to reservoir pressure and this phase is referred to as *build-up*.

Permeability can be calculated from both drawdown and build-up phases of the WFL pretest. In the case of the drawdown, fluid flow into the WFT probe is normally assumed to be spherical or hemispherical in nature and the interpretation is performed using a steady state simplified model that allows for a quick calculation of permeability at the wellsite. The interpretation is usually performed for both the pretest chambers, to test the consistency of the results.

It should be noted that WFL drawdown permeability represents the effective permeability to water in the invaded zone, since the fluid that is flowing into the probe is essentially mud-filtrate. In addition, the measurement has a very limited depth of investigation, typically 1 inch or 2, therefore the pressure data are heavily influenced by the formation damage. Because of these two factors, permeability calculated from drawdown is often lower than that expected and can be considered to be a lower limit for the actual formation permeability.

Permeability can also be estimated from the rate of build-up. The procedure requires more attention in this case, since the assumption of spherical flow has to be validated. However, the calculated values are less influenced by the borehole environment since the depth of investigation is a few feet rather than a few inches. For this reason, the measured permeability should reflect the uninvaded zone better and be closer to an effective permeability to hydrocarbon.

In addition to the pretest drawdown and build-up phases, permeability can also be estimated during the recovery of reservoir samples. In this case, longer periods of pressure

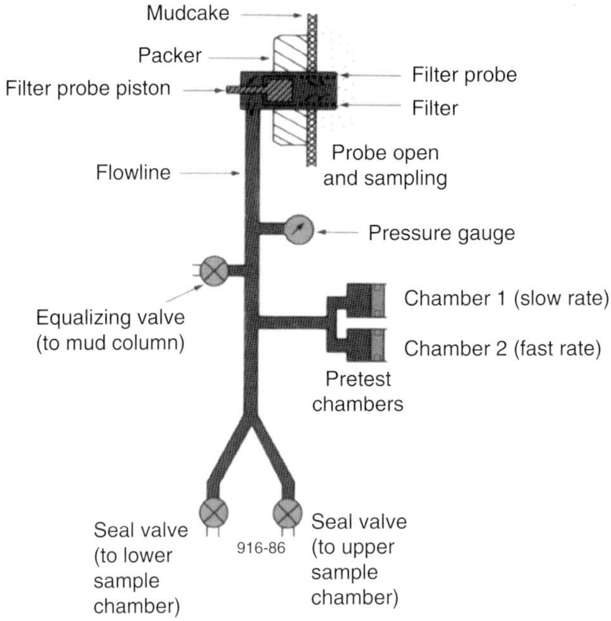

Figure 4.21 WFT tool basic configuration [16].

measurements are available and greater pressure differentials are measured, which allow for a more reliable permeability determination. However, this kind of test can only be performed once or twice during a single wireline operation.

Care should always be taken when dealing with WFT permeability. In fact, a number of important points must be checked before validating these data:

- Poor measurements can be obtained when the seal to the formation is not certain, when the formation permeability is very low and when gas is present.
- The pretest drawdown and build-up interpretation actually provide a mobility value, K/μ, rather than permeability alone. Since the fluid is usually a mixture of oil and mud filtrate in unknown percentages, the actual value for the viscosity μ is often uncertain.
- The WFT derived permeability, especially in the case of pretest drawdown, refers to a zone of the reservoir that has been altered by borehole operations. Formation damage may result in lower than expected permeability, while stimulation procedures may cause abnormally high values.
- When good quality data is available, WFT measurements provide information about effective permeability to the fluid saturations existing at the moment of sampling. Since in general these conditions are different compared to other measurement techniques, the computed permeability will also be different.

As a conclusion, WFT measurements provide permeability values that are difficult to compare with the other typical sources of data. With some exceptions, WFT data provide information that can be used more in a relative than in an absolute way and therefore its normal use is in completion decisions rather than in integrated studies.

B. Nuclear Magnetic Resonance Measurements (NMR)

Nuclear Magnetic Resonance measurements provide the only means to estimate a continuous vertical permeability profile in the borehole through wireline measurements. The principles of the tool have already been briefly presented in paragraph 4.1.3.2.B and can be found in Ref. [18].

The physical principle which is involved in the estimation of permeability is the so called *wall relaxation*. The relaxation of water saturating a pore is greatly shortened by contact with the grains of the rock at the adjacent pore walls. These wall effects provide the sensitivity of NMR to pore size and hence to permeability.

Many equations have been proposed that allow for the estimation of permeability from NMR measurements. From a general viewpoint, they can be classified in 2 broad categories:

1. Relationships that make use of the Irreducible Water Saturation estimation. The general form of these equations is:

$$K \propto \phi^4 \, Sw_{\mathrm{i}}^2$$

2. Relationships that make use of the T_2 (Transverse Relaxation Time) distribution. The general form of these equations is:

$$K \propto \phi^4 \, T_2^2$$

In the first case, the estimation of permeability is actually a two-step process. In the first step, a cut-off is imposed to the T_2 distribution, in order to distinguish between capillary bound (irreducible) and movable water. Then, the derived Sw_{i} is applied to compute permeability. Timur's equation [33] is the most classical example of this kind of derivation.

In the second case, the logarithmic mean of the T_2 distribution in milliseconds is used directly in the estimation. The relationship proposed by Sen et al. is among the most popular [34].

In fact, various authors have proposed slightly different forms of these relationships, as a function of the methods and the lithological types considered in their investigations. In all cases, whenever possible, the values of the constants and the exponents should be determined in the laboratory, utilising NMR measurements on core materials.

Another important point is the underlying assumption that permeability is related to pore size. While in most cases this is true, it should not be forgotten that permeability is actually dependent upon pore *throat* size. In some cases, diagenetic effects may alter the opening of the pore throats more than the pore volume, thus impairing permeability much more than porosity. Another example is given by vuggy carbonates, where permeability can be very low in spite of the large pore dimension, since vugs can be poorly connected. In these cases, the standard NMR equations may provide misleading results, since NMR is primarily sensitive to the size of pore bodies and is insensitive to the size of pore throats.

4.1.5.4 Well Testing

Well testing provides a reliable means of estimating reservoir permeability. When a well is put on production or injection, when its rate is changed or when the well is shut, the reservoir reacts with a pressure behaviour that is directly related to its flow potential and hence to permeability.

A number of types of well tests can be used to calculate reservoir permeability. All of them are based on the application of the basic flow equations to the interpretation of a recorded pressure and rate data set. Among these tests, we have the short Drill Stem Tests (DST), conventional well tests (Drawdown and Build-Up) and multiwell tests (Pulse and Interference). A number of popular monographs provide adequate descriptions of these kinds of test and their interpretation [35].

Well testing represents one of the most widely utilised techniques for determining reservoir permeability. The classical method is the interpretation of the Build-Up pressure data, in a plot of observed shut-in pressure data vs. the so-called superposition time. Fig. 4.22 shows one of these graphs, familiarly called the **Horner plot**.

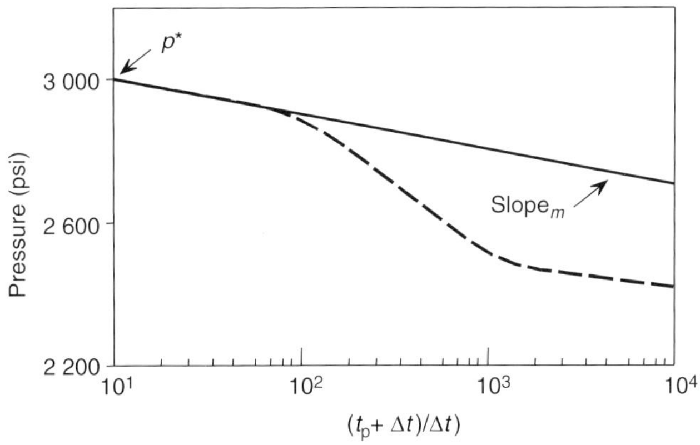

Figure 4.22 Horner plot for permeability estimation.

When the flow periods are well differentiated, the pressure data should have a straight-line portion, whose slope is equal to:

$$m = \frac{162.6 q B \mu}{KH} \qquad (4.14)$$

where q is the rate of the well before shut-in, B and μ are the oil formation volume factor and viscosity, K is the permeability and H the thickness of the producing interval (field units). This equation can be solved to estimate reservoir permeability.

An alternative technique to estimate permeability is by means of type-curves matching, whereby pressure and pressure derivative data are plotted against elapsed time, in a graph which is called **log-log diagnostic**. In this case, permeability can be estimated by the stabilisation value of the derivative, $\Delta p'_{st}$, as shown in Fig. 4.23. The equation in this case is the following (field units):

$$k = \frac{141.2 q B \mu}{\Delta p'_{st} H} \qquad (4.15)$$

where all the other terms are the same as in the previous equations. A clear stabilisation of the derivative is in this case a necessary condition to the interpretation.

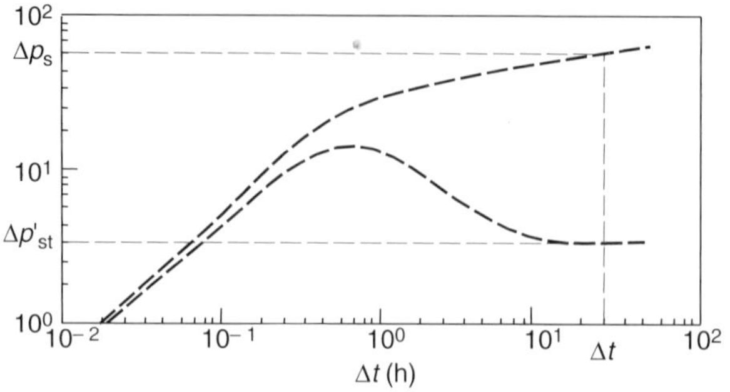

Figure 4.23 Log-log diagnostic for well test interpretation.

As with any other technique, the interpretation of well test permeability raises a number of issues, that have to be clarified in order to understand the nature and the correct utilisation of this information.

- **Quality of the data**. The reliability of the pressure and rate measurements is always a basic issue. The quality check stage is very important, because many factors can lead to erroneous interpretations. Typically, the lack of a precise identification of the flow periods, insufficient duration of the tests, low resolution pressure measurements and multiphase flow conditions are among the points that should be taken into account. Moreover, uncertainty often exists on the identification of the straight line or in the derivative stabilisation level [1].
- **Permeability derivation**. Whatever the interpretation procedure, the parameter that can be inferred from a well test interpretation is in fact the slope of a straight line or the stabilisation of a pressure derivative. As indicated by the above equations, the determination of permeability actually requires the knowledge of a number of other parameters, some of which carry an inherent uncertainty. Among them, the most critical parameter is usually H, the producing thickness. This is also the reason why reservoir engineers sometimes prefer to talk about producing thickness KH, rather than permeability only. In some cases, when the tests are carried out using a production tool, which measures at the same time bottom hole pressure and rates, this uncertainty can be reduced.
- **Effective vs. absolute permeability**. Well tests measure the effective permeability to oil (or gas), at the saturation, pressure and temperature conditions prevailing in the reservoir. This point cannot be overlooked, since most of the techniques discussed in this chapter provide an estimation of absolute permeability. When the reservoir is at irreducible water saturation, Sw_i, the effective permeability to oil is a fraction of the absolute permeability, whose value depends upon the petrophysical properties of the reservoir rock and the wettability conditions. In many cases this value, which corresponds to the maximum oil relative permeability, $Kr_{o\,max}$, can be close to 1 and there-

1. In some cases, even with good quality data, the straight line portion cannot be found at all.

fore a direct comparison of absolute and effective well test permeability can be attempted. However, the value of Kromax can frequently be as low as 0.6 and in this case, in the absence of other factors, the well test permeability can result in significantly lower values compared, for example, to core permeability. The situation is even worse in the case of multiphase tests, when the water saturation in the reservoir is higher than Sw_i and/or in the presence of gas. In these cases, the calculated effective permeability to oil is a function of the oil saturation in the reservoir and can be considerably lower compared to the absolute permeability.

- **Volume of investigation**. Another important point to take into consideration is that the radius of investigation in a typical well test is in the order of hundreds of feet, which represents a volume of reservoir thousands times bigger than any other estimation technique. That is, well tests provide information at megascopic scale, compared to the macroscopic scale data provided by cores, wireline measurements and empirical correlations (Fig. 1.2). The direct implication is that large scale reservoir heterogeneities are involved in the measurements, which have an important impact on the computed permeability (Fig. 4.24).

It is commonly assumed that, when good quality data are available, well tests provide the best estimation of the actual formation permeability. While this is possibly true, care must be exercised when these data are to be compared with other independent permeability sources, to make sure that a consistent interpretative procedure is performed. This issue will be discussed in detail in the final paragraph of this section (paragraph 4.1.5.8).

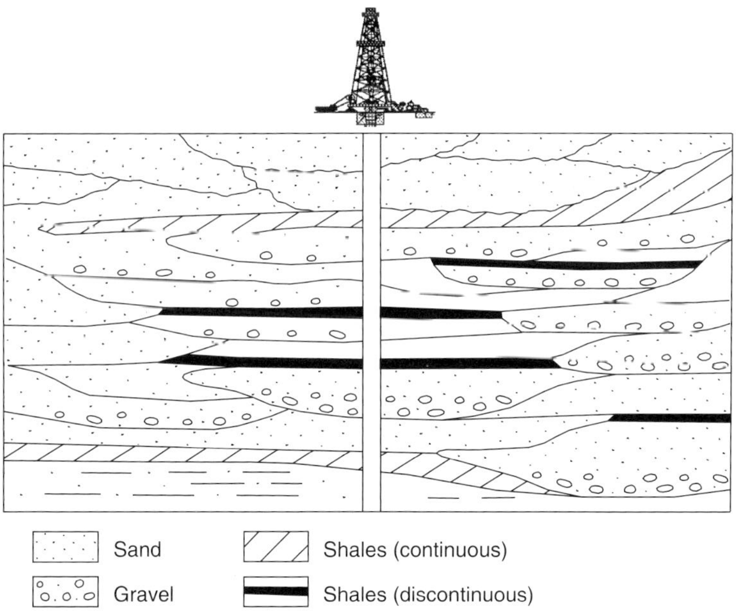

Figure 4.24 Volume of investigation of a well test and large scale heterogeneities.

4.1.5.5 Flowmeter Logging

Production flowmeter logging is used in many fields as a reservoir monitoring tool. The main feature of these measurements is that they provide a quantitative description of a number of parameters as a function of depth, the most important being the fractional flow rate.

In the framework of an integrated study, flowmeter results are of particular interest, since they provide the essential link between the static description and the actual dynamic performances of the well. In particular, a flowmeter profile can be converted from rates to permeability, when the total *KH* of the well is known [36]. The procedure consists in a vertical allocation of the total *KH* as a function of the measured flow rate. The average permeability k over an interval ΔZ along a well trajectory is calculated as:

$$k = \frac{\Delta F}{\Delta Z} KH \tag{4.16}$$

where ΔF is the fraction-change in total flow over interval ΔZ, while KH is the total permeability-thickness of the well determined from well test data. The procedure is illustrated in Fig. 4.25.

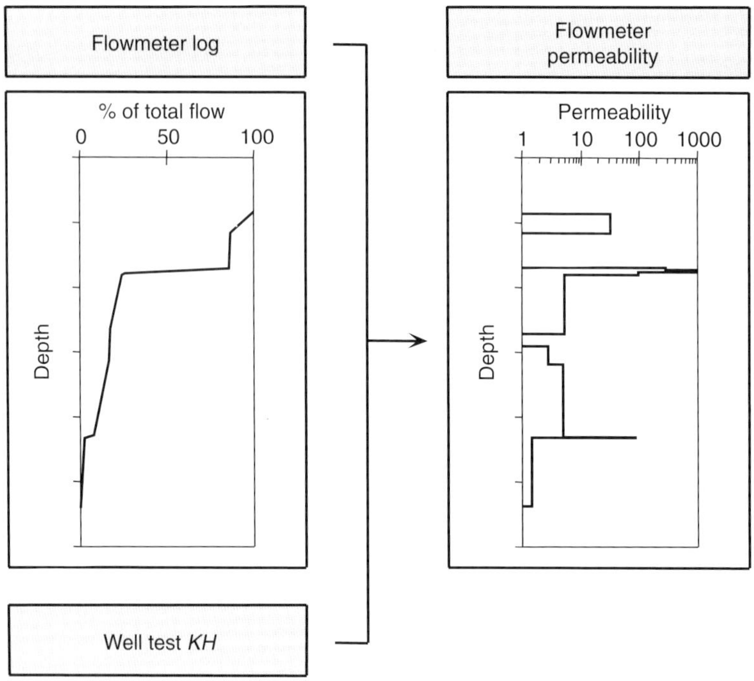

Figure 4.25 Procedure for deriving a permeability profile from flowmeter data.

The resulting permeability profiles can be compared with core measurements or other macroscale permeability profiles, thus providing a means to validate the available **static** petrophysical model. Fig. 4.26 displays an example of a comparison of flowmeter data with core measurements, where the data have been transformed into cumulative permeability-thickness profiles. In this case, a good agreement can be observed between the two types of data.

Attention must always be paid in these types of validation. Flowmeter data, for example, are influenced by near-borehole damage (skin), which could lead to an underestimation of the potential productivity of the affected intervals. In addition, the vertical resolution of the tool is generally lower with respect to conventional macroscale permeability estimations. Core plugs, on the other hand, must be evenly sampled, in order to avoid bias in the resulting profiles.

From this point of view, an unsatisfactory match between core and flowmeter data cannot be surprising. In fact, flowmeters provide an estimate of effective permeability to oil, while core data express absolute permeability. In addition, the kind of information provided by flowmeters is based on a much larger support volume compared to conventional macroscale data. The evidence of a mismatch between these data may actually indicate the presence of large scale heterogeneity, undetected in usual macroscale measurements, which have a significant impact on fluid flow characteristics.

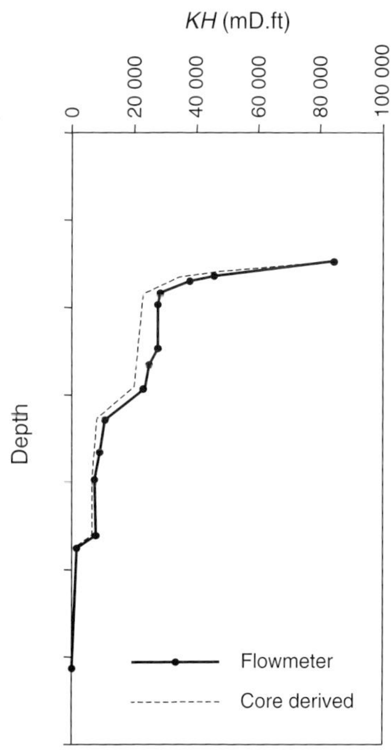

Figure 4.26 Permeability/thickness product comparison for core and flowmeter data.

4.1.5.6 Empirical Correlations

The most common way to estimate permeability profiles in uncored wells is through some permeability predictor, typically in the form of an empirical equation. This normally requires a calibration data set that is represented by one or more key wells where comprehensive information is available in terms of core and log data. This calibration data set is used to build the predictor and to test the reliability of the results.

In this paragraph, three types of permeability predictors will be reviewed, namely porosity-permeability relationships, multiple linear regressions and existing empirical equations. It is worth noting that all these techniques aim at measuring absolute permeability at in situ conditions, since the input parameters used to estimate permeability normally come from downhole wireline measurements.

A. Porosity-Permeability Relationships

By far the most used permeability predictor is the porosity-permeability relationship. It has long been recognised that most reservoir rocks show a reasonably linear relationship between these parameters in a semilog scale, which allows for the estimation of permeability when a porosity profile is available.

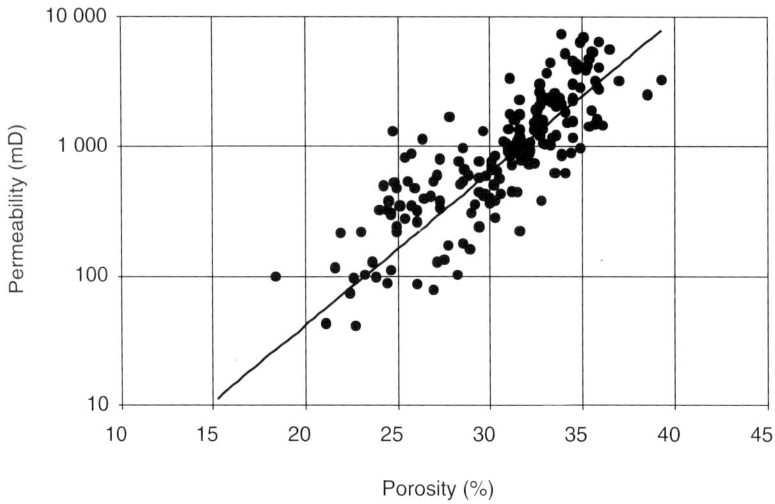

Figure 4.27 *K* vs. ϕ relationship for a fluvial reservoir.

Figure 4.27 shows a typical example of a K/ϕ plot relative to a fluvial siliciclastic reservoir. At first glance, it is evident that a clear porosity-permeability relationship exists for this reservoir and that a predictor could easily be constructed by means of a simple linear regression. In fact, the resulting equation shows a correlation coefficient of 0.81, which is reasonably good from a purely statistical point of view. However, a closer look at this graph reveals that the actual variability of permeability for a given porosity value is still significant and can be greater than one order of magnitude. This variability cannot be captured by the regression equation and any predicted permeability profile will inevitably average the actual permeability profile.

This smoothing effect, which in general will be inversely proportional to the correlation coefficient, can be observed in Fig. 4.28, where the histograms of the actual and predicted permeability values are compared. The dispersion parameters of the two distributions (variance and standard deviation) highlight the averaging effect of the regression procedure. What is important to note in this example is that the predicted permeability profile will be effective in estimating the average characteristics of the true profile, but will be ineffective in estimating the extreme values. These extreme values, from a fluid flow point of view, are the most interesting parts of the distribution, since they may represent either high permeability streaks or barriers.

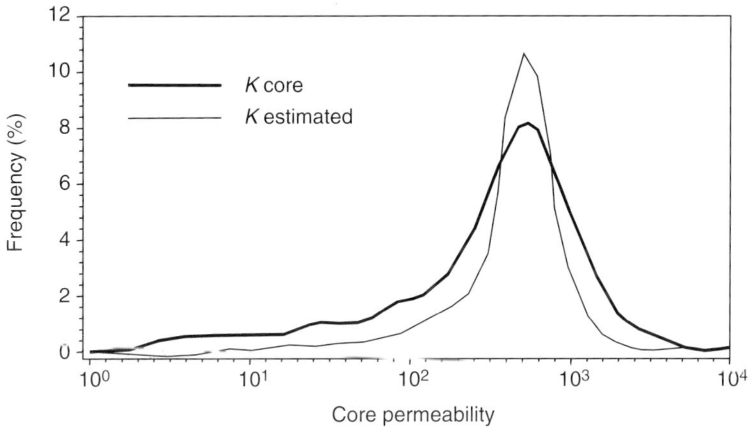

Figure 4.28 Histogram of predicted vs. actual permeability values.

Reservoir rocks show a wide spectrum of porosity-permeability relationships [37]. In some formations, like for example homogeneous clastic rocks, these relationships show very low dispersion and can be reasonably used for prediction purposes. In other cases, as it is frequently for carbonates, this relationship is very loose and does not allow any safe regression. The example of Fig. 4.27 can probably be considered as a limit example for a prediction exercise, since the smoothing effect is quite significant in this case. Distributions with a correlation coefficient lower than 0.8 should probably be handled through more sophisticated techniques.

A good practice in estimating permeability from these kinds of plots is to split the whole K/ϕ data set into subsets, which show a more homogeneous behaviour. The simplest way is of course to use different K/ϕ plots for different layers or areas of the field. Much better results can be obtained when a facies analysis has been performed on the reservoir under study, since the facies classification criteria are often related to petrophysical properties. It is not uncommon to observe a remarkable improvement in the correlation coefficient of the single facies K/ϕ relationships, as illustrated in Fig. 4.29.

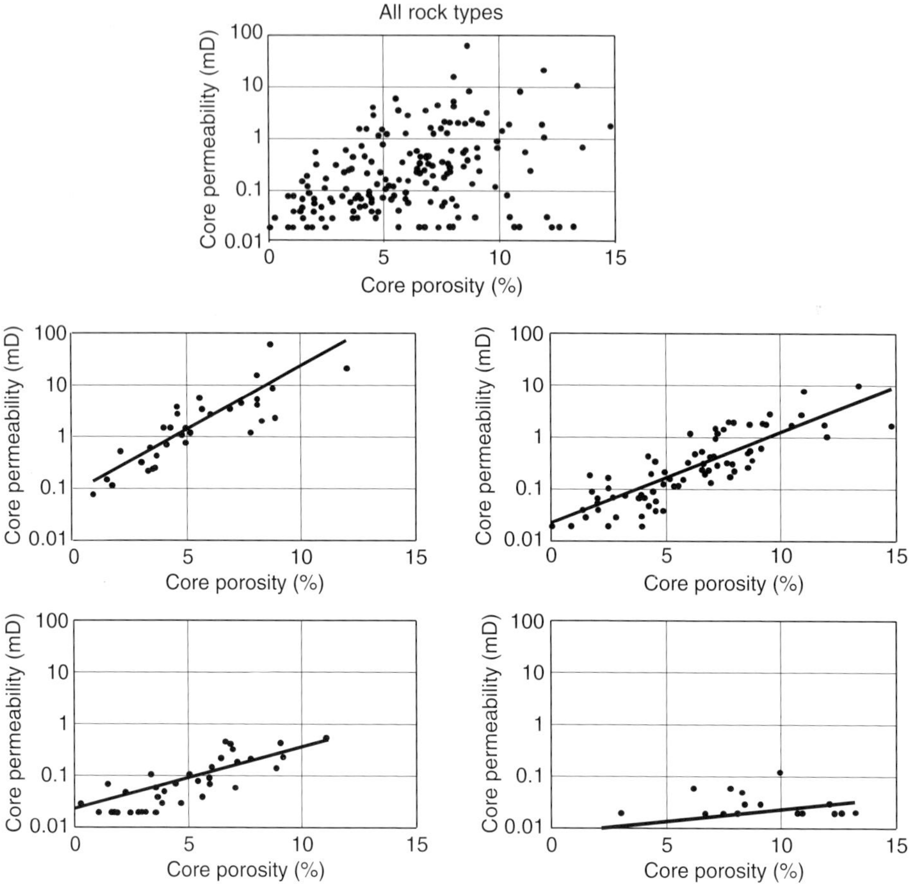

Figure 4.29 K vs. ϕ relationships for different rock facies.

B. Multiple Linear Regressions

In many cases, even in clastic reservoirs, the porosity-permeability relationship appears too scattered to allow for a reliable regression. This normally happens when the petrological variables (cementation, grain size distribution, diagenetic alteration, etc.) play an important role in the porous structure of the rock. In these case a more sophisticated approach is required that can take into account the influence of other parameters in addition to porosity.

Multiple linear regression represents a more complex estimation technique with respect to the simple K/ϕ plots and usually allows for a fast and reliable permeability estimation in most reservoirs.

The methodology is based on estimating the coefficients c of a predictive equation of the type:

$$\text{Log } K = c_0 + c_1x_1 + c_2x_2 + \ldots + c_nx_n$$

The estimation is performed using one or more key wells, where core permeability profiles are available. The resulting equation can then be applied to any other wells, where the independent variables x are known.

The independent variables should be selected from the parameters that have a strong impact on permeability. Typical examples are porosity, water saturation and volume of shale but also depth and geographic position could be used, to account for systematic vertical and lateral trends. Ref. [38] presents an exhaustive review of the application of multiple linear regressions for permeability estimation.

It should also be noted that the porosity-permeability regressions discussed in the previous section are just a particular case of a multiple linear regression, when the number of x variables is equal to 1. Therefore, multiple linear regressions suffer from the same type of limitations. In particular, this kind of estimation will result in a distribution of predicted permeability values which is narrower than the original data set, leaving the problem of estimating extreme values open to some extent. However, the correlation coefficient is usually better than a simple K/ϕ relationship, because more parameters are taken into account.

Figure 4.30 compares the actual permeability distribution of a cored wells with 2 predicted distributions, derived respectively from a multiple linear regression and from a K/ϕ relationships. The improvement obtained with the inclusion of more independent variables is evident, especially in the high permeability range.

As in the case of the K/ϕ relationships, multiple linear regression can be improved by working with distinct subsets of data, especially by facies, and also by applying some weighting scheme to the data, usually to the low and high permeability ranges. This allows for a reduction of the smoothing effect of the regression, but it also has the disadvantage of producing biased results, since the choice of the weighting scheme is subjective.

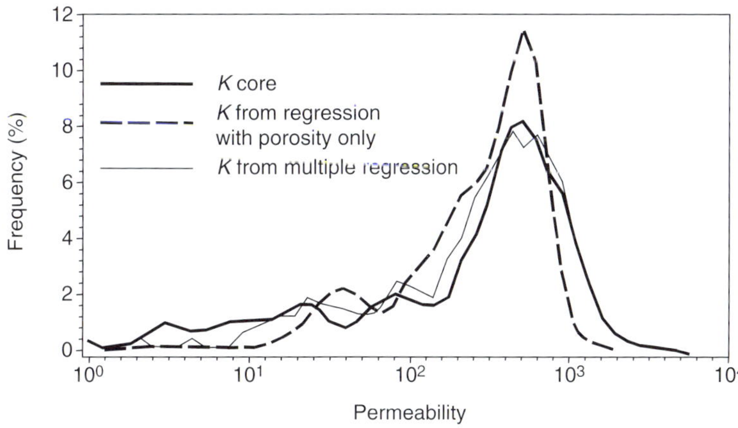

Figure 4.30 Actual and predicted permeability distributions.

C. Empirical Equations

Since 1927, when Kozeny originally presented his equation, a number of authors have been proposing empirical correlations to predict permeability. In general, these equations make use of more readily available information, like porosity or water saturation, to derive a permeability profile at the well locations.

In most cases, these equations can only provide rough estimates of permeability. The main problem of this approach is that, while permeability is dependent on the size and distribution of *pore spaces* within the rock framework, this parameter is always unknown. Therefore, alternative reservoir characteristics must be used.

On this basis, the proposed empirical correlations can be classified as follows:

- **Grain-based models**. These models are based on the application of petrological parameters like grain size and sorting, as measured for example from sieve analysis. The underlying assumption is that these petrological parameters control pore size and distribution. Two of these models (Berg and Van Baaren) have already been mentioned in paragraph 4.1.2.

- **Surface area models**. These are among the most reliable equations for estimating permeability. The concept in this case is that permeability will be impaired by large surface areas, as in the case of clays or other diagenetic alterations of the pore space. In the classic equation of Timur, the surface area is implicitly assumed proportional to water saturation.
  The whole family of equations based on magnetic resonance, discussed in paragraph 4.1.5.3.B, belong to this category, since the T_2 transverse relaxation time of the hydrogen is related to the surface area of the pores.

- **Pore size models**. Direct information on pore size can be obtained in the laboratory through capillary measurements. The threshold pressure and the shape of the capillary curves can be statistically related to pore size and distribution and, hence, to permeability.

Some of the most widely used equations are given in Table 4.1, with the corresponding reference publication.

Table 4.1 Empirical equations to compute permeability.

Model	Author	Reference
Grain-based	Berg	[4]
	Van Baaren	[5]
Surface area	Timur	[33]
	Sen	[34]
	Coates	[39]
Pore size	Kozeny-Carman	[3]
	Winland	[40]

The application of these empirical equations has to be made with caution, since they have been derived applying theoretical considerations to particular sets of samples. It is not uncommon to produce permeability profiles that do not match the available information at all. In general, with the possible exception of NMR methods, the use of these equations should be limited to the case where very little experimental permeability data is available. The choice of a particular one of these equations will depend on the reservoir under study and the available data. Also, whenever possible, it is highly recommended to test the reliability of the results against core permeability measurements.

4.1.5.7 Neural Networks

Neural networks provide an alternative to the traditional techniques of estimating permeability. The method has been introduced in the last years, following the widespread availability of powerful computing resources and has rapidly found a number of applications.

Neural networks can be programmed to recognise patterns, to store and retrieve database entries, to solve optimisation problems, to filter noise from experimental data and to estimate sampled functions when the analytical form of the function is unknown. The last feature is the one relevant to the problem of estimating permeability.

An artificial neural network is basically a system of several simple processing units known as nodes or neurons, associated with each other by simple connections. The strength of these connections may be changed by varying the weight attached to them. The process of adjusting the weight values is known as the *training process*. The training process is the fundamental procedure through which the neural network is calibrated to the particular estimation problem. Therefore, in order to apply this technique, a good data set of reference permeability information is needed. Typical input data are wireline log curves, depth and porosity.

The description of the structure of the neural networks and their use is beyond the scope of this review. Ref. [41, 42] provide detailed information on this matter. What is important to notice is that, in contrast to any other estimation method, neural networks do not make use of a pre-defined relationship, since the estimation function is built through experience during the training phase. In this respect, neural networks are model-free estimators.

Another interesting point is that the predicted permeability distribution does not obey any statistical rule, e.g., the preservation of the mean value. In fact, one of the drawbacks of the predictions made through regressions, i.e., the smoothing effect and the loss of the extreme values, is not a concern in the case of neural networks. In fact, the technique allows the actual variability of the data to be preserved.

The latter feature makes neural networks particularly suited to the case of heterogeneous formations. Fig. 4.31 (left) shows a porosity-permeability crossplot for a siliciclastic reservoir, where diagenetic effects have heavily altered the original pore network, resulting in a very heterogeneous system. A quick glance at this figure shows that traditional regression algorithms would give poor results in predicting permeability from porosity. Neural networks, on the other hand, provide an excellent fit to the experimental data (right).

Neural networks have some disadvantages, too. Firstly, the training process has to be done with caution and can be a lengthy process, which requires a good calibration data set. The excellent results shown in Fig. 4.31 have been obtained utilising a comprehensive training data set, which is not always available in real cases. On the other hand, failing in correctly calibrating the network may result in aberrant results. Another point to take into consideration is that the methodology is not yet an *off the shelf* application and requires expertise by the geoscientist.

Neural networks, like other techniques that make use of wireline measurements as input, aim at measuring absolute permeability at reservoir conditions.

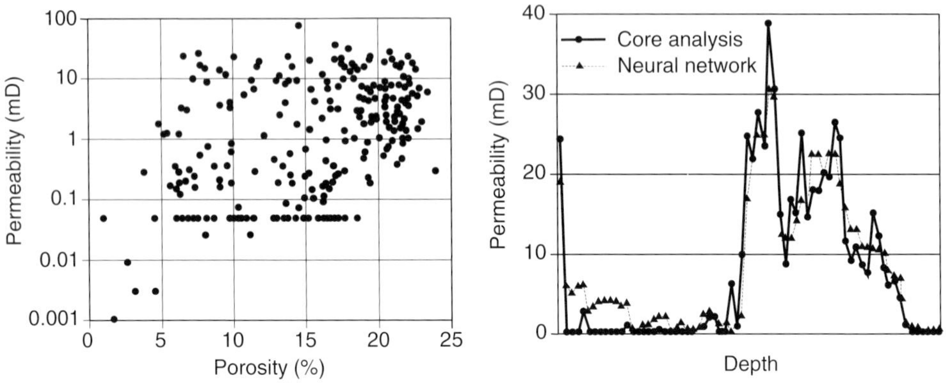

Figure 4.31 Permeability estimation from neural networks [41].

4.1.5.8 Integrating the Information

In the previous sections, we have described some of the most commonly used techniques to compute permeability at the well locations. However, it is evident that when we are to consolidate these different kinds of data in a single permeability model for the reservoir under study, distinctions have to be made concerning the type of information that each source provides.

In fact, the interrelationships between the different types of permeability estimation depend on three important factors [43]:

1. **The scale factor**. There is a scale factor attached to each type of measurement. In general, following the definitions given in Chapter 1, we have permeability information relevant to very different support volumes, from the macroscale (core analysis, WFT measurements, wireline logs and empirical correlations) to the megascale (flowmeters, well testing). The critical point is that each type of measurement encompasses different types of reservoir heterogeneity. Typically, for example, permeability from well testing is different from core permeability, since the former is affected from large scale heterogeneities that may reduce (e.g., presence of shales) or increase (e.g., by fractures) the estimated permeability value (Fig. 4.24).
2. **Measurement environment**. The various methods presented differ in that the measurement environment is not the same. Core analyses are usually performed at ambient temperature and pressure, while methods that use wireline logs provide estimated values at reservoir conditions. Another important parameter is fluid saturation: core per-

meability is measured on samples 100% saturated with a single fluid, thus providing values of absolute permeability, while flowmeter data or well test interpretation provide effective permeability values at the prevailing saturation conditions in the reservoir.

3. **Type of measurement**. Some of the listed techniques provide more or less direct measurements of permeability (e.g., core analysis), while others provide indirect estimation (e.g., regressions). The former will be, in principle, more reliable than the latter.

Table 4.2 summarises the relationships among the various permeability estimation techniques.

Table 4.2 Characteristics of various permeability estimation techniques.

Technique	Scale	Pressure & Temperature	Saturation	Method
Core analysis	Macro	Ambient/In-situ	Absolute	Direct
WFT	Macro	In-situ	Relative	Indirect
NMR	Macro	In-situ	Absolute	Indirect
Regressions	Macro	In-situ	Absolute	Indirect
Correlations	Macro	In-situ	Absolute	Indirect
Neural networks	Macro	In-situ	Absolute	Indirect
Flowmeter	Mega	In-situ	Relative	Indirect
Well testing	Mega	In-situ	Relative	Indirect

Before any integration is attempted, it is important to understand the nature of the information obtained from each type of permeability data. Most of the discrepancies that are frequently observed among the available data may find a justification in this table.

In general, reconciliation of the different available permeability data is not an easy job and there are no absolute rules. Even when the observed differences may be clearly related to some of the factors discussed above, the procedure to generate a single, consistent permeability model is not straightforward.

The best way to integrate permeability data is probably by a stepwise approach, aimed at progressively integrating data from the macro to the megascopic scale.

A. Macroscale Data Integration

At the macroscale, all types of indirect estimations should always be compared against corrected core data, when available. NMR logs, regression models, empirical equations and neural networks should always be calibrated to honour core measurements, unless doubts exist about the representativeness of these measurements (e.g. in the case of unconsolidated samples).

This procedure allows for the generation of synthetic permeability profiles for the largest number of wells, thus providing a consistent database for permeability modelling.

Additionally, when a facies description is available, average permeability values (or distribution functions) may be assigned to each individual facies. This procedure has the advantage of allowing a straightforward subsequent generation of a 3D permeability model, when a 3D model of facies is available.

B. Well Test Integration

The integration of macroscale permeability data and well test (megascale) permeability is a major issue in any reservoir study. Differences are often observed between these types of estimates.

In general, as already mentioned, these kind of measurements need not agree. Well tests measure an effective permeability to the prevailing saturation conditions in the reservoir and this value can be considerably different compared to the absolute permeability measured in the laboratory on core plugs. Also, the degree of reservoir heterogeneity involved in the two measurements can be very different.

One simple technique to integrate these estimates is to compare, for each well where both a macroscale permeability profile and an interpreted well test are available, the respective productivity thickness:

$$KH_{\mathrm{wt}}/K_{\mathrm{r}} = C\,\Sigma\,k_i h_i$$

where:

- KH_{wt} well test productivity thickness, mD/ft
- K_{r} relative permeability at the thickness-average saturation
- $k_i h_i$ macroscale (core or log/derived) productivity thickness, mD/ft
- C dimensionless correction factor

When using this equation, we assume that the well test result is the reference information and we tune the correction factor C until a satisfactory agreement is found between the 2 types of measurements. Note that in this case we are assuming a layered permeability model, where each i elementary reservoir interval of thickness h is contributing with its individual k permeability. This amounts to an arithmetic averaging of permeability values. It should be noted that, while this is the theoretical model to be applied for a layered reservoir, experience shows that in many instances this is not the case.

The correction factor C may actually account for a permeability model which is in some way different from the simple layered model. In fact, the correction factor C may include other uncertainty components, related for example to problems in defining the actual producing thickness in the well test interpretation, presence of heterogeneities or any other factor which can be related to the different scale domain of the 2 types of measurements.

The correction factor C has to be computed for all the wells and the tested intervals and the resulting values should be carefully analysed. When a reasonably constant value of C can be identified, this can be safely applied to all the wells having a vertical permeability profile but no well test data. This allows for conditioning the resulting permeability distribution to well test results. Note also that dividing the well test permeability thickness by the relative permeability value amounts to computing an absolute permeability value, which is the quantity required in the dynamic simulation phase.

In some cases, the values of the computed correction factor of C at various well locations may be too different to be considered of use. For example, when the values of C span between 0.2 and 0.9, this is too large a range to permit the identification of a characteristic value to apply to all the wells. These differences may actually suggest that there is a consistency problem in either of the log, core or well test data in those wells where the C factor presents extreme values. An accurate analysis of these wells should be undertaken, in order to correct or eliminate the spurious data.

A more complete integration of well test permeability can be obtained through direct conditioning of geostatistical models. This will be discussed in paragraph 4.2.4.2.A.

4.1.6 Net/Gross Ratio

Net/Gross ratio aims at representing the portion of reservoir rock which is considered to contribute to production. This ratio is computed by means of appropriate cut-offs on the log curves and then applied to the global thickness of the formation to compute the familiarly called Net Pay, the producible thickness. The concepts of Net/Gross ratio, cut-off study and Net Pay are therefore interrelated and in this context they will be used interchangeably.

The determination of the Net/Gross ratio is one of the typical steps in any reservoir study, which is practically always performed. Yet, it is also possibly the least documented and most subjective stage of the whole chain of a study. An exhaustive review of the published literature reveals that very little information is available on Net/Gross calculation methodologies. This is probably because no general rules exist and in most cases the choice of the cut-offs to be applied is an empirical procedure, which is left to the sense and the experience of the petrophysicist or the geologist.

In the following sections, after a general digression on the meaning of the cut-off concept, we will analyse the factors to be taken into account when selecting a cut-off value for a given reservoir and we will then review some of the applicable techniques to compute the Net/Gross ratio. In a final section, the real need for a Net Pay determination will be discussed and alternative strategies will be indicated.

4.1.6.1 The Cut-Off: a Dynamic Parameter

The cut-off is a threshold value applied to specific reservoir parameters, in order to split the formation into pay and non-pay sections. The implications of a proper selection of the type and value of the cut-off are important, in terms for example of volumetric estimates of the Oil in Place.

The main issue in discussing the choice of a particular cut-off is the understanding of its dynamic nature. This simple and rather obvious point is often neglected, as geologists tend to choose the cut-off values only on the basis of the lithological and petrophysical properties of the reservoir rock. To explain the dynamic nature of cut-off, let us consider the most typical approach to a cut-off selection, i.e., the core permeability/porosity cross-plots of the type shown in Fig. 4.32.

The usual approach to Net/Gross determination is the selection of a base permeability value (1 mD in this example) and the use of a regression function to derive the correspond-

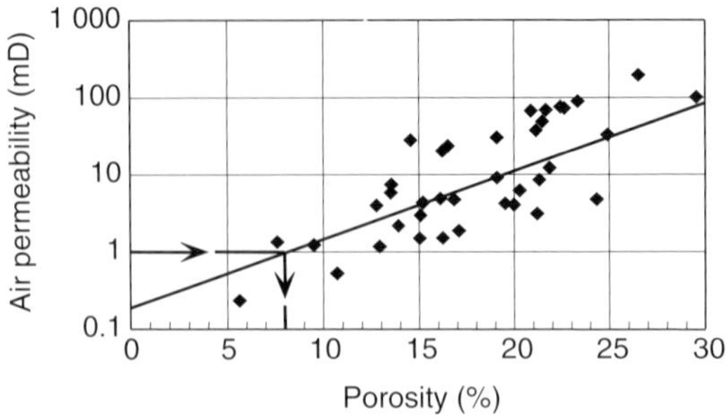

Figure 4.32 Typical permeability-porosity cross-plot.

ent porosity value. While the petrophysical properties of the rock are important in properly selecting the cut-off value, it should be appreciated that this value is also related to several other *dynamic* factors, like the saturating fluids, the depletion mechanisms and the recovery projects. The following points should therefore be taken into consideration:

- **Mobility vs. permeability**. The choice of the cut-off depends on mobility (K/μ), rather than on permeability only. This means that, for the same formation with a given permeability, the cut-off value may change considerably as a function of the reservoir fluid. Taking the extreme case of a dry gas reservoir and a heavy oil reservoir, where the difference in viscosity can be as high as 4 orders of magnitude (e.g., 0.02 against 200 cP), the cut-off value will follow the same rule. This is why many gas reservoirs with permeability in the range of the microdarcy are producing around the world, while much higher cut-offs, possibly in the 10 – 100 mD range, are usually applied to heavy oil reservoirs. Note that this also implies the somewhat surprising consequence that the same reservoir will have different Net/Gross and Net Pay maps depending on the reservoir fluid. Values usually adopted for the mobility cut-off are in the range of 0.5 – 1 mD/cP.
- **Air vs. oil permeability**. As already discussed in a previous section (4.1.5.2), core permeability should undergo all the necessary corrections (Klinkenberg, overburden) before being used in the cut-off determination. In addition to that, effective (oil) rather absolute (air) permeability should be used, since this is the actual saturation condition in the reservoir. Absolute permeability always being higher than effective oil permeability, its use will lead to overstated results of oil in place. Fig. 4.33 compares the regression functions determined from a set of air and oil permeability data. It can be noted that for the same entry permeability value of 10 mD, the difference in the porosity cut-off is around 3 p.u.
- **Dependence on pressure drawdown**. In a depletion drive reservoir, the cut-off depends on the pressure drawdown. It is somehow intuitive that a greater drawdown will allow for the drainage of relatively tight rocks, that would not produce under smaller pressure differentials. Therefore, during the depletion of a reservoir, the actual cut-off changes, since as higher drawdown are imposed on the formation, progressively tighter rocks will contribute to flow. This process is time dependent and will be

in general more severe in a heterogeneous, layered formation. The example highlights the inherent difficulty of choosing a single, *correct* cut-off value.

- **Dependence on depletion mechanism**. In a water drive reservoir, or during waterflooding operations, where the pressure is being mantained, the cut-off will depend, in general, upon the efficiency of the sweeping process. A saturation cut-off, based on a selected value of the residual oil saturation *Sor,* may be more appropriate in these cases. Alternative methodologies for calculating the net-pay in waterflooded formations have been proposed recently [44].
- **Dependence on the exploitation stage**. As a consequence of the last 2 points, the cut-off value in a given reservoir may be different during a primary and a secondary recovery project. For example, if primary production can be related to natural depletion and the secondary recovery is represented by a water injection project, then the cut-off value will be in general different in the 2 cases. This leads to the conclusion that the same reservoir, with the same saturating fluid, should be described with different net pay maps before and after the start of a recovery project.

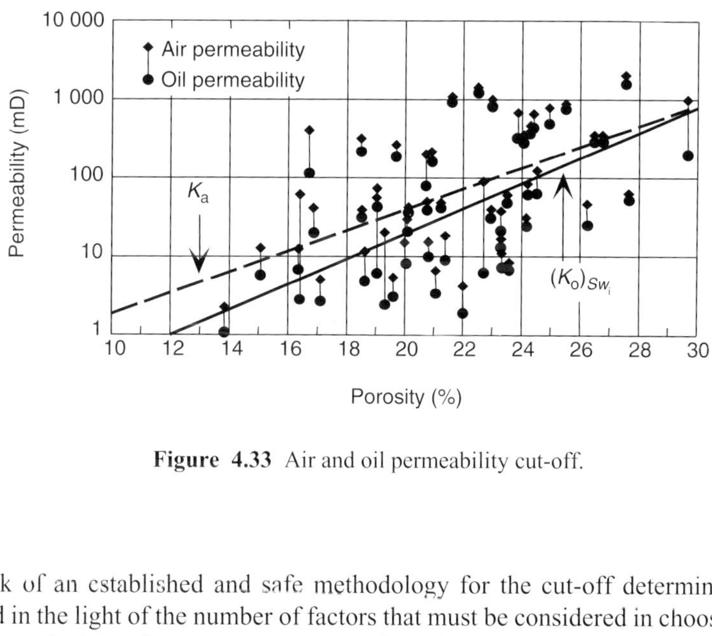

Figure 4.33 Air and oil permeability cut-off.

The lack of an established and safe methodology for the cut-off determination can be understood in the light of the number of factors that must be considered in choosing a cut-off value. In fact, the issue frustrates any systematic and analytical approach and in most cases the solution has to be found on a case by case basis. This in turn implies that there will always be a considerable degree of subjectivity in this process.

4.1.6.2 Defining Cut-Off Criteria

The main problem in a cut-off study is the definition of the starting (base) permeability value, that adequately represents the threshold between producible and non-producible rocks.

From a general point of view, the knowledge of the lithological type, the reservoir fluids and the producing mechanisms will allow for the determination of the likely range of the cut-off value. Other data like well testing results, WFT mobility and NMR free-fluid index profiles sometimes yield valuable additional information. In the case of old fields, it is often interesting to verify the different cut-off values adopted in previous studies and take advantage of the huge production information available. In some cases, for example, historical resistivity cut-off based on field experience prove to be very robust.

Once the base cut-off value has been defined, whether it be permeability, saturation, resistivity or any other variable, the determination of the values of the other petrophysical variables is normally straightforward. Typically, once a base permeability value is chosen, then porosity can be derived from a core permeability/porosity cross-plot, while the volume of shale can be found in the relevant Porosity/Vshale cross-plot. The same holds for water saturation. This procedure normally allows for the selection of a consistent set of petrophysical cut-offs. The methodology is schematically illustrated in Fig. 4.34.

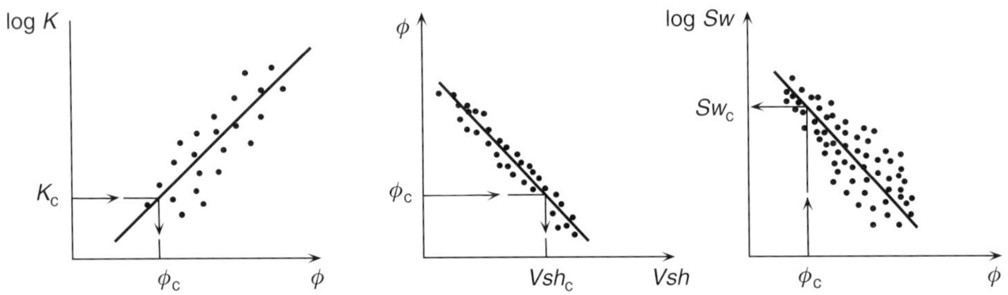

Figure 4.34 Typical procedure for the definition of a set of petrophysical cut-off values.

An alternative way of defining a Net Pay is through the concept of facies. The simplest procedure is the definition of a binary classification, let's say pay-non pay, or shale-sand. In this case, the ratio between the pay and the total thickness provides a simple, yet sometimes viable cut-off.

When a facies classification scheme is available, based for example on a cluster analysis study, then a more robust approach can be attempted. If the facies have also been characterised from a petrophysical point of view, then a combination of these facies can adequately describe the pay concept. In these cases, Net Pay or Net/Gross maps can be constructed as percentage of the pay facies thickness over the total thickness.

4.1.6.3 Notes on Cut-Off Application

The application of cut-offs and the determination of Net Pay maps is one of the oldest and most traditional works of reservoir geologists. But why do we apply cut-offs to rock formations?

In its essence, the cut-off concept aims at defining the *effective* petrophysical properties of a given geological unit, in the presence of poor reservoir zones. Considering for example a relatively thick sand-shale sequence, it is clear that different average porosity and permeability values will result depending on whether or not a cut-off is applied. While this is possibly unimportant for porosity, which is an additive variable, it is clear that calculating an average permeability that includes the low values could lead to a severe underestimation of the effective permeability of the layer [2]. Note that, from this point of view, the cut-off definition can be conceptually considered as an *upscaling* or *homogenisation* problem.

As with any upscaling issue, the cut-off is dependent on the support volume and its application should therefore be considered differently when working in 2D and in 3D.

When working in 2D, with a limited vertical resolution (number of layers), the cut-off study is an essential part of the work, since in most cases the individual geological layers include several lithological units with very different petrophysical properties. In this case, these geological layers represent a large support volume, which need a proper homogenisation procedure.

In 3D, on the other hand, the cut-off problem should be addressed in a different way. The possibility of working in greater detail makes the cut-off concept much less relevant, since in this case the average petrophysical properties of a given unit will be closer to the effective properties. Consider for example a fine scale facies stochastic model. In this case, it could be argued that the problem of defining a cut-off could be avoided completely. In fact, the total pore volume can simply be computed as the summation of individual facies, regardless of their pay or non-pay characteristics, while facies effective permeability can be determined through adequate upscaling procedure. In the simulation model, on the other hand, each facies (rock type) can be characterised with individual saturation functions (possibly derived from some pseudisation procedure). This is the reason why the cut-off is often neglected when working in 3D. The simulation model itself has all the information to decide whether or not a particular facies (or rock type) is pay, under all the possible static and dynamic circumstances. For example, if a facies has a very low porosity, capillary pressure will be very high and the facies will not contribute to flow, unless a very high drawdown is imposed. Therefore, from this point of view, we do not need an *a priori* judgement of whether or not that particular facies will contribute to flow.

Nevertheless, from a practical standpoint, a cut-off is often beneficial even in 3D. Eliminating the poorest facies from the simulation may help the numerical performance of the model and the running time, while the overall accuracy of the results will be unaffected. Fig. 4.35 shows an example, for a geostatistical reservoir description, where the non-pay facies have been removed. Note how these kinds of application software allows the geoscientist to quickly visualise the main interconnected paths through the reservoir under different cut-off hypotheses.

In conclusion, the cut-off is still a useful concept, provided that its dynamic nature is fully understood. From this point of view, it is interesting to note that in the SPE definition, Net Pay represents **the portion of the reservoir containing oil and gas reserves which are economically recoverable**. This definition is perhaps the most striking evidence that the concept of cut-off cannot and must not be considered a static parameter, as it is implicitly treated in most cases.

---

2. It should be noted that if a porosity-permeability relationship is used to generate the fieldwide permeability distribution, than an underestimation in the average porosity value will result in a correspondent underestimation of permeability, even when the volumetric computation is correct.

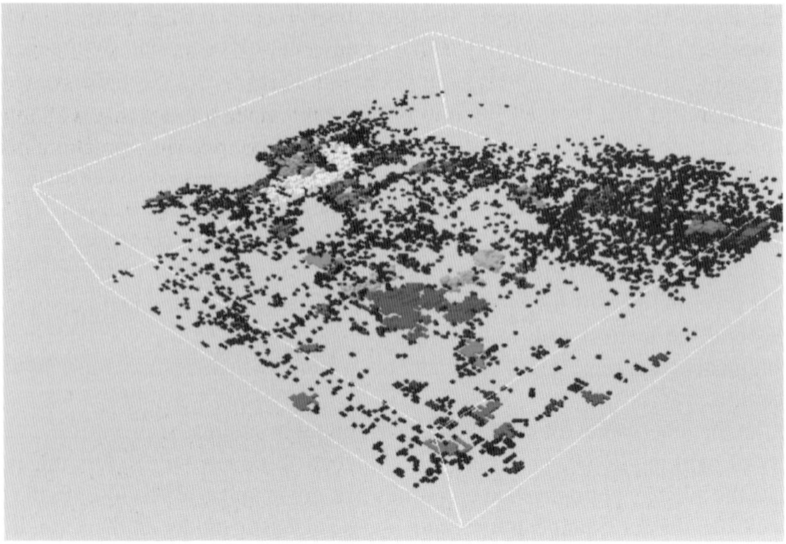

Figure 4.35 Facies Cut-off applied to a 3D description of the reservoir.

4.2 ROCK PROPERTIES DISTRIBUTION

The objective of the petrophysical interpretation, as discussed in the first part of this chapter, is to estimate the various petrophysical properties of the reservoir rock and their interrelationships, at the well locations.

The following step of a typical reservoir study is the analysis of the lateral variation of these properties, in order to generate 2D or 3D distributions at the scale of the reservoir. In this phase, the lack of information in the interwell regions represents the most significant challenge for the geoscientist.

For many years, interpolation was the only available technique for constructing reservoir maps. In fact, the conceptual geological model of the reservoir was the only guide to a sound interpolation of measured well values. This was performed either by hand contouring or, later, through more sophisticated computer mapping algorithms.

In the last 10 years the picture has changed, as new techniques which allow for more detailed reservoir descriptions have become increasingly common. Possibly the most noteworthy of these techniques is geostatistics. Geostatistics offers the possibility of studying the spatial variability of a given reservoir property (or a combination of properties) and to exploit these relationships in the definition of the reservoir model. In addition, geostatistics allows for the integration of different sources of information and in this respect it can be considered an effective integrated tool for reservoir description.

Another relatively young technique, which has considerably improved the capability to build accurate reservoir spatial models, is geophysics. Geophysics, particularly in recent years, has gained increasing importance in the field of reservoir characterization. The availability of improved computational power and new processing schemes have liberated seismology from its original primary purpose, i.e., structural mapping, and several new techniques, based on seismic attributes modelling, allow for a deeper insight in the interwell region.

In the next sections, we will see how these different techniques (interpolation, geophysics and geostatistics) can be utilised and interrelated to build a sound reservoir model in 2 or 3 dimensions. We will be dealing with the same properties already discussed in the previous sections, i.e., porosity, water saturation, net pay and permeability.

4.2.1 Porosity

Starting from the petrophysical interpretation available at the well locations, possibly averaged for given vertical sections (the reservoir layers), a spatial distribution of porosity can be obtained in different ways, depending on the amount and quality of the available data, the degree of desired accuracy and the time and the resources allocated to this phase.

4.2.1.1 2D Interpolation

The simplest way of obtaining a porosity map, for a given reservoir layer, is by simple interpolation of posted well values. This is usually done by means of computer mapping software, even though many geologists still prefer to generate hand contour maps and to digitise them afterwards. The procedure is in this case rather cumbersome, however it allows the geologist to keep maximum control over the generated porosity distribution, while holding to the conceptual geological model.

In general, the direct interpolation of well porosity values provides reasonable maps, when sufficient well density is available and when the spatial variations are related to large scale processes, like primary depositional patterns in clastic sediments. Of course, a consistent and reliable porosity database is mandatory in order to obtain meaningful results. Moreover, attention must be paid to uncontrolled extrapolation in unsampled areas.

An alternative to the largely available computer mapping software is provided by the simplest application of geostatistics, i.e., **ordinary kriging**. In this case, the correlation function of the variable being described, porosity in this case, is built with the available data, rather than assumed *a priori* by the gridding algorithm. This correlation function (the variogram or its inverse, the covariance), provides the spatial continuity of the variable, as well as its directional anisotropy. In addition, ordinary kriging provides a measure of local uncertainty (kriging variance).

When sufficient data are available for the construction of the experimental variogram, the use of kriging may result in improved results compared to the conventional interpolation algorithms. Fig. 4.36 compares 2 porosity maps obtained through conventional mapping (left) and ordinary kriging (right). The directional character and the range of continuity of the variable in unsampled areas represent the contribution of the correlation function (the variogram) to the resulting distribution.

It should be emphasised that any interpolation, including ordinary kriging, represents a smoothed image of the actual distribution of the variable under study. These algorithms

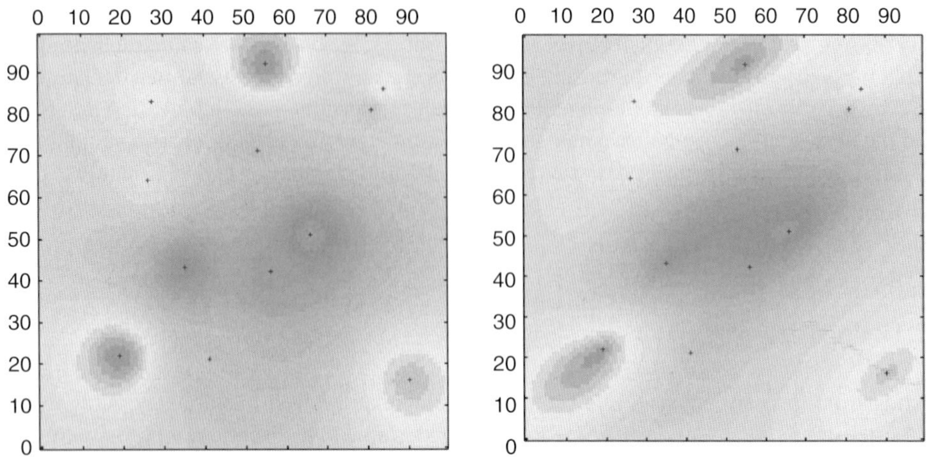

Figure 4.36 Porosity distributions: conventional (left) and kriging (right) maps.

actually work as a low-pass filter, which tends to smooth local details of the true spatial variability. However they have the advantage of providing a unique response, in terms of a single porosity distribution. In the case of ordinary kriging, this distribution minimises the estimation error and, in statistical terms, it is the *best* estimation according to the input model.

When reproducing the porosity heterogeneity is an issue, a different approach can be attempted, based on geostatistical simulation (or stochastic modelling, as it is often referred to). These algorithms allow for the preservation of the experimental (observed) variability, however they have the disadvantage of providing multiple equiprobable realisations of the spatial distribution under study. Therefore, the application of stochastic techniques to static parameters like porosity or Net Pay is generally limited to those cases when a probabilistic approach to the Oil in Place estimate is sought and when the overall uncertainty is to be quantified. In all other circumstances, the unique, average porosity (or Net Pay) distributions obtained by interpolation or kriging normally provide reliable estimates for volumetric computations.

4.2.1.2 Seismic Data Integration

Seismic data represent the only direct source of information in the interwell region of a reservoir. Until a few years ago, the use of this information for the sake of internal reservoir description was very limited. In fact, although it was recognised that variations in the lithological and petrophysical characteristics and fluid content of the reservoir rock had an impact over the recorded seismic signal, the data resolution and the processing capability were not sufficient to for detailed reservoir applications.

Nowadays the situation is different, and in most cases seismic information can provide valuable input to the reservoir characterization phase.

The basic idea behind seismic data integration is the use of some attribute of the recorded seismic volume to guide the generation of the distribution of the variable of interest, porosity in this case. From a general viewpoint, the procedure can be summarised in the following points:

1. **Calibration phase**. This is the first basic step of the integration. Wells provide high frequency, depth related information, while seismic data have low frequency, time related information. These data must therefore be calibrated, both vertically and areally. Sonic logs, checkshot velocity surveys and Vertical Seismic Profiles are typically used in this phase.
2. **Seismic attribute identification**. Modern interpretation packages offer the possibility of computing a large number of seismic attributes of different nature: punctual or averaged over time windows, based on amplitude, time or on the complex seismic trace. In addition, sophisticated seismic inversion techniques allow for the generation of volumes of acoustic impedance, another powerful attribute. The main objective of this phase is to identify the attribute that works better as a predictor for the variable of interest.
3. **Prediction**. The areal distribution of the variable of interest can be generated by integrating the well and seismic information. This phase is normally performed using linear or non-linear regression models or geostatistics.
4. **Cross-validation**. This step is performed by sequentially removing all the wells from the conditioning dataset one by one and estimating their values from the other wells. Although not always performed, this procedure provides a means to quantify the improvement related to the use of the seismic attribute.

As an example of a typical integration of seismic data, consider the porosity map shown in Fig. 4.37. This map has been obtained by simple contouring of well data, without any additional constraints and it displays quite a smooth appearance.

When an amplitude map is available, a relationship can be sought between this parameter and porosity, with the objective of utilising the denser seismic information to improve the modelling in the interwell area. Fig. 4.38 shows the seismic amplitude map relevant to the previous example, together with the observed relationship between porosity and amplitude derived at the well locations. The latter comprises 12 wells and shows an acceptable correlation factor of 0.82.

The simplest seismic data integration is to use such a correlation to transform the amplitude map directly into a porosity map. This map should then be adjusted to honour the measured porosity values at the well locations. In this case, as it is a simple linear transformation of the amplitude distribution, the resulting porosity map will look very similar to the input amplitude map.

In more sophisticated applications, seismic information can be integrated through a geostatistical approach. The basic principle in this case is to estimate the variable of interest by means of a correlation function as defined by well data (the variogram), plus a cross-correlation function that quantifies the relationship with another, related variable, i.e., the seismic information. This procedure is called **cokriging**.

From a theoretical viewpoint, the correct way to operate is to solve the full cokriging system of the 2 variables, porosity and seismic attribute. However, from a practical standpoint,

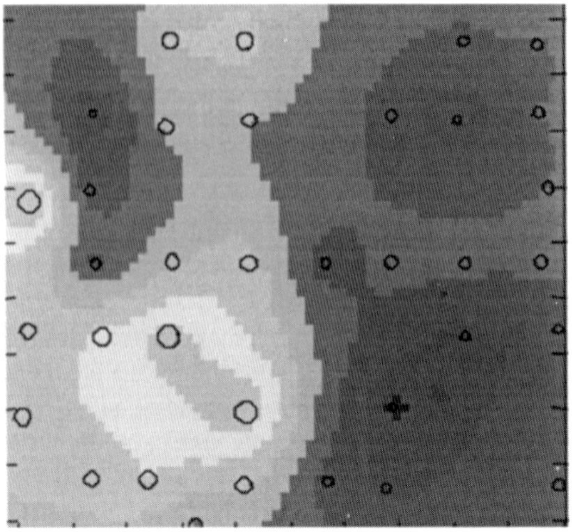

Figure 4.37 Porosity distribution (wells only).

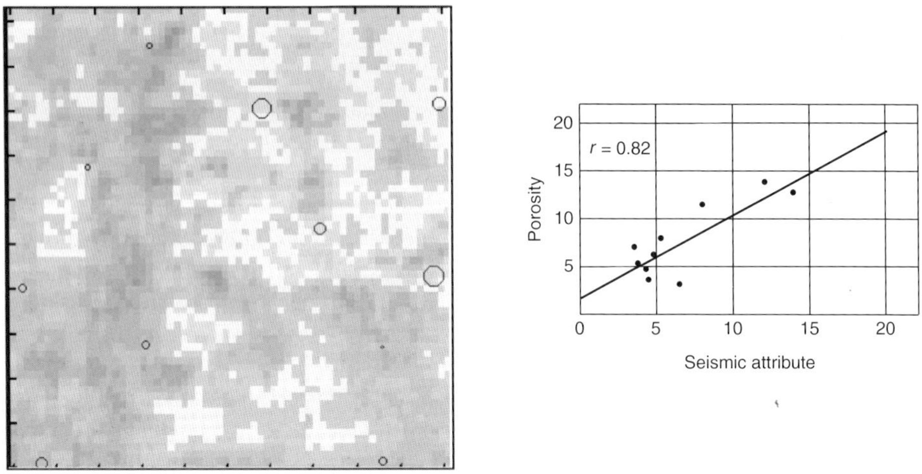

Figure 4.38 Amplitude map and relevant amplitude-porosity relationship.

cokriging is a cumbersome procedure and moreover, when applied to the dense population of seismic data, often gives problems of matrix instability in the solution. Therefore, simplified forms of cokriging, called **collocated cokriging**, are normally utilised [45]. The advantage in this case is that the implementation of collocated cokriging does not require the construction of the cross-variograms, but only the knowledge of the correlation coefficient between the 2 variables [3].

An example of application of collocated cokriging is shown in Fig. 4.39, relevant to the same porosity and amplitude datasets already shown previously. Comparing this image with Fig. 4.37, the higher degree of detail and the overall more realistic appearance of the resulting map can be appreciated. This kind of application in fact constitutes a more complete integration of the well and seismic information.

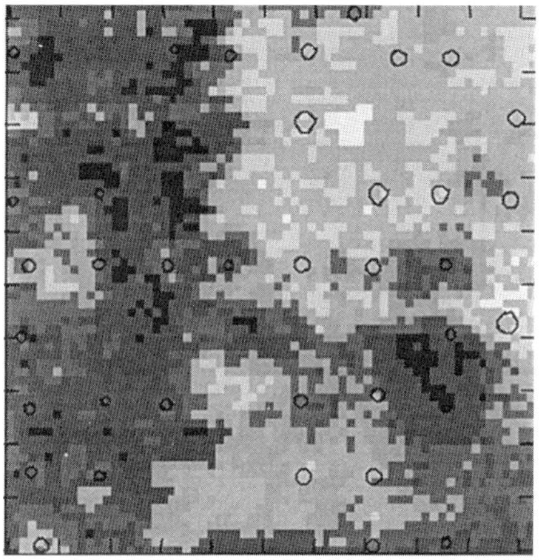

Figure 4.39 Porosity map obtained by collocated cokriging.

In conclusion, the integration of seismic information in reservoir description can lead to a considerable improvement in the quality of the final distribution. However, a number of points must be considered, in order to make sure that a proper integration is performed.

- **Choice of the seismic attribute**. Provided that the calibration phase is properly carried out, the choice of the seismic attribute to use is the critical step of the whole procedure. This choice is usually carried out by observing the existing relationship between porosity measured at the wells and the values of the various seismic attributes at the same locations. Care must be exercised in this work, since it is not unusual to observe a spurious correlation, i.e., a false good correlation coefficient that does not reflect any physical relationship between the variables. The probability of finding a spurious correlation increases with the number of seismic attributes considered and is inversely proportional to the number of observation points [46].

3. Another alternative approach to the full-cokriging system is the so-called **external drift** model, whereby the distribution of a secondary variable (the seismic attribute in this case) is assumed to be related to local trends in the primary variable. The shape of the (known) distribution of the secondary variable will therefore be imposed in this case on the (unknown) distribution of the primary variable.

- **Influence of other rock and fluid parameters**. The seismic signal is a complex response that is influenced by a number of parameters, like lithology, fluid content and overburden formations, in addition to porosity. Once a useful relationship is found between any seismic attribute and porosity, we must make sure that lateral variations in those parameters are not influencing the seismic response (and hence porosity estimation) away from the wells. An interesting example in this respect is provided by the giant Norwegian Ekofisk Field. Here, seismic impedance has been utilised to estimate porosity in the flanks of the structure, based on the favourable factor that lithology does not change throughout the field. In this case, variations in the seismic amplitude have been related to changes in the petrophysical properties of the reservoir rock. In the crestal part of the field, on the contrary, the presence of gas-saturated overburden sediments alters the seismic signal and does not allow for the use of any seismic derived attribute [47].
- **Validation**. Any estimation by means of seismic data has to be validated using well data, which is always the reference information. The cross-validation phase is an important part of the work and should not be neglected. In some cases, for example, good correlation coefficients may be found between seismic impedance and well porosity data in the calibration wells, while other wells may show poor correlation in the validation phase. These cases, when not detected, may lead to the generation of unreliable distributions.

The application of seismic data to reservoir description is still a growing technology, which requires good technical expertise and some prudence. However, its use is quickly expanding and more and more geoscientists are learning its use and understanding its advantages.

Most importantly, the potential of this technique is significant. There is little doubt that, in a few years, integrating seismic and well data will be routine practice in reservoir description.

4.2.1.3 3D Modelling

Many geoscientists still have a 2D perception of petroleum reservoirs. This is because the traditional way of studying and visualising petroleum reservoirs is through a series of 2D maps, which illustrate the geometrical and petrophysical characteristics of the reservoir at various stratigraphic levels.

In recent years, however, the evolution of computing capabilities and the introduction of new modelling theories, have made it possible for the geoscientists to work in a 3D space, with a resolution previously unknown. Two techniques will be mentioned here, geocellular modelling and geostatistics.

Geocellular modelling is a general term that describes the possibility of generating detailed 3D geological models starting from sparse well data. As already mentioned in Chapter 3, several applications are currently available in the market, which allow for the generation of 3D distributions of any reservoir property, starting from well vertical profiles. These packages usually offer flexible visualisation options and the reservoir can be modelled by means of fine scale, three dimensional grids, rather than by a series of maps.

On the other hand, it should be noted that the algorithms used in the generation of these 3D distributions are deterministic and not substantially different from those existing in the conventional 2D mapping software.

As a matter of fact, 3D geological modelling can be considered the domain of geostatistics. Since the beginning, the extension of the geostatistical theories to the 3D space appeared straightforward and all the currently available software for geostatistical modelling work in 3D. In fact, the popularity of geocellular packages is often related to their use as post-processors of external geostatistical algorithms.

In general, porosity can be modelled by means of geostatistics through one of the 2 following approaches:

- **Direct estimation**. Porosity is modelled directly in the entire volume of the reservoir, starting from a number of vertical porosity profiles at the well locations and a spatial correlation function.
- **Two-stage model**. A 3D lithological model of the reservoir is generated first, using a selected number of facies. Porosity is modelled afterwards, within each lithological facies.

The direct estimation of porosity gives satisfactory results in simple lithological environments and in general in all those cases where lithology is not the controlling parameter in terms of petrophysical properties.

The two-stage model, although more demanding, gives much better results in complex lithology reservoirs and in general in all those cases where lithology is the main controlling factor. Note that this approach is much more flexible than the previous one. In fact, when sufficient information is available (and when it is considered geologically correct) a different correlation function can be used for porosity within each facies, or group of facies. For example, a different model for the variogram could be chosen in overbank and channel deposits of a deltaic system.

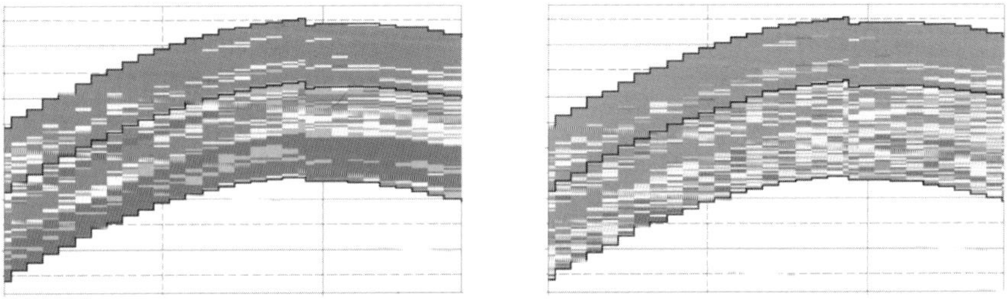

Figure 4.40 Two-stage geostatistical modelling: facies (left) and porosity (right).

Figure 4.40 (right) shows a cross-section of porosity values obtained from a two-stage model. In the same figure, the lithological model which has been used as a basis is shown on the left. Poor reservoir facies (light grey on the left figure) clearly correlate to low porosity zones in the correspondent porosity simulation, thus highlighting the dependence of the final porosity distribution on the prior lithological model.

Finally, it should be noted that the direct integration of the seismic information in 3D is a difficult problem, since the vertical resolution of any seismic acquisition is far poorer than the typical vertical scale of the geological description. Research is ongoing on this subject, but for the time being no satisfactory method exists to perform such integration.

4.2.2 Water Saturation Distribution

Water saturation is a complex petrophysical property. As already discussed in paragraph 4.1.4, the definition of a reliable water saturation profile at the well locations is in itself a difficult task, since several parameters are necessary for the estimation process.

When dealing with lateral (2D) or spatial (3D) distributions of water saturation, we are faced with the additional problem of finding a model that can correctly represent the spatial variability of this property when starting from a sparse set of well data.

In reservoir geology practice a number of methodologies are usually applied that allow for the generation of water saturation distributions. While in themselves these techniques are all applicable, when the objective of the work is an integrated study the perspective is different and some techniques are more useful than others.

In the next sections, we will discuss some of these techniques, their advantages, their limitations and their applicability.

4.2.2.1 Direct Mapping of Water Saturation Values

This is the oldest and possibly still the most used methodology for water saturation modelling. The technique is simple, in that it requires a data set of interpreted wells where a water saturation curve has been interpreted. Areal distributions can then be generated by means of automatic contouring packages, by interpolating the average water saturation values at each well location. Water saturation maps have to be generated for each geological layer.

This technique works well in the cases of dense well populations, where many points are available for a safe contouring application, and it also has the advantage of honouring well values. However, a number of points have to be addressed:

- No explicit correlation is considered with the other petrophysical properties, i.e. porosity and permeability.
- In water drive reservoirs or in the case of water injection projects, the number of wells available for the mapping can be greatly reduced, since many of the wells may have been invaded at the time of the logging. Since the objective is to generate a distribution of the original water saturation, invaded or partially invaded wells cannot be utilised.
- When using conventional interpolation algorithms, the shape of the transition zone cannot be correctly reproduced. Actually, the transition zone, or capillary fringe, shows a non-linear relationship between depth and water saturation, that cannot be properly handled by the interpolation algorithms. These procedures may lead to an overestimate of the quantity of water in the transition zone.
- When the available water saturation data set is not consistent, the interpreted distribution may carry a significant amount of random error, which is actually related to a lack of calibration of the original logs or differences in the interpretative parameters. Often,

some of the shapes that we are used to seeing in this kind of maps are actually related to consistency problems of the input data.

- In the dynamic simulation model, the original water saturation distribution is calculated by means of capillary pressure functions. This may not be consistent with the map obtained by interpolating water saturation data. In turn, this may generate inconsistency between the hydrocarbon in place calculated in the simulation model and the volumetric estimate.

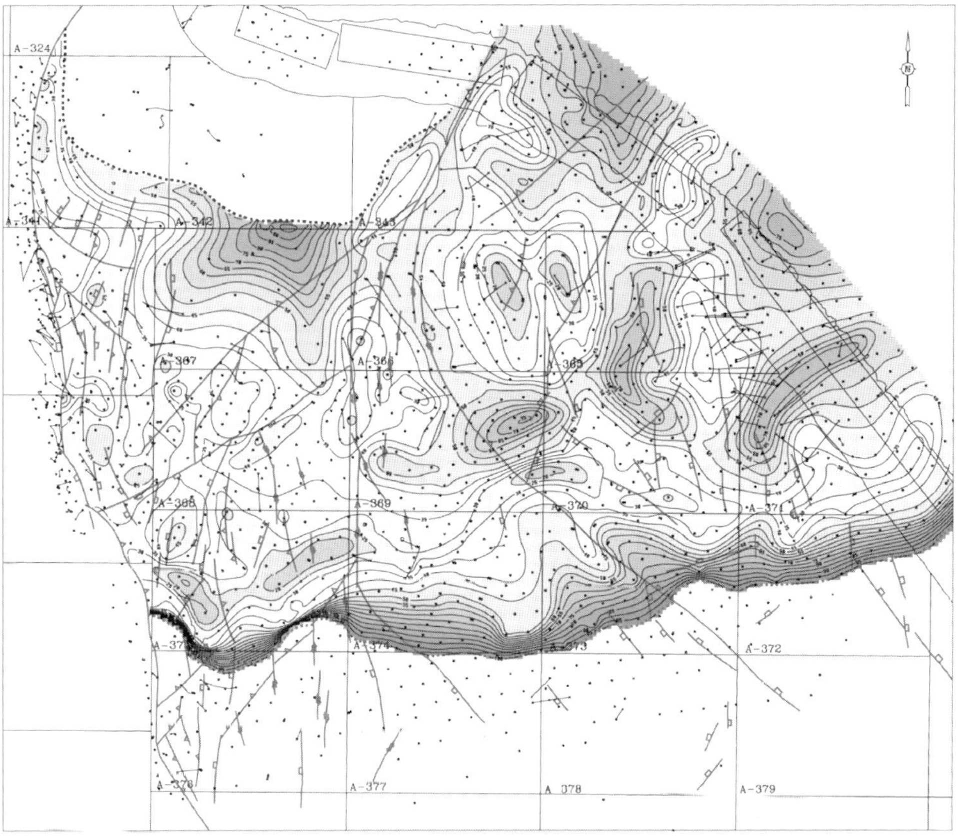

Figure 4.41 Water saturation map by direct well values interpolation (Courtesy of PDVSA E&P).

Figure 4.41 shows a water saturation map generated by direct contouring of well values. The example is related to a giant Venezuelan field, where a water-flooding project has been active for almost 30 years in the southern edge of the field. A number of peculiar features can be noticed, like the presence of spots of high water saturation values in the middle of the reservoir, which are actually related to extrapolations of the algorithm in poorly controlled areas. Also, the transition zone appears to be too extended which could in fact be related to the improper inclusion in the data set of wells already partially invaded by the injected water.

Among the advantages of this method is the possibility of dealing with laterally changing reservoir properties. Since the technique honours the well values, it accounts for lateral changes in the pore system characteristics. Of course, to capture such variability, a consistent petrophysical interpretation is mandatory.

4.2.2.2 Water Saturation/Porosity Relationship

Another popular technique for estimating water saturation distributions is the direct application of a water saturation/porosity relationship. In those reservoirs where the geometry of the pore system is relatively simple, a linear relationship is often observed between these parameters in semilog scale. This can be exploited in the modelling procedure.

The method is simple and fast and it also allows the generation of a consistent water saturation distribution, when unambiguous relationships can be found. Fig. 4.42 shows an example of a coastal, fine-grained sandstone reservoir. The cross-plot has been constructed using 10 wells that are far from the transition zone. The use of such a relationship, which shows an acceptable correlation function, will result in a reliable saturation map.

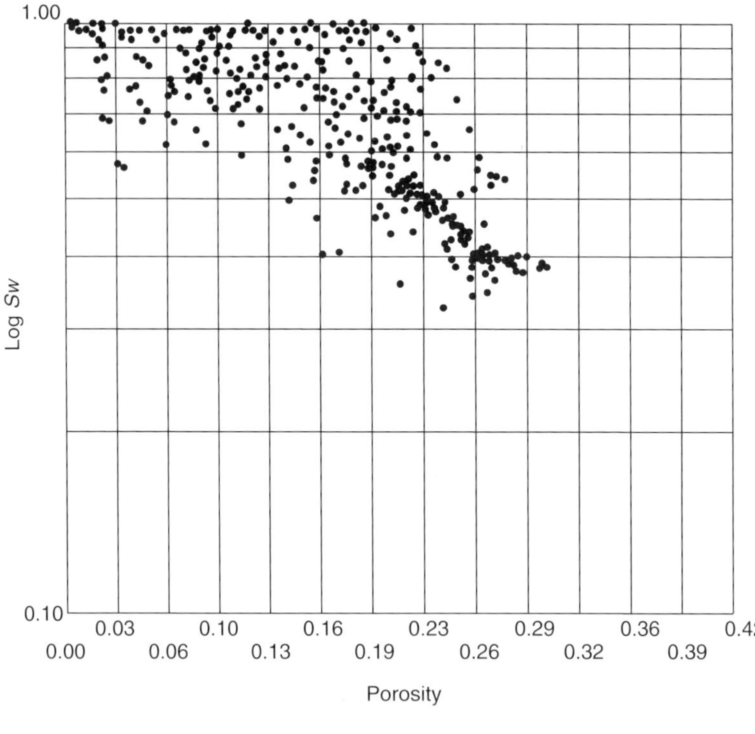

Figure 4.42 Water saturation/porosity relationship for a coastal sand.

The main disadvantage of this method is that the resulting water saturation distribution is independent of the height above the contact. In other words, the physical phenomenon of the decrease of water saturation with the distance from the water free level cannot be taken into account and the transition zone cannot be modelled. The technique is therefore suited for those reservoirs where a hydrocarbon-water contact has not been identified or where the height of the transition zone is negligible, for example in gas reservoirs.

4.2.2.3 Capillary Pressure Functions

Water saturation distributions can be calculated through the application of a capillary pressure function, estimated by means of multiple linear regression techniques. The concept in this case is that water saturation is dependent on the petrophysical properties of the reservoir rock, as well as the height above the Oil-Water Contact (OWC). Therefore, if a relationship can be established among these parameters at the well locations, it can then be applied throughout the field using the grid operation options of a common mapping software.

The general procedure for the determination of such a relationship is the following:

1. Select a number of wells representative of the different petrophysical ranges observed in the reservoir and located at different depths above the contact. Some of these wells should penetrate the OWC.
2. Perform a multiple linear regression to compute the m coefficients of an equation of the type:

$$\log Sw = m_1 \phi + m_2 \log h + \ldots + m_n$$

or, alternatively:

$$\log Sw = m_1 \phi + m_2 1/\log h + \ldots + m_n$$

where ϕ is porosity and $\log h$ is the logarithm of the height above the OWC. Other terms can be added if available, like for example the volume of shale. The regression can be performed using either the log curves or the capillary pressure functions derived from special core analysis. Alternative analytical expressions can also be considered. The use of a particular equation depends in general on the water saturation behaviour above the transition zone and should always be checked against the calibration data.
3. Validation of the computed coefficients by controlling the statistical parameters of the regression. Visual comparison of the simulated and interpreted water saturation curves for the input wells (Fig. 4.43).
4. Application of the computed equation, using a standard mapping package. Input grids will be the porosity distribution and a grid of the height above the OWC of the mid-point of the reservoir layer under study.

This methodology is more comprehensive than the previous ones, since most of the parameters that influence water saturation can be taken into account. In addition, the possibility of integrating core and log data allows for a consistent use of all the available information. The capillary fringe can usually be reproduced accurately, as well as the irreducible water saturation zone.

Figure 4.44 shows the water saturation map, estimated by this method, for the same geological unit of Fig. 4.41. The difference is evident, both in the definition of the transition

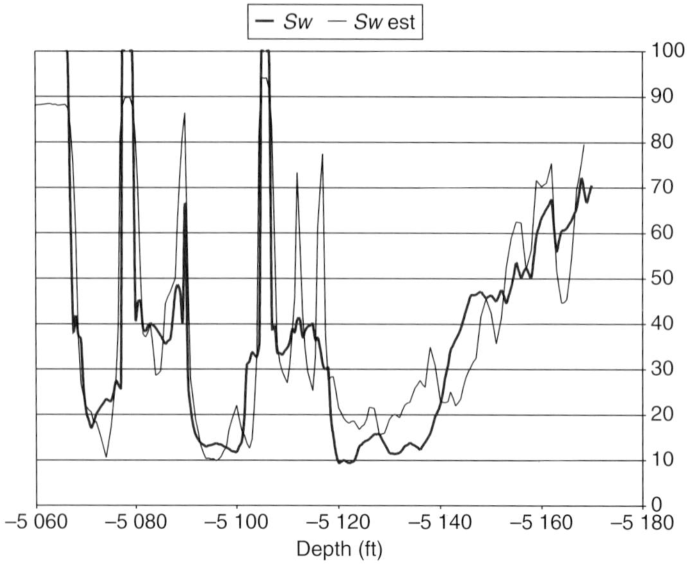

Figure 4.43 Interpreted vs. estimated *Sw* curves.

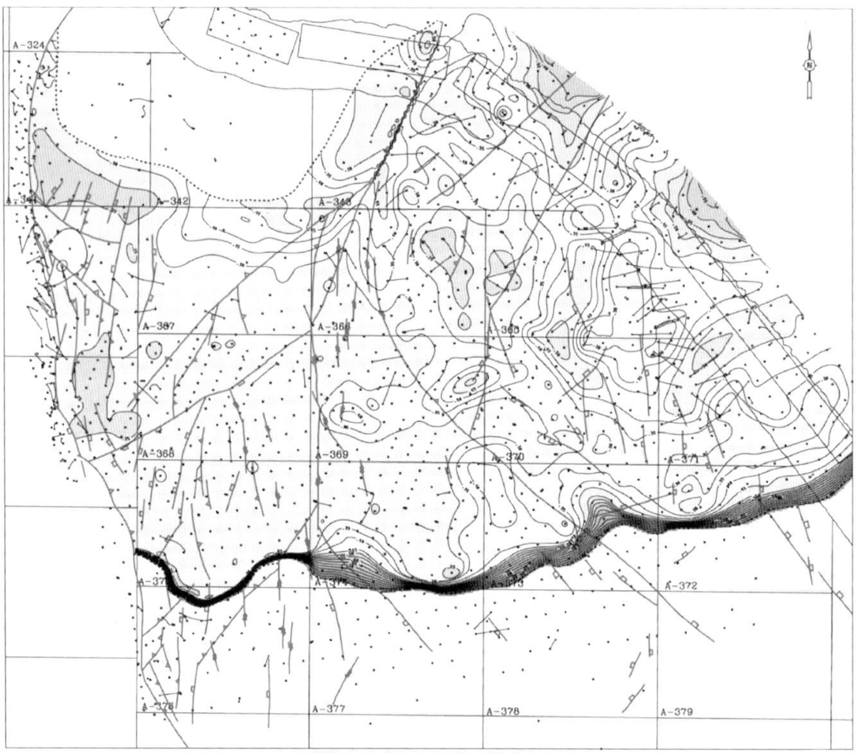

Figure 4.44 Water saturation distribution through capillary functions (Courtesy of PDVSA E&P).

zone and in the irreducible water saturation distribution. In this case, the use of the water saturation map of Fig. 4.41 resulted in an underestimation of the OOIP by 20%.

The maps constructed by capillary pressure functions have the advantage of generating distributions of porosity and water saturation that are mutually consistent. Also, these same functions can be utilised in the dynamic simulation phase, thus minimising coherency problems between the static and the dynamic models of the reservoir.

This technique also has some disadvantages however. When lateral variations in the pore geometry system exist, which result in a different porosity/water saturation relationship, these cannot be taken into account, since one single equation is applied to each geological horizon. Other problems may arise from the difficulty of identifying a satisfactory regression model, due to lack and/or inconsistency in the input data.

However, this is the recommended technique for water saturation distribution, especially in those cases where the final objective of the study is a simulation model.

4.2.2.4 3D Water Saturation Distributions

Water saturation, as other petrophysical properties, can be estimated in 3 dimensions. Some of the commercial G&G packages allow for the generation of such distributions, however it is important to verify which algorithm is used in this operation. Actually, in contrast with other petrophysical properties, water saturation has a distinct, non-linear dependence on depth (distance from the free water level), which makes it unsuitable for conventional, least-square type estimation algorithms.

Whenever possible, it is recommended to independently generate water saturation vs. depth curves per facies or porosity classes, through an accurate integration of the available information (Fig. 3.14). These curves should then be input to the geocellular modelling software. From this point of view, it should be noted that capillary pressure functions prove to be particularly suited for 3D distributions generation, since depth (the third dimension) is properly accounted for.

For the same reason (non-linearity with depth), the use of geostatistics for water saturation modelling would prove particularly complex. While it would be theoretically possible to apply some form of cokriging to the 2 variables water saturation and porosity (or permeability), it is probably simpler and more effective to model the relationships between the various petrophysical properties through a regression analysis. In addition, it should be remembered that what could be simulated in this case is a distribution of *irreducible* water saturation, since depth would not be accounted for.

4.2.3 Net Pay

Net Pay values are normally available for all the wells, for each reservoir layer, as a result of the application of some type of cut-off to the log curves or to the petrophysical interpretations (paragraph 4.1.6). The study of the areal distribution of the pay sections within the reservoir is important, in that it determines where the bulk of the reserves are located.

It should be emphasised that most of the techniques that have already been discussed in the porosity section are applicable also in the case of Net Pay and therefore will not be discussed here in detail.

4.2.3.1 2D Interpolation

In most cases, reliable Net Pay maps can be generated by simple contouring of the posted well values. Compared to porosity, the interpolation exercise is in this case more robust, since usually all wells can be taken into account, not just those with an available petrophysical interpretation. In fact, in some old fields, the number of wells with a porosity profile can represent only 10 or 20% of the total number of wells, while the net pay computation can also be performed by means of old suites of logs like SP or old electrical logs. Of course, it is imperative in this case to make sure that the Net Pay calculation performed in these old wells is consistent with the cut-off study carried out in the wells with a petrophysical interpretation.

Another point that makes the Net Pay interpolation easier compared to porosity is that the conceptual depositional model of the reservoir can be used as a guide in the mapping work. This conceptual model can also be used in the definition of the experimental variogram, when geostatistics is applied. Fig. 4.45 (left) shows a Net Pay map obtained by ordinary kriging, where a distinct directional pattern NW-SE can be observed. This map has been obtained using only the values of 18 wells and a variogram model.

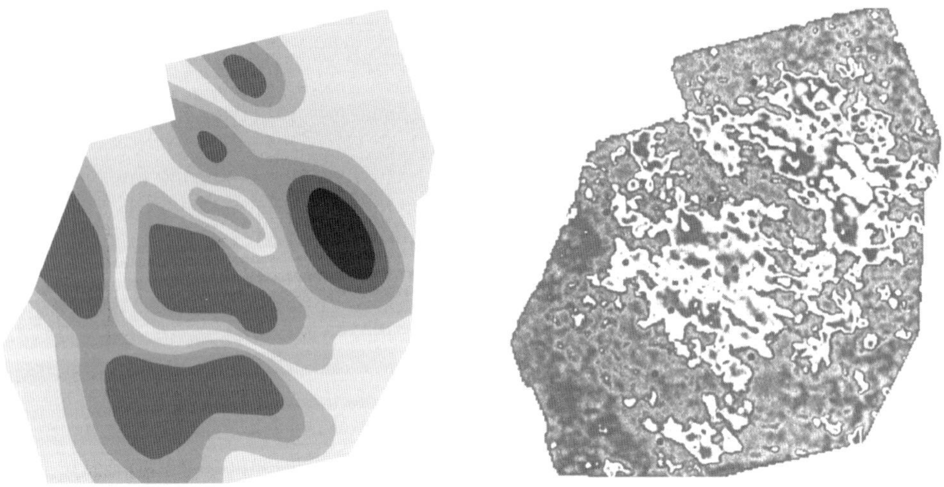

Figure 4.45 Net Pay map obtained by ordinary kriging (left) and collocated cokriging (right) [48].

As has been discussed in paragraph 4.2.1.1, these contoured maps provide a smoothed version of the actual Net Pay distribution. A stochastic approach can be attempted when the uncertainty in the Net Pay distribution is to be addressed, for example in a probabilistic determination of the OHIP.

4.2.3.2 Seismic Data Integration

The integration of seismic data can provide a valuable input to the estimation of a Net Pay distribution map. The available methodologies have already been discussed in paragraph 4.2.1.2. Two example of seismic data integration will be discussed in this section.

The first example concerns the same reservoir presented in Fig. 4.45. In this case the Net Pay distribution can be improved by utilising the seismic information, in the form of an amplitude map. The integration has been obtained by collocated cokriging and the results are shown in the same figure (right). Note the higher degree of detail obtained through the utilisation of highly sampled seismic data and the overall more realistic appearance of this map with respect to the ordinary kriging version.

The improvement in the overall quality of the results can also be appreciated by analysing the cross-validation results (Fig. 4.46). Here, it can be observed that the introduction of the seismic information has resulted in a substantial reduction in the estimation error compared with the ordinary kriging map.

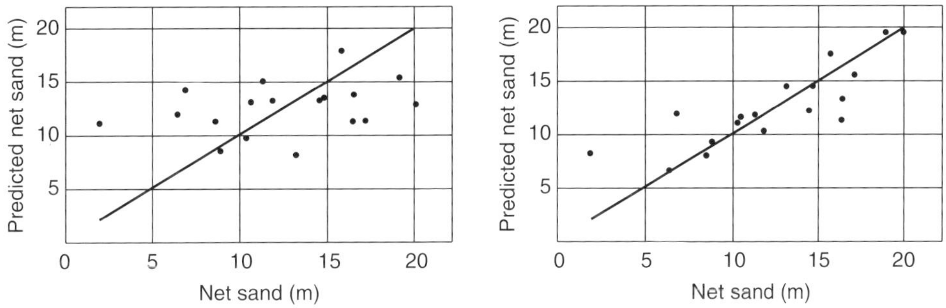

Figure 4.46 Cross-validation results: ordinary kriging (left) vs. collocated cokriging (right) [48].

The second example concerns seismic inversion, which represents another potential source of data for integration. Fig. 4.47 shows a cross-section obtained through the inversion of a 2D seismic line, based on one single calibration well. Low impedance (light grey) is related in this case to the presence of sand bodies and the visual comparison with the log profiles of the wells along the cross section shows a remarkably good correlation between seismic and well data. In fact, these wells have not been used in the inversion and have simply been depth adjusted and posted over the seismic section.

The quality of the results, as observed in these wells, gives confidence in the reliability of the inversion data in undrilled areas and confirm that this additional information can be safely used to construct maps of Net Pay distribution.

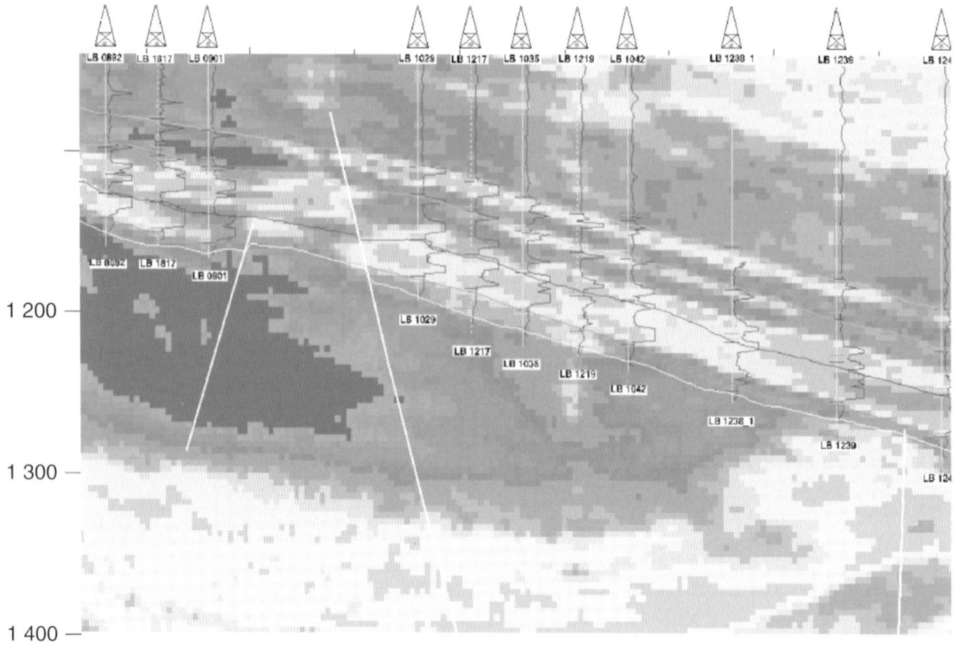

Figure 4.47 Inverted seismic section and well correlation (Courtesy of PDVSA E&P).

4.2.3.3 3D Modelling

As in the case of porosity, a 3D distribution of Net Pay can be generated by means of geocellular modelling packages or geostatistical techniques.

Note, however, that while porosity is a *continuous* variable, Net Pay is better considered as a discrete variable. That is, each point in space can be considered either pay or non-pay and the resulting Net Pay distribution is the sum of all the pay sections of the reservoir.

As a matter of fact, 3D Net Pay distributions are seldom *explicitly* generated. Rather, they can be visualised by selecting the pay facies in an existing lithological stochastic realisation or putting some defined cut-off on the correspondent 3D porosity distribution (Fig. 4.35).

4.2.4 Permeability Distribution

The derivation of a reliable 2D or 3D distribution of permeability is one of the main issues of an integrated reservoir study, since the characteristics of the fluid flow in the reservoir simulator depends on the spatial structure attributed to permeability.

First of all, it is important to note that a single, recommended methodology for permeability distribution modelling cannot be defined. As already discussed in paragraph 4.1.5.8, many sources of information may be available and it is not an infrequent occurrence that the geoscientist, when starting such a study, has to rely on just a couple of cored wells with relevant permeability profiles, a handful of synthetic profiles derived from good quality log data and perhaps a few well tests. How to integrate this information to derive a reliable permeability distribution?

In the next sections, we will try to focus on some of the most used techniques for deriving permeability distributions in 2 and 3 dimensions. It should be kept in mind, however, that other techniques can be applied, depending on the availability of the basic data, the technical and professional level of the resources and the objective of the study.

4.2.4.1 2D Interpolation

When the study is based on a traditional, 2-dimensional approach (maps and sections), a permeability map can be derived in different ways. The most straightforward method is a simple contouring of well-averaged values. This exercise can be based either on calibrated macro-scale (core or log) data, on well test derived permeability or on a combination of the 2 types of data, following the approaches indicated in paragraph 4.1.5.8. Both simple interpolation software or more sophisticated geostatistical algorithms can be used. In all cases, a number of points must be addressed first:

- **Data availability**. As in any other interpolation exercise, direct mapping of permeability values requires a sufficient number of data points, in order to minimise uncontrolled extrapolations. Since the number of cored wells is usually too low for such an operation, this in turn means that there needs to be available a number of permeability values derived from log calibration or other techniques at uncored well locations.
- **Spatial variability**. All interpolation algorithms, geostatistical or not, utilise an underlying spatial correlation function. Commercial mapping software does not allow the geoscientist to calculate the correlation function relevant to the variable under study, while the default options usually imply rather continuous distributions. Such options may prove to be inadequate for permeability, due to its high spatial variability, and may lead to unrealistically smooth maps. The spatial variability of permeability is often much greater than the density of the available wells. Any geoscientist that has attempted the construction of a horizontal variogram of permeability has suffered the frustration of finding what geostatisticians call pure **nugget effect**, i.e., the apparent lack of a spatial correlation function. When the correlation length of permeability (the range of the variogram) is lower than the well spacing, assumptions (educated guesses) must be made about the actual permeability distribution function.
- **Vertical averaging**. When considering a vertical absolute permeability profile, measured or estimated, we are faced with the problem of averaging these values for a given thickness, typically the reservoir layer. Theoretically, the correct operator for such an operation is the arithmetic average, since fluid flow takes place in parallel along the horizontal beds of a reservoir layer. However, the issue eludes many theories and in fact many geoscientists prefer to use alternative types of averaging (harmonic, geometric or power averaging), which give less optimistic results. The issue will be discussed in more in detail in the upscaling section (paragraph 7.5.2.1).

Whenever enough quality data are available that allow for the definition of a robust spatial model and a correct operator for averaging the values, a permeability map can be generated. In all cases, the results should be accurately checked, in order to make sure that the resulting distribution shows the expected spatial structure and looks geologically sound.

An alternative approach to permeability mapping is represented by a porosity/permeability transform, based on a regression function on core data. In this case, of course, the resulting permeability distribution will strongly resemble the primary porosity map. In this case, the following points must be considered:

- **Consistency of the relationship**. The first and obvious point is that a representative porosity/permeability relationship is required for such an operation. While in siliciclastic reservoir this is not uncommon, in carbonate reservoirs this relationship is often too scattered to be of use. The utilisation of different functions for different rock facies usually provide improved results (Fig. 4.29).
- **Spatial distribution**. When using a porosity/permeability transform, the underlying assumption is that the two properties have the same spatial correlation function, which is in general very unlikely. If we are incapable of deriving an experimental correlation function for permeability from the available data, we may have to accept this assumption, but it is important to understand that, in the majority of cases, the generated permeability distribution is likely to be far too smooth compared with the actual distribution.
- **Type of averaging**. When using this method, we do not perform any explicit averaging operation of permeability values at the well locations. We do perform, however, an averaging operation on porosity, to derive the representative value for each layer. It should be noted that to perform an arithmetic average on porosity values and then to utilise a porosity/permeability relationship amounts to performing an implicit geometric averaging of permeability (since the logarithm of an average is not equal to the average of logarithms). This is another underlying assumption that has to be evaluated, insofar as it could result in underestimate permeability values being generated in stratified reservoirs.

From a general viewpoint, this method for deriving permeability distributions gives more satisfactory results compared to the straight interpolation of well averaged permeability values. One main reason for this is that usually the porosity map can be strongly conditioned from a geological viewpoint, so that this input information is directly transferred to permeability, which in fact is also a geologically dependent variable.

As an example of 2D interpolation, Fig. 4.48 shows a permeability map obtained by kriging a well test data set. In this particular field, the density of wells and the number of available well tests allowed for the building of a reliable correlation function that showed a marked NE-SW anisotropy.

When good quality data are available, interpolating well test permeability will produce reliable results and in fact most engineers tend to think that these maps are the best representation of the actual permeability distribution in the reservoir. However, it should be kept in mind that in many cases the quality of the available data is variable and may jeopardise the reliability of the resulting distribution. Uncertainties in the producing thickness *h*, partially penetrating wells, commingled tests or other factors may lead to a permeability dataset

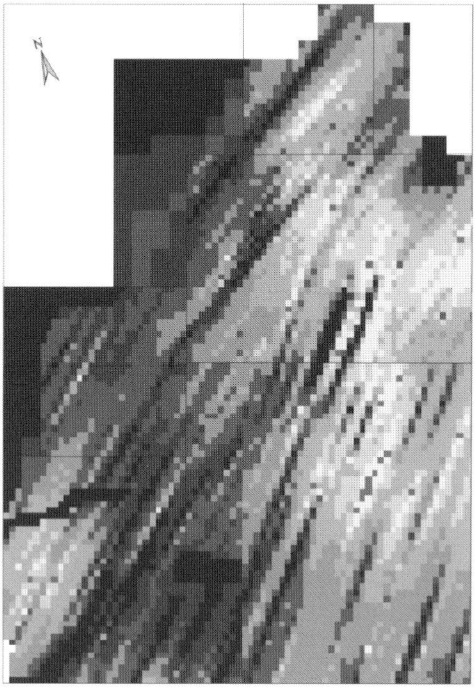

Figure 4.48 Permeability map obtained by kriging well test interpretation results.

which may be partially inconsistent. Attention must be paid to retain only those tests that give results of comparable quality [4].

One domain where well test permeability represents the basic information is the case of fractured reservoirs. In fact, in these reservoirs core data normally provide useful information only concerning matrix permeability, while the global permeability of the formation is governed by fractures and can be significantly larger. In these cases, well tests represent the only source for permeability estimation and distribution. As usual, the availability of a consistent interpreted well test dataset is a mandatory requisite.

4.2.4.2 3D Permeability Distributions

Starting from a number of vertical permeability profiles at the well locations, a 3D distribution can be generated by deterministic interpolation using standard geocellular modelling

4. In favourable circumstances, the permeability map can be integrated with maps of well PI's (Productivity Index). Under the assumption that the skin is constant, these maps mimic the underlying permeability trend and can be used as a conditioning (soft) information.

packages. Of course, the same limitations already discussed for the 2D case still apply in 3D, in terms of data availability and correlation functions.

Alternatively, 3D distributions of permeability can be generated through geostatistical techniques. In general, single stage and two-stage models can be applied, following similar procedures to the ones described in paragraph 4.2.1.3.

When performing a direct geostatistical modelling of permeability (single stage models), attention must be paid, as in 2D, to the correlation range of the selected distribution function. Such modelling is often frustrated by the absence of a clear variogram in the horizontal direction (the apparent nugget effect).

Another important point is that, as opposed to porosity, permeability is better handled with **simulation** rather then **estimation** algorithms. In fact, in the case of permeability, extreme values are essential components in the fluid flow process, since they represent either barriers or preferential paths to fluids. The simulation approach (stochastic modelling) honours the observed experimental variability of permeability, thus preserving the extreme values. A conventional estimation procedure, on the other hand, would generate an averaged, smoothed representation of permeability, which would be less effective in predicting the actual fluid flow performance.

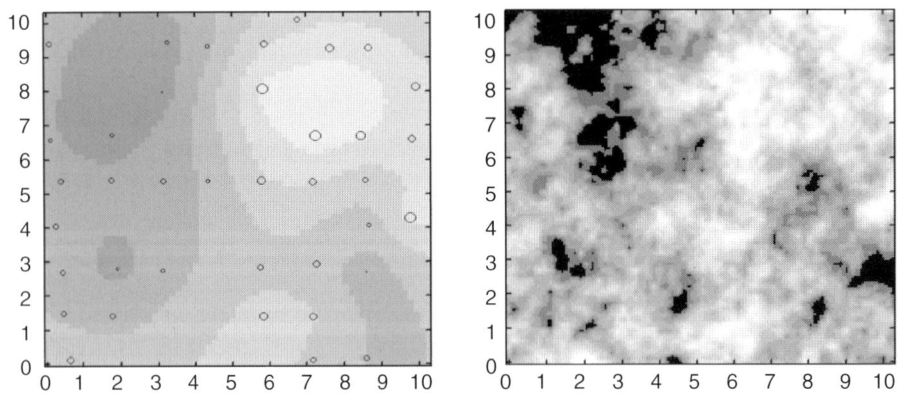

Figure 4.49 Estimated vs. simulated permeability maps.

Figure 4.49 compares two permeability maps, obtained by estimation and simulation, respectively. The difference between the maps is striking, in spite of being generated by means of the same conditioning wells and the same covariance function. In the simulated map (right image), the degree of heterogeneity is preserved, which provides a more realistic image of the resulting permeability distribution.

Simulation algorithms (or stochastic models), on the other hand, have the disadvantage of generating infinite equiprobable realisations of the permeability distribution and the geoscientist is left with the uncertainty of choosing one representative image.

Theoretically, no *a priori* method exists to choose one realisation over the others. In the routine practice, therefore, one image is often randomly selected as representative of the actual permeability distribution. While this is acceptable in most operational studies, it is always important to fully understand the implications related to such a decision.

Two stage models often represent an interesting alternative to direct permeability modelling. The underlying facies distribution represents in this case the template for generating a permeability model. Fig. 4.50 shows a 3D permeability distribution, corresponding to the same cross-section shown in Fig. 4.40. As in the case of porosity, the strong dependence of the resulting permeability field on the input facies modelling is evident.

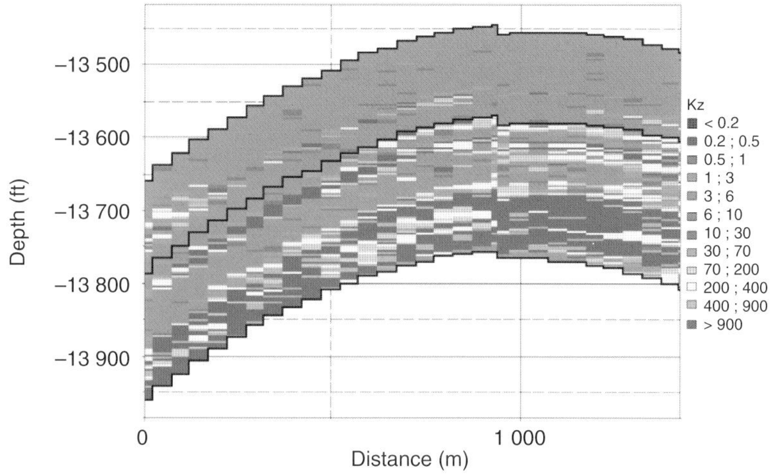

Figure 4.50 3D permeability modelling based on stochastic facies distribution.

A. Conditioning Stochastic Models to Dynamic Data

Stochastic models of permeability are usually generated on the basis of static (geological) data only. This is often a problem in the reservoir simulation phase, as modifications in the original petrophysical distribution would then be required for history match purposes.

The conditioning of stochastic models with well test permeability and more generally with production data has been the focus of a great deal of research in the last years, because of the perceived potential of this type of integration.

From a general point of view, two main streams of research have been investigated:

- ***A priori* constraint**. This represents the *traditional* approach to the integration of well tests into geostatistical models. In this case, the well test results enter directly in the numerical processing as an input information and the objective is to generate a random permeability field that honours the average well test permeability in the region around the well. Ref. [49] and [50] provide two examples of an *a priori* process.
- ***A posteriori* constraints**. In this case, the geostatistical model is modified after its generation, to honour the well test average permeability value. This is therefore an inverse problem, where a known petrophysical field is perturbed in order to satisfy some posterior constraint. Ref. [51] and [52] present two examples of this approach.

In general, the objective of these techniques is to provide average permeability values for each facies or geological object which honour not only the static information, but also the available dynamic data.

The use of flowmeter data, in this respect, is of particular interest. Flowmeter profiles not only provide basic information on total well productivity (or injectivity), but they also give the vertical distribution of such productivity. Since the static (geologic) information is always available along the vertical well profiles, flowmeter data potentially provide the crucial link between the static and the dynamic models of the reservoir.

As we have seen in paragraph 4.1.5.5, permeability profiles can be generated starting from flowmeter data, when an interpreted well test is available. These profiles can be compared with calibrated macroscale permeability data at the well locations, thus providing in themselves a key validation step.

In general, when good quality data are available, flowmeter data should be considered the reference information, since they reflect the actual production behaviour of the field. When inconsistencies are found, macroscale data should then be modified in order to honour flowmeter data. The problem, in this case, is to find a consistent way to modify the macroscale data, which could then be applied in a predictive mode, i.e., in those wells where no flowmeter data is available.

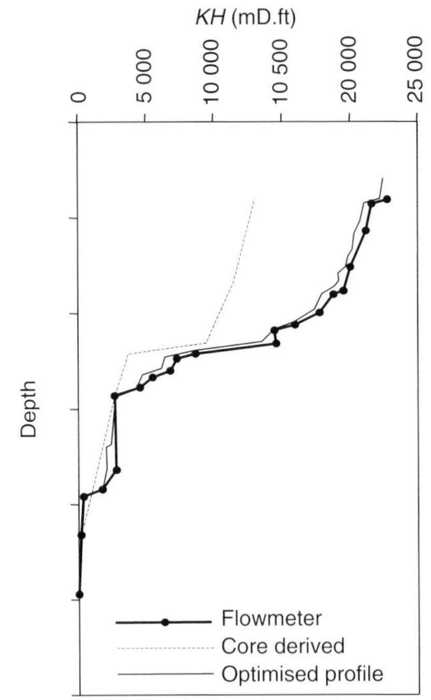

Figure 4.51 Macroscale, flowmeter and optimised permeability profiles.

A powerful method to achieve such consistency is the use of optimisation techniques [36]. These techniques are based on an inversion loop that minimises the observed difference between a reference value (flowmeter permeability) and an estimated value (macroscale permeability). The optimisation loop globally adjusts the average permeability values of each facies, in order to honour the reference flowmeter data.

Note that, in order to be predictive, the inversion has to be performed on several wells at a time. This allows modified facies values to be obtained that best honour, on average, all the available flowmeter results. Note also that, since flowmeter permeability profiles are obtained imposing the total *KH* from well tests, honouring the flowmeter data implicitly means honouring well test data.

This technique may provide excellent results. Fig. 4.51 compares the permeability thickness profile obtained through macroscale *static* data, based on a facies classification, and the optimised profile, obtained after the inversion. The reference flowmeter data are also shown. The improvement in the match for this well is striking.

However, as always when dealing with optimisation procedures, care must be exercised in their application, since visually excellent results may sometimes be obtained at the expenses of the geological reality, e.g., facies with unrealistically high or low permeability. The geoscientist has the responsibility of supervising their application, for example putting constraints on the values that each facies can be given.

In conclusion, production log data represent an essential piece of information for permeability characterization, since they provide the critical link between the static characterization and the dynamic behaviour of the field. For this reason, in all reservoir study projects, the existence and the applicability of these types of data should always be checked carefully.

References

1 Neasham JW, The morphology of dispersed clay in sandstone reservoirs and its effect on sandstones shaliness, pore space and fluid flow properties. SPE paper 6858.

2 Sneider RM, Erickson JW (1997) Rock types, depositional history and diagenetic effects, Ivishak reservoir, Prudhoe Bay Fields. SPE-RE, Febr.

3 Carman PC (1956) The Flow of Gases through Porous Media. Academic Press.

4 Berg R (1970) Method for determining permeability from reservoir rock properties. Transactions, Gulf Coast Association of Geological Societies.

5 Van Baaren JP (1979) Quick look permeability estimates using sidewall samples and porosity logs. 6th Annual European Logging Symposium Transaction.

6 Beard DC, Weyl PK, Influence of texture porosity and permeability of unconsolidated sand. AAPG Bulletin 52, 2, 249-369.

7 Lucia FJ, Petrophysical parameters estimated from visual descriptions of carbonate rocks – A field classification of pore space. JPT 35, 2, 629-635.

8 Begg SH, Gustason ER, Deacon MW, Characterization of a fluvial dominated delta: zone 1 of the Prudhoe Bay Field. SPE paper 24698.

9 Clark NJ (1960) Elements of petroleum reservoirs, SPE-AIME.

10 Choquette PW, Pray LC, Geologic nomenclature and classification of porosity in sedimentary carbonates. AAPG Bulletin 4, 2, 207-250.

11 Amyx JW, Bass DM, Whiting RL (1960) Petroleum Reservoir Engineering - Physical Properties. McGraw-Hill.

12 Dotson BJ, Slobod RL, McCreery PN, Spurlock JW (1951) Porosity measurements comparison by five laboratories. Trans. AIME, 341.

13 Ruth D, PohJoisrinne T (1993) The precision of grain volume porosimeters. The Log Analyst, Nov.-Dec.

14 Reiss LH (1980) The Reservoir Engineering Aspects of Fractured Formations. Ed. Technip, Paris.

15 Worthington PF, Daines JM, Bratli RK, Nicolaysen R (1997) Comparative evaluation of core compaction corrections for clastic reservoirs. The Log Analyst, Sept.-Oct.

16 Dewan JT (1983) Essentials of Modern Open Hole Log Interpretation. PennWell Books.

17 Marett G, Kimminau S (1990) Logs, charts and computers: the history of log interpretation modeling. The Log Analyst, Nov.-Dec.

18 Kenyon WE (1997) Petrophysical principles of application of NMR logging. The Log Analyst, March-April.

19 Prammer MG, Drack ED, Bouton JC, Gardner JS (1996) Measurements of clay-bound water and total porosity by magnetic resonance logging. The Log Analyst, Nov.-Dec.

20 Epping WJ, Eggenkamp IM, Reid I, Added value from NMR measurements to characterise gas reservoirs in the UK Southern North Sea. SPE paper 56944.

21 Chang D, Vinegar H, Morriss C, Straley C (1997) Effective porosity, producible fluids and permeability in carbonates from NMR Logging. The Log Analyst, March-April.

22 Solanet F, Khatchikian A, Breda E (1998) NMR as standard porosity tool in the San Jorge Basin: log responses and applications. SPE-REE, Dec.

23 Woodhouse R (1988) Accurate reservoir water saturation from oil-mud cores: questions and answers from Prudhoe Bay and beyond. The Log Analyst, May-June.

24 Leverett MC (1941) Capillary behaviour in porous solids. Trans. AIME.

25 Anderson WG (1986) Wettability literature survey. Part 1: Rock/oil/brine interactions and the effect of core handling on wettability. JPT, Oct.

26 Archie GE (1942) The electrical resistivity log as an aid in determining some reservoir characteristics. Trans. AIME.

27 McCoy DD, Warner Jr. HR, Fisher TE (1997) Water salinity variations in the Ivishak and Sag Rivers reservoirs at Prudhoe Bay. SPE-RE, Febr.

28 Pirson SJ (1963) Handbook of Well Log Analysis. Prentice-Hall Inc.

29 Myers MT, Pore combination modeling: a technique for modeling the permeability and resistivity properties of complex pore systems. SPE paper 22662.

30 Klinkenberg LJ (1941) The permeability of porous media to liquids and gases. Drill and prod. Prac.

31 Jones SC, The profile permeameter: a new, fast, accurate minipermeameter. SPE paper 24757.

32 Georgi DT, Jones SC, Application of pressure decay profile permeametry to reservoir description. SPENC paper 9212.

33 Timur A (1968) An investigation of permeability, porosity and residual water saturation relationships for sandstones reservoirs. The Log Analyst 9, 4.

34 Sen PN, Straley C, Kenyon WE, Whittingham MS (1990) Surface to volume ratio, charge density, nuclear magnetic relaxation and permeability in clay bearing sandstones. Geophysics 55, 1.

35 Earlougher RC, Advances in well test analysis. SPE Monograph 5.

36 Mezghani M, Van Lingen P, Cosentino L, Sengul M, Conditioning geostatistical models to flowmeter data. SPE paper 65122.

37 Nelson PH (1994) Permeability-porosity relationships in sedimentary rocks. The Log Analyst, May-June.

38 Wendt WA, Sakurai S, Nelson PH (1986) Permeability prediction from well logs using multiple regression. *In:* Reservoir Characterization. Academic Press, 181-221.

39 Coates GR, Dumanoir JL (1973) A new approach to improved log derived permeability. SPWLA 14th Annual Logging Symposium, Lafayette.

40 Kolodzie S, Analysis of pore throat size and use of the Waxman Smits equation to determine OOIP in Spindle Field, Colorado. SPE paper 9832.

41 Mohaghegh S, Ameri S, Arefi R (1996) Virtual measurement of heterogeneous formation permeability using geophysical well log responses. The Log Analyst, March-April.

42 Rogers SJ, Chen HC, Kopaska-Merkel DC, Fang JH (1995) Predicting permeability from porosity using artificial neural network. AAPG Bulletin, Dec.

43 Ahmed U, Crary SF, Coates GR (1991) Permeability estimation: the various sources and their interrelationships. JPT, May.

44 Cobb WM, Marek FJ, Net pay determination for primary and waterflood depletion mechanisms. Paper SPE 48952.

45 Wenlong Xu, Tran TT, Srivastava RM, Journel AG, Integrating seismic data in reservoir modeling: the collocated cokriging alternative. SPE paper 24742.

46 Kalkomey CT (1997) Potential risks when using seismic attributes as predictors of reservoir properties. The Leading Edge, March.

47 Doyen PM, den Boer LD, Pillet WR, Seismic porosity mapping in the Ekofisk field using a new form of collocated cokriging. SPE paper 36498.

48 Gastaldi C, Biguenet JP, De Pazzis L (1997) Reservoir characterization from seismic attributes: an example from the Peciko Field (Indonesia). The Leading Edge, March.

49 Deutsch CV (1993) Conditioning reservoir models to well test information. Transaction of the 1992 Convention on Geostatistics, Troia. Kluwer Academic Pub.

50 Holden L, Madsen R, Jakobsen KA, Tiolsen CB, Vik S, Use of well test data in stochastic reservoir modelling. SPE paper 30591.

51 Blanc G, Guérillot D, Rahon D, Roggero F, Building geostatistical models constrained by dynamic data – A posteriori constraints. SPE paper 35478.

52 Hu LY, Le Ravalec M, Blanc G, Roggero F, Noetinger B, Haas A, Corre B, Reducing uncertainties in production forecasts by constraining geological modeling to dynamic data. SPE paper 56703.

Chapter 5

Hydrocarbon in Place Determination

The determination of the Original Hydrocarbon In Place (OHIP) [1] is typically the concluding phase of the geological study. It is here that the reservoir description, in terms of internal and external architecture and rock properties characterisation, is finally quantified in a number, which represents the amounts of oil and/or gas initially in place in the reservoir.

The OHIP figure is one of the most important parameters of any oil field. Even though the economic importance of a project is obviously much more closely related to the reserves of a given field (i.e., the producible part of the OHIP), the OHIP is the parameter that gives the clearest view of the importance of the hydrocarbon accumulation and consequently of the foreseeable exploitation projects.

In the framework of an integrated reservoir study, the importance of an accurate determination of the OHIP value is also related to the potential reservoir energy that the hydrocarbon volume represents, which is dependent on the compressibility of the oil and gas phases. As will be discussed in more detail in Chapters 6 and 7 (dedicated to reservoir dynamic model and numerical simulation, respectively), the OHIP figure is an essential parameter in establishing the various sources of reservoir energy and their relative contribution. An incorrect OHIP estimation would have to be compensated by a correspondingly incorrect estimation of other reservoir parameters (e.g., pore volume compressibility), in order to reconcile the available dynamic information. This would lead, of course, to errors in the prediction phase.

The basic procedure for determining the OHIP is the volumetric, geologically based calculation. Material balance techniques may also offer very reliable estimations in some cases, but do tend to suffer from some limitations, usually related to the quality of the available dynamic data. Reservoir simulation, on the other hand, does not provide an independent assessment of the OHIP, since the value computed by the model is just a reflection of the input geological parameters.

In the next sections, the traditional volumetric methodologies for computing the OHIP will be discussed, both on a deterministic and probabilistic basis. Material balance-based techniques will also be reviewed briefly.

---

1. The notations OHIP and N (as the oil in place is often referred to in reservoir engineering applications) will be used interchangeably throughout the text.

5.1 VOLUMETRIC ESTIMATES

The estimation of the hydrocarbons originally in place at reservoir conditions (N_R) is based on the well-known equation (field units):

$$N_R = GBV\ N/G\phi(1 - Sw) \tag{5.1}$$

where:

GBV Gross Bulk Volume of the reservoir (bbl)
N/G Net to Gross ratio
ϕ porosity, fraction
Sw water saturation, fraction

When, for a given reservoir, average values of these parameters are known, the determination of the oil in place is straightforward.

In the majority of cases, however, the reservoir will be described by means of maps (or 3D grids) of the relevant geological parameters, therefore the above equation should be applied to all the nodes of the support grid used to build the static model.

The resulting map (or 3D grid) is often called the **equivalent hydrocarbon thickness** map, since it represents the (theoretical) hydrocarbon thickness throughout the reservoir. This map provides an image of the spatial distribution of the hydrocarbons, while its mathematical integration will generate the total OHIP value. Fig. 5. 1 shows an example of such maps.

As mentioned, Eq. (5.1) provides a value of OHIP at reservoir conditions. In the case of oil, the relationship that allows for the determination of the relevant figure at surface conditions (N) is the following:

$$N = \frac{N_R}{B_o}$$

where B_o is the oil formation volume factor, expressed as *rb/stb*. The oil formation volume factor is defined as the ratio between the volume of oil at the prevailing reservoir conditions to the stock tank volume and is a strong function of pressure.

It should be appreciated that the application of such an elementary formula is not as straightforward as it appears, because several values can be defined for the formation volume factor. As will be shown in more detail in paragraph 6.2.4, different measurements of B_o are normally available to the reservoir engineer, depending on the type of experiment performed on the reservoir fluid samples. In the majority of cases, we have a **differential** B_o, a **flash** B_o and few **separator test** B_o's and all of them are different, especially in the case of light and volatile oils. In addition, a **composite** B_o can be determined, through the combination of the differential B_o and the separator tests results.

It should also be appreciated that all of these values of B_o will give *wrong* results to a certain degree, as the actual idea of *surface* OHIP is a hypothetical figure, which is virtually impossible to achieve (unless a theoretical, 100% recovery factor is considered). On the other hand, any of these values could be used when some qualitative estimation of the surface OHIP is sought.

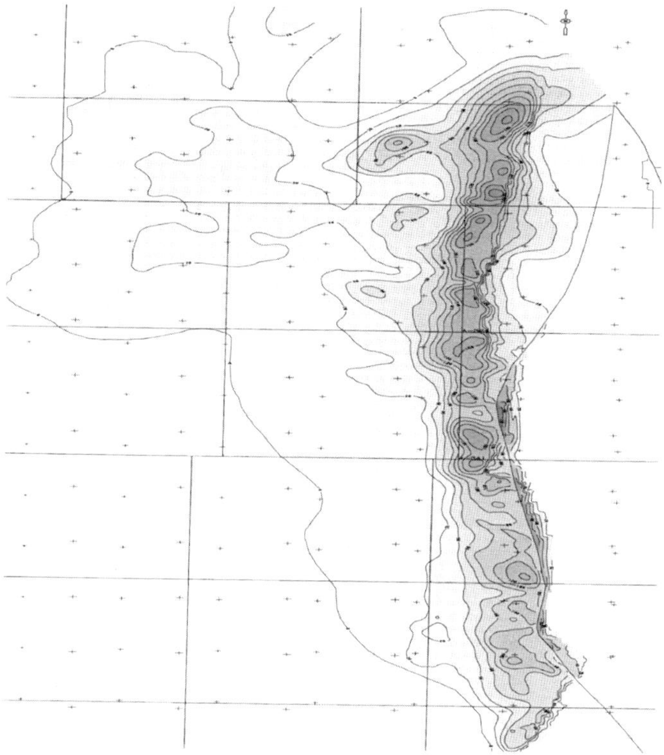

Figure 5.1 Equivalent hydrocarbon thickness map (Courtesy of PDVSA E&P).

Anyway, the use of the composite B_0 is recommended, since it takes into account both the reservoir depletion process and the actual surface separator conditions, thus best approximating the global oil volumetric behaviour. Additionally, as the composite formation volume factor is normally used in material balance calculations, its use in the OHIP determination allows for the comparison of the 2 methods and guarantees the consistency of the results.

Note, however, that the use of the reservoir condition OHIP (Eq. 5.1) would eliminate the ambiguity related to the choice of B_0 to be used in the determination of the surface figure, while providing a sound and consistent estimation of the quantity of hydrocarbon actually present in the reservoir.

5.1.1 Deterministic Evaluation

A deterministic estimate of the OHIP is obtained by simply combining the 2D or 3D grids of the various geometrical and petrophysical parameters that appear in Eq. 5.1, following the methodology that has been outlined before. Note that, when the various distributions are available in grid format, such OHIP estimation reduces to a quick, straightforward mathematical operation.

The deterministic evaluation is the technique that has traditionally been applied for the computation of the OHIP since the beginning of the oil industry. In this methodology, all the various input parameters are calculated deterministically and no allowance is given for any related uncertainty. In other words, the distributions of the geological parameters are considered free of error, even if this is obviously not true.

The process of building a geological model starting from a few scattered data (wells) is characterised by the presence of uncertainties at all stages, which are related to random and systematic errors in the measurements, lack of representative data, interpretation problems and so on. We may choose to ignore this uncertainty but we must be aware that the resulting deterministic estimates can be severely biased.

In fact, the final deterministic OHIP figure is just one particular outcome of a given interpretative process. Should a different approach have chosen, say a different log interpretative model or even a different gridding algorithm, a different OHIP would have been obtained.

Of course, we cannot understate the psychological relevance of the single outcome, **the** OHIP value. Geoscientists have always found it easier to search for one single value of OHIP, possibly because this represents a simpler and faster approach. Also, managers often feel uncomfortable with uncertainty and probability and, from this point of view, there is little doubt that a certain *culture of determinism* predominates in the petroleum industry.

On the other hand, a probabilistic approach to the evaluation of the hydrocarbon reserves provides a much closer insight into the uncertainty related to the estimation process, as well as a good feeling for the accuracy of the results.

Ultimately, it is the geoscientist's responsibility to understand the advantages related to such an approach and to educate management in the correct handling of the probabilistic distribution of the OHIP figures.

5.1.2 Probabilistic Evaluation

In general the probabilistic approach represents much better practice in the problem of computing the OHIP.

The basic idea behind a probabilistic computation is to take into account the uncertainties related to the various parameters involved in the computation. The simplest approach is therefore to treat the variable of Eq. 5.1 in a probabilistic way, by assigning them distribution functions, rather than a single, deterministic value.

This is the so-called Monte Carlo approach. In its simplest, adimensional application, it amounts to randomly sampling the input parameters distributions, in order to generate a probability distribution function of the variable of interest, the OHIP in this case. The procedure is illustrated in Fig. 5.2.

In the framework of an integrated reservoir study, the simple Monte Carlo approach does not represent a standard procedure, since in most cases the input reservoir parameters are described with spatial distributions (2D or 3D grids), rather then modelled with adimensional distribution functions.

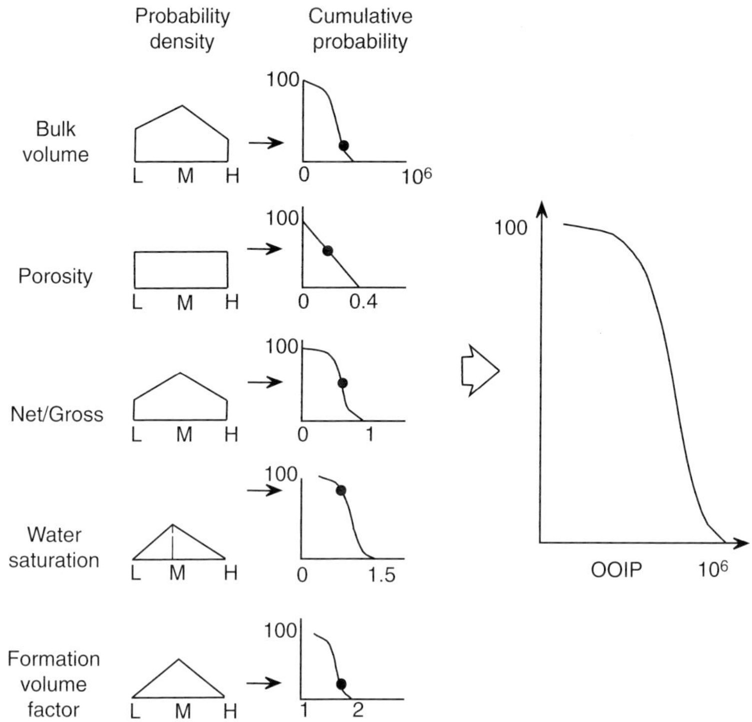

Figure 5.2 Monte Carlo technique for probabilistic OHIP estimation.

Under these conditions, stochastic modelling provides a powerful technique to model the uncertainty in the OHIP. The technique allows for the definition of the uncertainty related to each input parameter, through the generation of sets of equiprobable realisations. Later, the combination of the various parameter realisations, provides an estimate of the uncertainty associated to the OHIP figure. A more complete treatment of uncertainty assessment through stochastic modelling can be found in paragraph 3.3.3.1, while typical results are illustrated in Fig. 3.20.

The probabilistic evaluation of the OHIP based on stochastic modelling, however, can be a lengthy process, especially if all the parameters are treated as random variables.

A faster approach is to model the less uncertain properties in a deterministic way, while applying the stochastic approach to the most important parameters. Alternatively, methods based on experimental design allow for the exploration of the uncertainty space through the use of a limited set of stochastic images [1]. A further method to speed up the probabilistic estimation is by means of the combined use of Monte Carlo methods and stochastic simulation [2].

Finally, it should be noted that the assessment of uncertainty related to the estimated value of OHIP will give the geoscientist a much better sense of the possible sources of inconsistencies that may arise in the following phases of the study. If, for example, in the numerical model the simulated pressure is too low, this could be related to a bad choice of the geological input in the model, resulting in an underestimated value of OHIP. A significant uncertainty in OHIP, shown by a highly dispersed distribution of the computed values, will provide the rationale for selecting an alternative geological model, rather than modifying other reservoir parameters.

5.2 MATERIAL BALANCE ESTIMATES

The material balance equation expresses the law of conservation of matter applied to petroleum reservoirs. Simply stated, it relates the net reservoir voidage due to production to the expansion of reservoir fluids.

The material balance equation has been used for many years in the calculation of the hydrocarbon originally in place. The results of such calculations are significant because they are largely independent from the parameters that are used in the volumetric estimation, thus providing a completely independent assessment of the OHIP.

A more detailed treatment of the technique is presented in paragraph 6.6 and Appendix 1. Here, it will suffice to mention that the possibility of estimating the OHIP from the material balance equation is related to some underlying assumptions, which are summarised here:

- **Pressure equilibrium**. A uniform pressure is assumed to exist throughout the reservoir at a given time. This is a critical assumption, since the expansion properties of the rock and fluids are stated in terms of prevailing reservoir pressure. The definition of a representative pressure decline as a function of time is probably the most important factor in the material balance approach. Moreover, a significant pressure drop is necessary, in order to obtain meaningful results.
- **Reliable production data**. In the material balance equation, the reservoir voidage is expressed in terms of oil, gas and water production. Therefore, the reliability of these data is critical to the accuracy of the results. Care must be exercised in verifying the quality of the available data and in applying the necessary corrections.
- **Representative PVT data**. The PVT properties of the reservoir fluids also have also a considerable impact on the final results. Composite data (see paragraph 6.2.4.2) are generally considered to be the best approximation to the reservoir-well-separator system.

In the next sections, the material balance technique applied to gas and oil reservoir will be illustrated briefly.

5.2.1 Gas Reservoirs

The material balance equation for gas reservoir is considerably simpler than that for oil reservoirs. Due to the very high compressibility of gas, some of the terms that appear in the general formulation, e.g., rock and connate water expansions, may be safely neglected and the final form of the equation becomes:

$$\frac{P}{Z} = \frac{P_i}{Z_i}\left(1 - \frac{G_p}{G}\right) \tag{5.2}$$

where P is pressure, Z is the gas deviation factor, G is the gas initially in place and G_p is the produced gas. The subscript i refers to initial conditions.

Note that this represents the equation of a straight line, whose intercept with the X axis yields the value of the initial gas in place, G. Fig. 5.3 provides a typical example of such plots.

Due to the high mobility of gas, the evidence of a reasonably unambiguous straight line is quite common in gas reservoirs, especially when formation permeability is medium to high. However, downward or upward curvatures are not uncommon and may be related to fluid exit or entry in the system, respectively. Typical such examples are interference with other pools or aquifer influx. Compaction can also cause a non-linearity of the P/Z plot.

When a clear P/Z trend is found, the material balance approach to the determination of the OGIP is possibly the best available method. Volumetric estimates should be double checked for inconsistencies, when significant differences are noticed. One typical source of difference, for example, is related to a possible compartmentalisation of the reservoir, due to the presence of sealing faults. In any case, effort should be paid to try to find some reasonable explanation for the observed differences.

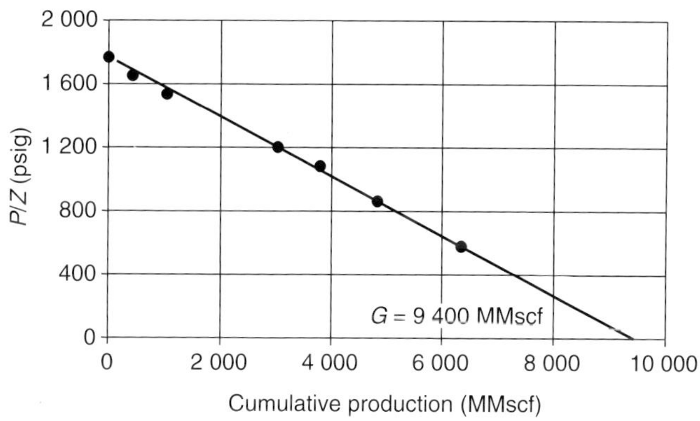

Figure 5.3 P/Z plot for a gas reservoir.

5.2.2 Oil Reservoirs

For many years in the past, material balance represented the main approach to OOIP computation. Different formulations of the general material balance equation were used, depending on the prevailing type of reservoir mechanism. Therefore, undersaturated oil reservoirs, solution gas-drive reservoirs, water drive reservoirs, gas-cap drive reservoirs and compaction drive reservoirs were normally analysed using simplified expressions of the general equation (Eq. A.1 of the Appendix).

Nowadays, despite the widespread application of numerical methods, material balance still represents an important basic technique for oil in place computation.

From a general viewpoint, the OOIP can be computed either directly, by expressing the material balance equation as a function of the only unknown N, or by means of graphical techniques, which allow for the simultaneous determination of N plus a secondary parameter, related to an additional source of energy, usually the gas-cap volume or the water influx. The well-known techniques proposed by Havlena and Odeh [3] are the methods used mainly for this type of evaluation. Fig. 5.4 illustrates one example, for the OOIP computation of the relatively simple case of an undersaturated oil reservoir with an active aquifer influx.

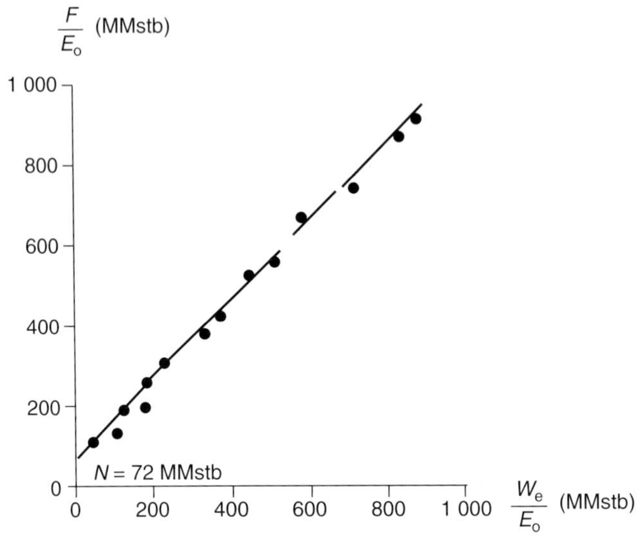

Figure 5.4 OOIP computation from material balance.

In all cases, the N value determined from material balance computation must be validated against the volumetric OOIP from the geological study. The two estimations will never agree exactly and any difference greater than, say, 10% should be investigated. When flaws in either technique are ruled out and when robust material balance solutions are available, two cases may arise:

- **The material balance gives lower OOIP than the volumetric calculation**. In this case, the inconsistency may be related to differences in the reservoir volume being investigated, for example in the presence of faulted reservoirs, where some of the fault blocks are not in communication with the main producing part of the reservoir.
- **The material balance gives higher OOIP than the volumetric calculation**. Since the material balance provides an estimation of what Schilthuis called *active oil* [4], it is possible that too strong a cut-off has been applied in the volumetric calculation and that some of the oil trapped in the low porosity rocks actually contributes to the global expansion.

Other situations may be invoked, of course, that may explain the observed differences between volumetric and material balance results. However, the two techniques rely on completely different approaches to the estimation problem and, once possible causes of differences have been investigated, agreement between the results should not be forced. This in turn implies that, in general, the similarity of the results should not be considered a measure of the accuracy of either technique.

References

1 Damsleth E, Hage A, Volden R (1992) Maximum information at minimum cost: A North Sea development study with experimental design. JPT, Dec.

2 Linjordet A, Nielsen PE, Siring E (1997) Heterogeneities modelling and uncertainty quantification of the Gullfaks S_{or} Brent formation in place hydrocarbon volume. SPE-FE, Sept.

3 Havlena D, Odeh AS (1963) The material balance as the equation of a straight line. JPT, August.

4 Schilthuis RJ (1936) Active oil and reservoir energy. Trans. AIME.

Chapter 6

Basic Reservoir Engineering

Any reservoir study involves a review of the basic reservoir engineering parameters. This work is usually done by collecting and analysing the basic dynamic data of the reservoir, in terms of rock and fluid parameters, pressure data and production and injection performance. These data will then form the typical input for the numerical simulation model.

In the framework of an integrated reservoir study, however, the basic reservoir engineering work should not be considered as a phase of mere data collection and review. There are at least two important points that should be noted in this respect:

- **Integration with the geological model**. Every single reservoir engineering task is somehow related to the geological model. Pressure data, production and injection performance, special core analysis and fluid properties must be considered in the framework of the available geological architecture, while these data provide an invaluable feedback to the geological model itself. Most of these integration opportunities have been widely discussed in Chapter 3. The collection and the analysis of the reservoir engineering data should then be performed in close agreement with the geologists working on the project.
- **Integration with the simulation study**. The basic reservoir engineering can provide invaluable information on the **dynamic model** of the reservoir, i.e. the combination of drive mechanisms that govern its level of energy at each stage of the field exploitation. In turn, the proper understanding and definition of the dynamic model is a mandatory requisite for the subsequent simulation phase. In this respect, basic reservoir engineering techniques should be used to define the relative impact of each parameter in the global production performance, through a proper dynamic characterisation of the reservoir.

The objective of this chapter is therefore to review the reservoir engineering tasks that should be performed in the framework of a typical integrated reservoir study, as well as their interrelationships with other phases of the study itself.

The first part of this chapter will describe qualitatively the main reservoir drive mechanisms. The following sections will focus on the fluid properties (PVT), rock-fluid characteristics and pressure and saturation modelling.

Finally, the last part will be dedicated to some of the most useful reservoir engineering techniques that can be used to explore and characterise the reservoir mechanics, i.e. material balance and streamline simulations.

6.1 RESERVOIR NATURAL DRIVE MECHANISMS

Natural drive mechanisms refer to the energy in the reservoir that allows the fluids to flow through the porous network and into the wells. In its simplest definition, reservoir energy is always related to some kind of **expansion**.

Several types of expansions take place inside and outside the reservoir, as a consequence of fluid withdrawal. Inside the reservoir, the expansion of hydrocarbons, connate water and the rock itself provides energy for the fluid to flow. Outside the producing zone, the expansion of a gas cap and/or of an aquifer may also supply a significant amount of energy to the reservoir. In this case, the expansion of an external phase causes its influx into the reservoir and will ultimately result in a **displacement** process.

There are five basic types of mechanisms that are commonly used to classify the dynamic behaviour of a reservoir:

1. Fluid Expansion
2. Solution Gas Drive
3. Water Drive
4. Gas Cap Drive
5. Compaction Drive

Very few reservoirs belong completely to one of these categories. In fact, in most cases the main producing mechanism for a given reservoir may change during the exploitation of the field. Typically, for example, an undersaturated oil reservoir produces under fluid expansion conditions in the initial period of the exploitation, until the reservoir pressure falls below the bubble point pressure. At this stage, the solution gas drive mechanism becomes predominant. In addition to that, in the majority of cases, more than one mechanism is active at any time for a given reservoir, therefore the most common producing mechanism could be referred to as *Combination Drive*.

The understanding of the main energy resources of a reservoir during the different stages of a field life is essential to any simulation exercise. As a matter of fact, the impact of each reservoir energy component should be quantified beforehand and explicitly input into the reservoir simulator.

The analysis of the production data of the field, together with a correct characterisation of the fluid and rock properties of the reservoir, usually provide a good insight into the energy mechanisms.

In the next sections, the main characteristics of the basic drive mechanisms will be reviewed and, for each of them, the most influential parameters will be indicated.

6.1.1 Fluid Expansion

Fluid expansion occurs as the reservoir undergoes a pressure depletion. In such conditions, when no external influx is present, the reservoir fluid essentially displaces itself.

In gas and gas condensate reservoirs, fluid expansion is often the predominant drive mechanism and accounts for the recovery of a significant part of the hydrocarbon originally in place. On the contrary, in the case of undersaturated oil reservoirs, the liquid phase expansion contributes only a little to oil recovery, since oil compressibility is usually very low, especially in medium to heavy gravity oils.

In undersaturated oil reservoirs producing by fluid expansion, the pressure declines very rapidly, while the producing Gas-Oil-Ratio (GOR) remains constant and equal to the original solubility ratio, Rs_i.

Typical recovery figures for these reservoirs ranges between 1-2% of the OOIP. Higher recovery, but normally still below 5%, can be obtained in the case of volatile oils, or when the expansion of the rock and connate water plays an important role. However, in most cases, the initial liquid expansion is followed by the liberation of gas, which allows for a substantial reduction in the pressure decline gradient.

In conclusion, fluid expansion is inherently the least efficient drive mechanism, and usually needs to be supplemented with external energy sources.

6.1.2 Solution Gas Drive

When the reservoir pressure falls below the saturation pressure, gas is liberated from the hydrocarbon liquid phase. Solution gas drive, or dissolved gas drive, indicates the process of expansion of the gas phase, which contributes to the displacement of the residual liquid phase.

Initially, the liberated gas will expand but not flow, until its saturation reaches a threshold value, called critical gas saturation. Typical values of the critical saturation ranges between 2 and 10%. When this value is reached, gas starts to flow with a velocity which is proportional to its saturation. The more the pressure drops, the faster the gas is liberated and produced, thus lowering further the pressure, in a sort of chain reaction that quickly leads to the depletion of the reservoir.

At the surface, solution gas drive reservoirs are characterised in general by rapidly increasing GOR's and decreasing oil rates, while the pressure decline tends to be less severe than in the liquid expansion phase. Generally no or little water is produced. The ideal behaviour of a field under dissolved gas drive depletion is illustrated in Fig. 6.1.

As can be observed, the GOR curve has a peculiar shape, in that it tends to remain constant and equal to the initial Rs_i while the reservoir pressure is below the bubble point; then it tends to decline slightly, until the critical gas saturation is reached. This decline corresponds to the existence of some gas in the reservoir, that cannot be mobilised. After the critical saturation is reached, the GOR increases rapidly and finally declines towards the end of the field life, when the reservoir approaches the depletion pressure. The final recovery factor in this kind of reservoirs is normally rather low, ranging approximately from 7 to 35% of the OOIP.

The most important parameter in solution gas drive reservoirs is gas-oil relative permeability. Actually, the increase in the GOR curve is related to the increased gas permeability with respect to oil, as its saturation increases. The lower the critical gas saturation, the more rapidly the gas will be mobilised and produced, thus accelerating the depletion and impairing the final recovery. Fig. 6.2 shows 2 sets of gas-oil relative permeability data, corresponding to ideal extreme cases. The curve to the right represents a more favourable mobility ratio as far as the oil recovery is concerned.

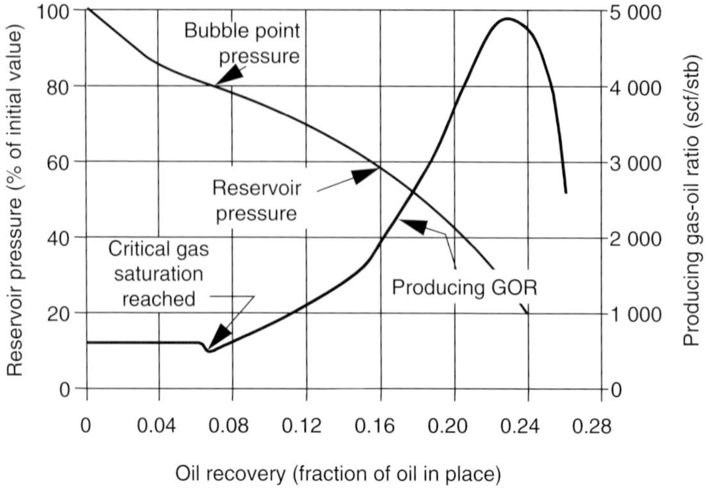

Figure 6.1 Ideal production behaviour of a solution gas drive reservoir.

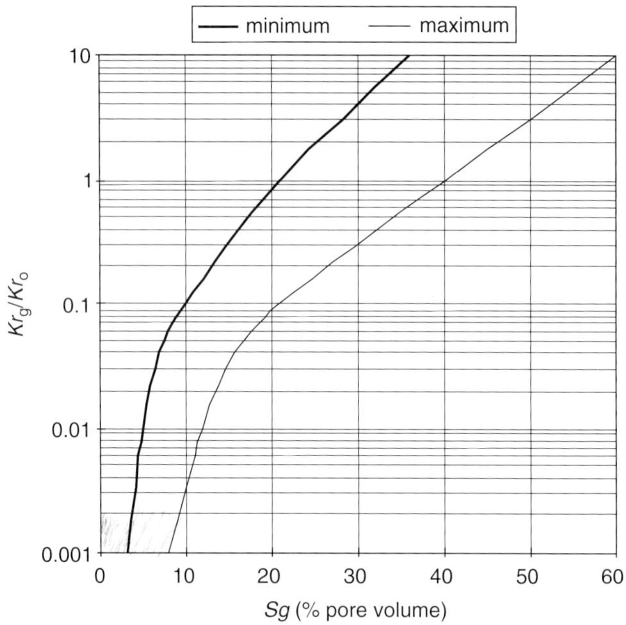

Figure 6.2 Gas-oil relative permeability curves.

For this reason, the availability of a reliable set of gas-oil relative permeability curve is mandatory when evaluating a solution gas drive reservoir. Whenever possible, the curves determined in the laboratory should also be compared with field derived Kr_g/Kr_o values (see paragraph 6.3.2.1).

Other rock and fluid parameters influence the performances of a solution gas drive reservoir to a lesser extent.

- **Gravity Segregation**

One important point when dealing with solution gas reservoirs is the influence of gravity. It is generally assumed that in these kinds of reservoirs the impact of gravity is negligible. The important implication is that the recovery is not rate-sensitive and that these reservoirs can normally be produced at the highest sustainable rates. However, in some favourable cases, the effect of gravity may act so as to induce the segregation of phases in the reservoir. When this happens, gas is not produced but rather migrates towards the top of the structure to form a secondary gas cap, which in turn acts to maintain the pressure in the reservoir.

The occurrence of a gravitational segregation in the reservoir can often be inferred from the production behaviour of the field. For example, when gravity is acting, wells located updip in the structure will produce preferentially with higher GOR with respect to the wells located downdip. Another important information comes from the rate of success of workovers aimed at reducing the producing GOR by shutting off the highest perforations. However, the most important sign that gravity is acting is a stable or only slightly increasing GOR.

Figure 6.3 shows the production profiles relative to two oil fields, both producing by solution gas drive. The practically constant GOR profile indicates that a gravitational segregation process is taking place in one of the two reservoirs (solid line). The two fields have roughly the same oil in place, but, as it can be noted, the cumulative production and consequently the total recovery are very different in the two cases. In fact, the different behaviour is in line with the different geological characteristics: one is composed of lenticular sand bodies of medium to low permeability, while the other consists of stacked marine sand bodies of great continuity and high vertical permeability. The latter conditions are favourable to gravity segregation.

Gravitational segregation, when present, plays an important role in the performance of the field. Recovery figures can be significantly higher and for this reason, when favourable conditions are deemed to exist for the gravity to act, wells are often operated at low rates, in order to permit the segregation of phases in the reservoir.

6.1.3 Water Drive

Many hydrocarbon reservoirs are connected down-structure with natural aquifers, which can provide an important source of producing energy. As oil is produced, pressure declines in the reservoir and when the pressure disturbance reaches the aquifer, water starts to expand and to flow into the reservoir. Therefore, in this case, the producing mechanism is related to a displacement process, since the expansion takes place mostly outside the reservoir.

From a geometrical point of view, water drive fields may be described as bottom or flank drive, depending on the relative configuration of the aquifer and the reservoir. In bottom drive reservoirs, the oil zone is completely underlain by water, while in flank drive reservoirs, the oil is in contact with the aquifer only in the peripheral parts of the field. These different configurations pose distinct problems, when production is concerned, since the former

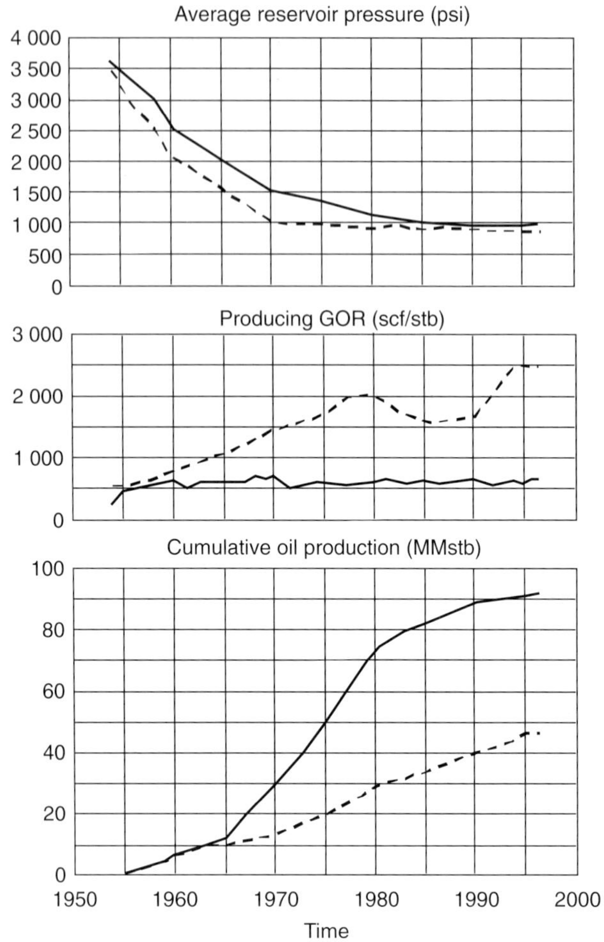

Figure 6.3 Impact of gravity in the performance of 2 solution gas drive reservoirs.

are more prone to water coning problems, while the latter will in general experience water fingering or under-running (paragraph 6.1.3.1).

The efficiency of a water drive mechanism depends in the first place on the volume of interconnected water. In fact, since water compressibility is very low, in the order of 5 10^{-6} vol/vol/psi, several thousands barrels of water must be present and able to expand, in order to produce a single barrel of oil. The dimension of a natural aquifer are usually referred to by a dimensionless ratio:

$$r_D = r_A / r_R$$

which represents the aquifer radius to the reservoir radius. It is generally assumed that a value of around 50 represents a strong natural water drive. In some cases, when the areal extension of the geologic formation is huge, the pressure disturbance may also not reach the aquifer boundary within the producing life of the field. These types of reservoirs are called **infinite acting**.

Another important parameter of water drive systems is aquifer permeability. High permeability is essential for a water drive to be effective, since the pressure gradients must propagate relatively rapidly, in order to allow for a sufficient volume of water to be involved. Very big aquifers may prove to be completely inefficient, if their permeability is insufficient to guarantee a rapid pressure propagation. Conversely, relatively small aquifers may respond fairly well, when the transmissibility of the system is high. This is the case, for example, of some fractured formations. It could be argued that a lower threshold of around 100 mD is necessary, for a water drive system to behave efficiently.

The production performance of a water drive reservoir are quite different from those observed for solution gas drive. When the aquifer volume is large enough, the reservoir will in general show a fairly low pressure decline and furthermore, this decline may become smaller with time, since the aquifer response is often delayed. When the water influx rate equals the fluid production rate, the reservoir pressure may eventually stabilise to a constant value, which is somewhat lower than the initial pressure. In extreme but not infrequent examples, for very large aquifers with high transmissibility, pressure does not show any decline with time and remains equal to the initial pressure, even in the presence of high withdrawal rates.

The producing GOR will remain constant and equal to the initial solution GOR, as long as the pressure remains above the bubble point.

As far as the oil production is concerned, a slow but steady decline is generally observed, which is related to the progressive water invasion of the structurally lowest wells. Therefore, unless wells are abandoned and worked over, water drive fields are characterised by a progressive increase of the water production. Moreover, if water and oil have approximately the same mobility, the total fluid production of the field remains fairly constant. The ideal production performance of a water drive field is illustrated in Fig. 6.4.

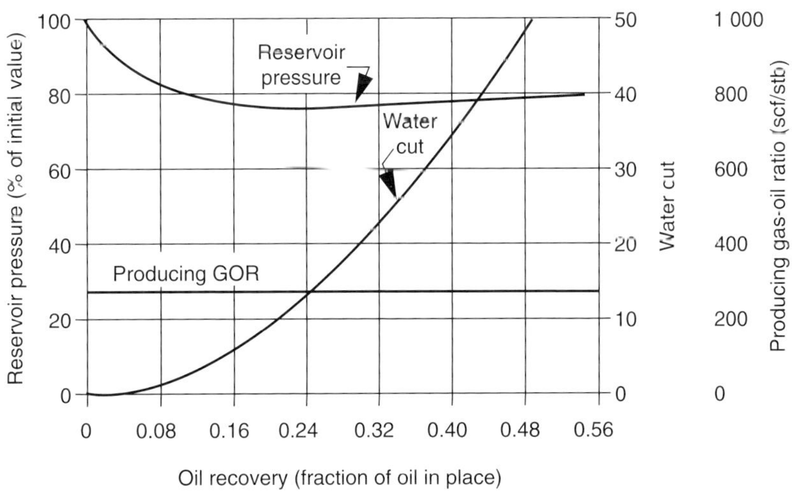

Figure 6.4 Ideal behaviour of a water drive reservoir.

In those cases where the aquifer supply is not strong enough to maintain the pressure above the bubble point at the desired production rates, the pressure may decline below the saturation pressure and some gas is liberated in the reservoir. To some extent, this may be beneficial, since the expansion of the gas phase provides the reservoir with an additional source of energy, which may reduce the rate of pressure decline. However, if the pressure keeps declining, the producing GOR may rise significantly and the solution gas drive process may prevail, thus impairing the final recovery.

Water drive reservoirs usually exhibit the highest recovery efficiency. Reported values of recovery factors in these reservoirs ranges from 30 to 80% of the original OOIP, with an average in the vicinity of 50%. Significantly, the highest figures have been reported for high permeability reservoirs.

In all cases, the final recovery of a water drive reservoir depends upon the efficiency of the displacement process. The overall or global efficiency E_g of a displacement process is defined as the product of three independent components:

$$E_g = E_m E_a E_v$$

where E_m, E_a and E_v are the microscopic, areal and vertical displacement efficiency, respectively.

At the microscopic scale, the water-oil relative permeability is by far the most important factor, since it defines the relative mobility of the 2 phases at various saturation conditions, as well as the residual oil saturation for a waterflooding process. Important properties like rock wettability are also strongly related to the microscopic displacement process. The microscopic efficiency also defines the highest recovery factor attainable for a water-oil system, in the case of a 100% volumetric displacement efficiency.

At the macroscopic scale, represented by the areal and vertical displacement efficiency, the most influencing factor is reservoir heterogeneity. Large scale reservoir features like shale streaks, faults, fractures and in general all those features that represent barriers or, conversely, high conductivity paths to fluid flow, impair the homogeneity of the displacement process. Oil may remain trapped behind a local geologic trap, while water may quickly reach the producing wells through some high permeability streaks. A good description of reservoir heterogeneity, as discussed in paragraph 3.4, is therefore essential for a wise exploitation of water drive reservoirs.

- **Unstable Displacements**

The recovery of a water drive reservoir may or may not be rate sensitive, depending on the stability of the displacement process. The existence of a stable process, in turn, depends upon a number of reservoir and fluid properties.

In general, two types of unstable displacements occur in water drive reservoirs, i.e., **coning** and **fingering**. Although the two processes may occur in the same reservoir, the former is typical of bottom aquifers and the latter of flank aquifers.

The first of these processes, water coning (or *cusping*, as it is often called for horizontal wells) defines the movement of the fluid contact surface (Water-Oil Contact, in this case) towards a producing well, due to the presence of viscous pressure gradients established around the wellbore by the production itself. This movement is counterbalanced by gravity forces, therefore the existence and the magnitude of a coning process can be defined, at any time, by the resulting potential gradients around the wellbore. Fig. 6.5 shows a simplified example of a water coning in a vertical well.

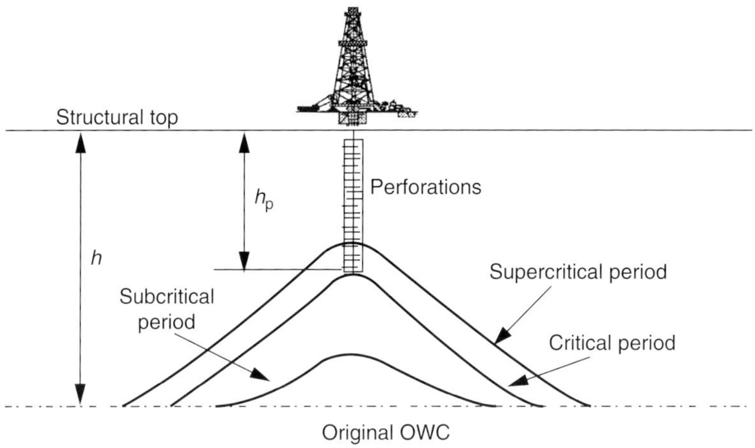

Figure 6.5 Water coning in a vertical well.

A number of parameters influence water coning: geometric factors (reservoir thickness, perforation position), rock and fluid properties (e.g., horizontal and vertical permeability, fluid densities and viscosities), production rates and the degree of reservoir heterogeneity.

Numerous analytical solutions have been proposed in the literature to describe the coning process, both in terms of breakthrough time and critical rate computation [1]. Alternatively, the process can be analysed by means of ad-hoc numerical simulation studies.

In all cases, it is important to understand and to quantify the impact of coning before embarking in the simulation study: the numerical field model in fact, due to the large dimension of the grid cells, is not suited to reproducing such process correctly.

Water fingering, on the other hand, defines the formation of water tongues that under-run the oil zone, as a result of excess production off-takes. This is a highly unwanted situation in all cases, since it not only causes early water breakthrough in structurally high wells, but it also poses serious problems to the recovery of the oil left above or below the water tongue.

Similarly to coning, water fingering is influenced by a number of rock and fluid parameters, probably the most important being the presence of reservoir heterogeneity and the fluid mobility ratio, M. The occurrence of this type of unstable displacement is therefore more likely in heavy oil reservoirs, where an unfavourable mobility ratio exists ($M > 1$). Like coning, water fingering is a result of the interaction of viscous and gravity forces in the reservoir, acting in opposite directions. Therefore, another significant feature is the reservoir dip, as it defines the influence of gravity in the process: the higher the dip, the more gravity will be acting, thus helping a stable displacement.

The formation of water fingering is related to the concept of critical rate, i.e., the maximum fluid withdrawal rate that causes the fluid interface to deform and to move updip through the formation. The situation of stable and unstable fronts is depicted in Fig. 6.6. The study of the critical rate for a given rock-fluid system can be made by means of analytical equations, the most commonly applied being the equation of Dietz [2]. Again, more detailed results can be obtained by means of numerical simulations.

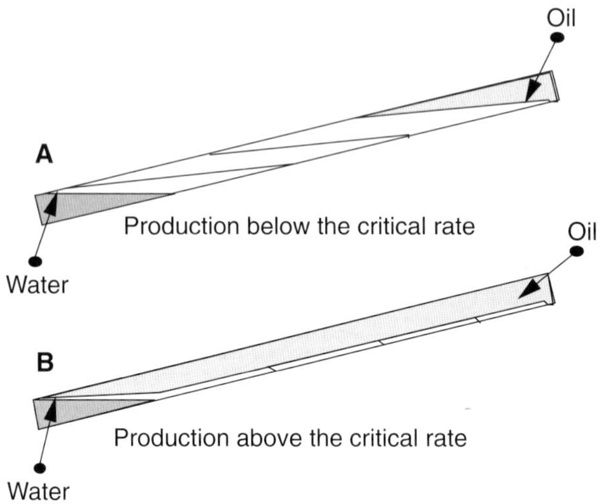

Figure 6.6 Stable displacement (**A**) and water fingering (**B**).

As in the case of coning, the presence and the impact of fingering must be investigated in the basic reservoir engineering phase, since this will influence the geometry of the simulation grid. Actually, high permeability paths have to be explicitly represented as individual layers, in order to correctly reproduce the fluid flow in the model.

6.1.4 Gas Cap Drive

Gas cap drive is the producing mechanism whereby a volume of free gas in the upper part of the structure of a reservoir expands into the oil zone to displace oil downdip, towards the producing wells.

Where an original (primary) gas cap exists, the oil phase is saturated and the pressure at the Gas-Oil Contact is equal to the saturation pressure. Therefore, a small pressure drop in the reservoir, related to fluid withdrawal, causes some solution gas to evolve from the oil phase. In other words, a gas cap drive is always accompanied by some degree of solution gas drive.

The relative impact of the two mechanisms basically depends on the size of the gas cap. The larger the gas cap (with respect to the oil volume), the smaller the pressure drop in the reservoir necessary for the gas cap to expand. Therefore, the larger the relative size of the gas cap, the smaller will be the impact of the solution gas drive process in the reservoir.

A gas cap can be either primary or secondary. In the latter case, a gravitational segregation mechanism must work in the reservoir, which allows for the gas to migrate upstructure.

The rock and fluid characteristics necessary for this process to happen are the same as those mentioned in the previous section: high vertical permeability, favourable oil mobility, low flow velocity. Also, the reservoir must either be thick or have appreciable dip, in order to provide a high vertical closure.

From a production point of view, a gas cap drive reservoir is usually characterised by a slow but fairly constant pressure decline with cumulative production. It may also be characterised by the production of substantial and increasing quantities of gas, especially from the updip wells. However, this is an unwanted situation, therefore wells are progressively worked over or shut-in, in order to prevent withdrawal of gas from the gas cap and preserve reservoir energy. Water production depends upon the presence and the activity of a natural aquifer, but is generally negligible. Fig. 6.7 shows the ideal production behaviour of a gas-cap drive reservoir.

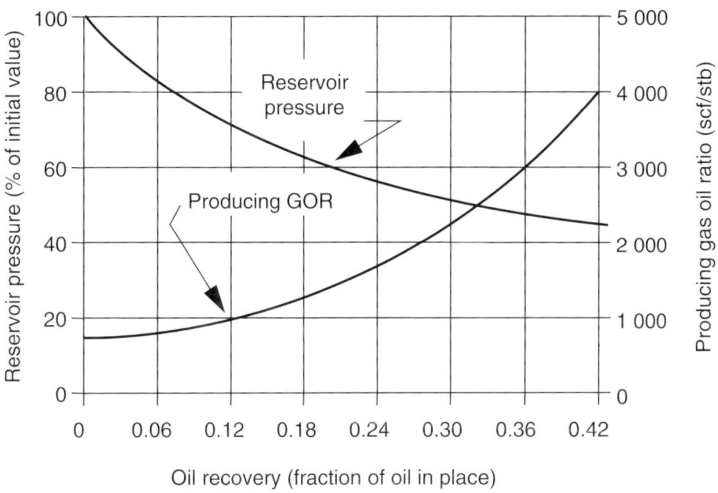

Figure 6.7 Ideal behaviour of a gas cap drive reservoir.

The recovery of gas cap drive reservoirs can be quite different, depending upon the dimension of the gas cap, the effectiveness of gravity segregation and the efficiency of the gas displacement process. When these factors combine in a positive direction, oil recovery can be as high as 70% of the original OOIP. However, when the size of the gas cap is small relative to the oil volume and in thin, heterogeneous reservoirs that prevent phases separation, the recovery can be as low as a solution gas drive reservoir, i.e., below 30%. The recovery efficiency of a gas-cap drive reservoir is also significantly influenced by the field production rate, since low rates may induce gravitational segregation and prevent the generation of unstable fronts.

The critical parameter in a gas cap drive reservoir is in the majority of cases the efficiency of the displacement process. As for solution gas drive reservoirs, gas-oil relative permeability is very important, in that it defines the relative mobility of the 2 phases. In particular, the value of the residual oil saturation to gas, Sor_g, defines the microscopic flow efficiency of the displacement process (drainage cycle). Likewise, reservoir heterogeneity is an essential factor in the final recovery of such reservoirs. The presence and the nature of these heterogeneities influence the stability and the uniformity of the displacement front, which in turn affect the volumetric sweep efficiency.

Under unfavourable circumstances, as in the case of water drive reservoirs, unstable processes like gas overrunning or coning may develop, which may eventually lead to the premature closure of wells and hence cause low recovery factors. The detection of unstable fronts and possibly the analytical study of the associated critical rates often provides useful insights into the mechanics of the reservoir, as well as useful indications about the simulation grid building.

6.1.5 Compaction Drive

Compaction drive is the producing mechanism related to the decrease in pore volume that in some reservoirs is a consequence of fluid withdrawal. To understand the mechanism, consider that the effective pressure P_E acting on the rock at a given depth, corresponds to the difference between the total pressure P_T (corresponding to the weight of the overburden formation) minus the fluid pressure P_F (what we ultimately refer to as formation pressure). Therefore, the following simple relationship holds:

$$P_E = P_T - P_F$$

Whenever pressure depletion is observed in the reservoir, this implicitly generates an increase in the effective pressure acting over the rock framework. Depending on the compressibility of the formation, this increase may induce a decrease in the pore volume and therefore provide some energy to the system.

In most cases, the **pore volume** compressibility of the reservoir formations (not to be confused with the **bulk** volume compressibility) is of the order of magnitude of $5\ 10^{-6}$ 1/psi, which is a small value compared to fluid compressibility, especially gas. For this reason, in most cases the contribution of compaction to the total recovery is small and often negligible. In addition, it is often considered that compressibility is constant with pressure.

However, a number of significant exceptions have been documented in the literature, where abnormally high values of compressibility provide the reservoirs with an important source of energy. The most well-known examples are the oil fields along the eastern flank of the Maracaibo lake, Venezuela, the offshore fields near Long Beach, California and most importantly the Ekofisk Field in the Norwegian sector of the North Sea. In all these fields, compaction of the reservoir formation is associated with a significant subsidence at the surface, which is the most spectacular expression of the underground rock compaction.

Investigations performed in the Bachaquero Field [3] showed that rock compressibility may account for up to 50% of the reservoir energy. Fig. 6.8 shows a typical relationship between cumulative oil production and surface subsidence in the area. Furthermore, there appears to exist a distinct threshold pressure drop, below which no significant compaction is observed. Nevertheless, when the threshold is exceeded, compaction increases rapidly and in a non-linear mode, thus providing the reservoir with a significant additional drive (Fig. 6.9).

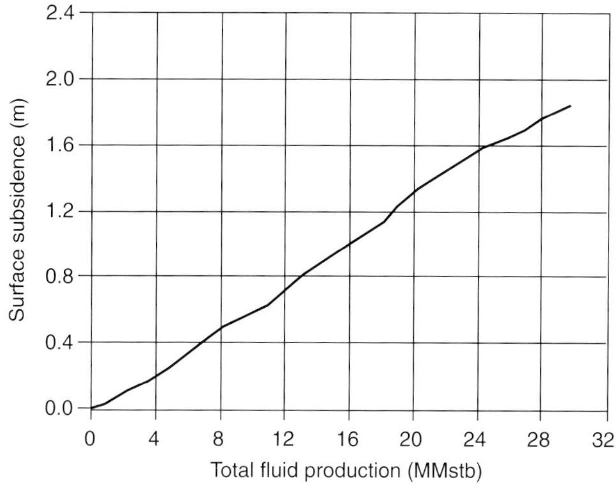

Figure 6.8 Cumulative oil production vs. surface subsidence.

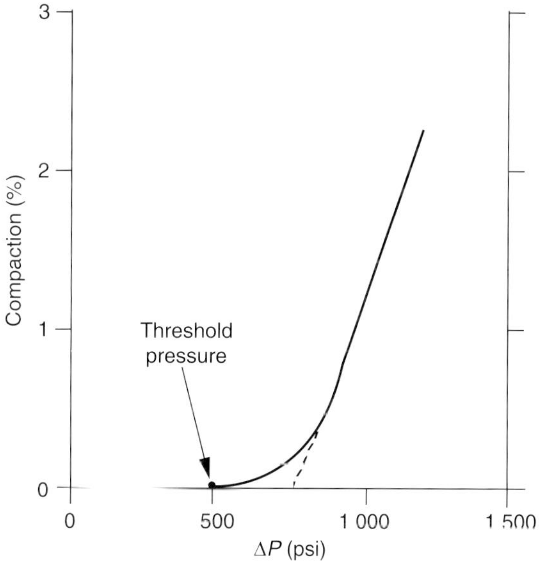

Figure 6.9 Formation compaction vs. pressure.

Compaction is also one of the main producing mechanisms in most of the North Sea Chalk reservoirs. In the Valhall Field, for example, the average contribution of compaction drive has been estimated to be around 50%, with value as high as 70% in the crestal, high porosity area of the field [4].

In the Ekofisk field, the best known example, the abnormal formation compressibility has been attributed to early migration of hydrocarbon and overpressuring of the formation, which processes led to the preservation of abnormally high porosity. Field exploitation and the consequent pressure depletion, caused the collapse of the pore structure and a significant reduction in the original pore volume. Currently, it can be estimated that more than 30% of the OOIP has been recovered by means of compaction drive. Ref. [5] contains a good summary of the experience gained in this field.

The production behaviour of a typical compaction drive reservoir is difficult to define, since it depends, among other factors, upon the particular rock framework and its mechanical properties. In general, however, the pressure and production behaviour is similar to that of water drive reservoirs, with the difference that only little water is produced, related in this case to the expulsion of some connate water or water trapped in shaly layers. In fact, as they are related to overpressure, typical compaction drive reservoirs are often isolated systems, disconnected from regional aquifers.

When the existence of a compaction drive is known or suspected, it becomes imperative to define its impact as a natural energy source. Material balance calculations are often a useful tool in this respect. When this is not taken into account, the excess energy in the reservoir can be misinterpreted as a higher OOIP or a stronger water influx, which in turn would lead to erroneous predictions.

Likewise, in the simulation model, compaction must be properly modelled, either as an adequate compressibility table as a function of pressure, or with a more sophisticated coupling with a geomechanical model.

6.2 FLUID PROPERTIES

Fluid properties are just as important to the reservoir engineer as rock properties are important to the reservoir geologist. Actually, the type of reservoir fluid is one of the main factors that influence the production behaviour of a reservoir and, consequently, the choice of the most appropriate exploitation plan and surface separation infrastructure.

As a matter of fact, all reservoir engineering applications, from the simplest to the most complex, require some assumptions concerning the Pressure-Volume-Temperature (PVT) behaviour of the reservoir fluids. The whole importance of these parameters is that they allow the observed volumes of gas and liquid production at the surface to be related to the corresponding underground withdrawal and, from this point of view, they can be considered to be the link between the subsurface reservoir and the surface production facilities.

More than other dynamic parameters, the properties of reservoir hydrocarbons have a strong impact over any engineering calculations. Furthermore, the PVT characteristics often carry a high degree of inherent uncertainty, which can be related in turn to sampling problems (or more in general to the actual representativeness of the available fluid samples), to the existence of vertical and/or lateral variations of the reservoir fluid composition, to the influence of post-migration processes of alteration or biodegradation and so on [1].

1. From this point of view, it is interesting to note how the whole subject of uncertainty assessment has been in recent years a widely discussed issue among the static modelling disciplines, while apparently little attention has been paid to the uncertainty related to dynamic parameters, like production data, measured pressures and, indeed, PVT properties. A recent paper addresses the problem of estimating such uncertainty [6].

The engineer must therefore apply a great deal of attention to the definition of the PVT model of the reservoir fluids and, in this respect, the integration of all the available sources of information may provide a means to reduce the uncertainty related to this phase.

In general, three types of PVT data sources are available in a typical reservoir study:

1. Experimental laboratory analyses on bottom hole or surface recombined fluid samples.
2. Field production data.
3. Generalised correlations.

In the next sections, after a quick review of the main PVT concepts and parameters, these sources of data will be analysed, to try to highlight the applicability of each one and the common pitfalls. In the last section, the importance of the integration of the different sources of data as a general procedure for PVT modelling studies will be discussed.

Finally, it should be noted that a complete discussion about PVT properties of the different types of reservoir fluids is a prohibitively wide task, which in fact has been the subject of a number of excellent textbooks [7, 8]. In this context, the attention will mainly concentrate on the relatively simple case of a crude oil reservoir, since what we are really interested in is the discussion of integration procedures.

6.2.1 Reservoir Hydrocarbon Fluids

Natural occurring petroleums are made up of extremely complex mixtures of hydrocarbon molecules and in general the resulting combination of these compounds may vary, in a reservoir deposit, from completely dry gas to heavy oils or tars.

The properties of a reservoir fluid depend on the chemical composition of the hydrocarbons and the reservoir temperature and pressure. These conditions determine the physical state of the hydrocarbon itself in the reservoir, i.e. liquid or gaseous.

The hydrocarbon state behaviour in the reservoir is usually described with phase diagrams, which relate the fluid state to the reservoir pressure and temperature. Fig. 6.10 shows the phase envelopes for the 4 most commonly found types of hydrocarbon deposits, i.e. crude oil, volatile oil, gas-condensate and dry gas (wet gas can be considered as a particular case of dry gas, when some condensate occurs at the separators). For each fluid type, the upper line of the envelope represents the lower pressure and temperature limits for the existence of a liquid phase and is called **bubble point line**. Likewise, the lower line represents the upper pressure and temperature limits for the existence of a gaseous phase, and is called **dew point line**. The area within the phase envelope, on the contrary, represents the pressure and temperature conditions at which both the liquid and the gas phases are present simultaneously. The point called C, where the bubble point and the dew point lines join, is called **critical point** and refers to a particular pressure and temperature condition where the liquid and gas properties are identical.

The physical state of the hydrocarbon at the moment of the discovery depends on the initial pressure and temperature of the reservoir, indicated as P_i and T_i in Fig. 6.10. It can be noted that crude oils and volatile oils are always liquid at initial reservoir conditions, since the critical temperature is higher than the reservoir temperature. For the gas and gas-condensate, on the contrary, the critical temperature is lower than the initial reservoir temperature, therefore these accumulations are initially in gaseous state.

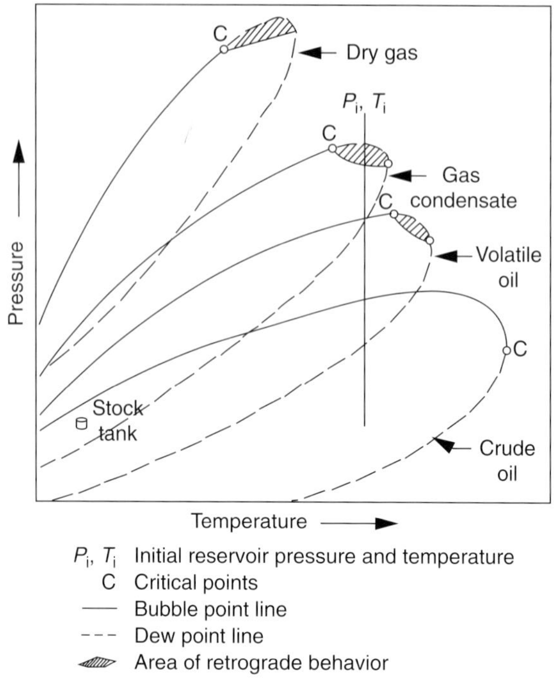

Figure 6.10 Phase behaviour of the main types of hydrocarbon reservoirs.

The vertical line in the figure shows the typical changes in pressure and temperature that happen in the reservoir as a consequence of field exploitation. In fact, it is generally assumed that the underground withdrawal process can be described by an isothermal depletion, since a significant reduction in fluid temperature happens only at the surface. As a reference, stock tank pressure and temperature conditions are also shown in the figure.

Let's now consider the typical behaviour of the various types of reservoir fluids, when production is started at the surface, following the ideal isothermal depletion line:

- **Crude oil**. In this case, the phase envelope and the initial pressure and temperature conditions shown in Fig. 6.1 indicate that the reservoir fluid is **undersaturated**. As fluids are produced at the surface, the pressure in the reservoir drops below the bubble point line and some gas is liberated, whose amount and composition is dependent upon the chemical composition of the hydrocarbon mixture.
- **Volatile oil**. The phase behaviour of volatile oils is qualitatively very similar to that of crude oils, however in this case, a much larger quantity of gas is liberated. This behaviour is related to the greater amount of intermediate components in the hydrocarbon mixture, which tend to escape the liquid phase together with the lighter components. For this reason, these fluids are also called **high shrinkage oils**.
- **Gas-condensate**. In this case, the reservoir temperature is higher than the critical temperature, however the two-phase region extends to the right of the critical point and

therefore, when pressure is reduced, the dew-point line is crossed. A small amount of the initial gas condenses in a liquid phase, following an inverse behaviour with respect to oils (which tends to vaporise with the depletion). For this reason, these fluids are often called retrograde.

- **Dry gas**. The reservoir temperature is always higher than the critical temperature, even during the isothermal depletion. Therefore, no liquid is formed in the reservoir. The P-T separator conditions lie outside the 2-phase region, therefore no liquid is condensed at the surface (with the exception of some water of condensation). These gases are normally composed of large percentages of the lighter hydrocarbon compounds, methane and ethane.
- **Wet gas**. The situation is very similar to that depicted for dry gas reservoirs, since no liquid is formed in the reservoir during the depletion. However, the pressure-temperature separator conditions fall inside the 2-phase region, therefore in this case some liquid is formed at the surface.

An approximate distinction of the different types of hydrocarbon reservoirs can also be made by means of surface production parameters. The following table (Table 6.1) summarises, for each reservoir type, the commonly accepted ranges of the main surface parameters.

Table 6.1 Reservoir type distinction based on surface production parameters.

	Crude oil	Volatile oil	Gas-condensate	Wet gas	Dry gas
Initial GOR, scf/stb	< 2 000	2 000–6 000	6 000 –20 000	20 000–100 000	> 100 000
Colour	Black to light green	Dark straw	Straw to colourless	Colourless	Colourless
API gravity	10–45	40–50	45–65	n/a	n/a
Composition	C_{7+}: > 40%	C_{7+}: 12.5 to 40%	C_{7+}: 2 to 12.5%	primarily $C_1 - C_2$	primarily C_1
Oil FVF, rb/stb	< 2.0	> 2.0	n/a	n/a	n/a

6.2.2 Main Oil and Gas PVT Parameters

The main PVT properties of oil and gas are needed to relate the observed surface volumes to reservoir volumes. These parameters are usually determined in the laboratory through appropriate PVT tests (see paragraph 6.2.4).

1. **Oil formation volume factor, B_o**. This is the ratio between the volume of oil at the prevailing reservoir conditions and the volume at surface (**stock tank**) conditions. It is an adimensional parameter and it is normally expressed as rb/stb.
2. **Gas formation volume factor, B_g**. This is the ratio between the volume of free gas at the prevailing reservoir conditions and the volume at surface (**stock tank**) conditions. It is also an adimensional parameter and it is expressed as rb/scf. (This parameter should not be confused with the gas expansion factor *E*, which is commonly used in gas reservoir engineering and has dimension scf/rcf).

3. **Solubility ratio, *Rs***, also called **solution Gas/Oil Ratio**. This is defined as the quantity of surface gas that dissolves in one stock tank barrel of oil, at the prevailing reservoir conditions. It is expressed as scf/stb.

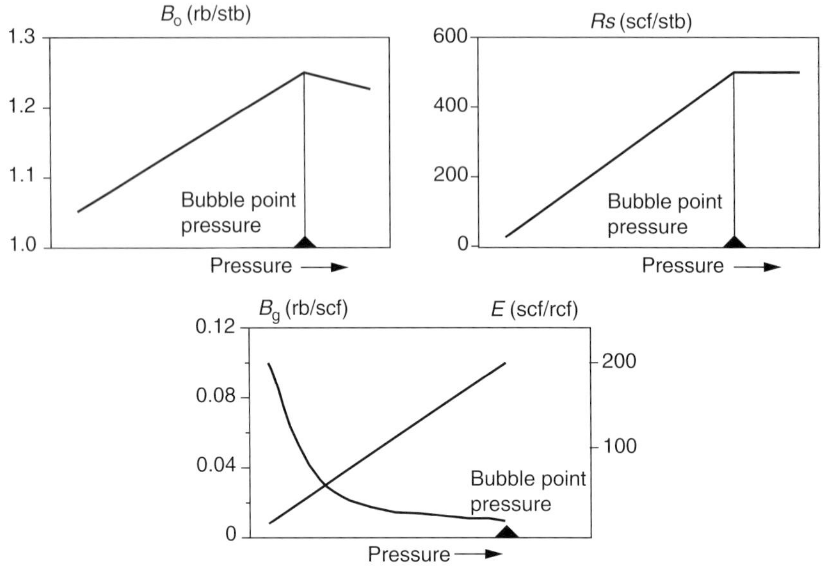

Figure 6.11 Main PVT parameters as a function of pressure.

These parameters are strictly functions of pressure and their characteristic behaviour is shown in Fig. 6.11.

The oil formation volume factor, B_o, is always greater than 1, since the volume of oil at reservoir conditions is larger than the corresponding volume at the surface, due to the liberation of some of the dissolved gas. Below the bubble point, it has a typical decreasing trend as a function of pressure depletion, which is related to the lower quantity of gas which evolves from the oil at reduced pressures. Above the bubble point, on the contrary, the oil volume factor decreases slightly, as a consequence of the liquid phase compressibility.

The behaviour of the gas formation volume factor, B_g, indicates a rapid, non linear increase with the depletion, which is related to the large compressibility of the gas. The behaviour of the gas expansion factor E is also shown, by comparison.

The solubility ratio, Rs, shows a similar trend to that observed for the oil volume factor. As pressure declines, less gas will be able to dissolve in the liquid phase at the prevailing reduced conditions, therefore the value of Rs will be lower. Above the bubble point, the values are constant and equal to the initial value Rs_i. In this region, the oil is undersaturated with gas, which implies that it would dissolve more gas if it were available.

In later sections, we will see how these parameters can be computed and eventually rearranged for practical reservoir engineering applications.

In addition to the aforementioned, other PVT parameters are usually needed in typical reservoir engineering applications. These parameters describe the volumetric and physical behaviour of the fluids in the reservoir, rather than relating underground to surface volumes. Some of them refer to the liquid phase and some to the gas phase.

1. **Oil and gas compressibility, c_o, c_g.** Oil compressibility expresses the expansion of the fluid phase above the bubble point and has units 1/psi. The fractional oil recovery above the bubble point is closely related to this parameter. Gas compressibility is normally derived from correlations.
2. **Oil and gas viscosity, μ_o and μ_g.** These parameters are needed to describe the fluid flow in the reservoir and are expressed in centipoises, cP. Oil viscosity is usually determined from PVT tests, while gas viscosity is readily available from existing correlations (see paragraph 6.2.6).
3. **Oil and gas densities, ρ_o, ρ_g.** Densities as a function of pressure are used to compute the vertical gradients of the fluids in the reservoir. They can be computed from PVT measurements or obtained from existing correlations.

The complete analyses for oil and gas, provided by most laboratories, may actually include much more information. Of particular interest are the compositional data, which can be utilised to model the PVT behaviour of the hydrocarbons through appropriate Equations of State (EOS). We refer the reader to specialised texts for more details on this issue.

6.2.3 Fluid Sampling Procedures

Reservoir fluids are usually sampled early in the life of a field, in order to gain information about the initial state of the hydrocarbon accumulation. The objective is of course to obtain a representative proportion of oil and gas at the moment of sampling.

The availability of reliable reservoir fluid samples is the main requisite for the correct modelling of fluid properties and their distribution within the field. In this context, the importance of sampling procedures cannot be overemphasised since the quality of the available fluid information is often related more to the representativeness of the fluid sample than to the laboratory measurements. In fact, it is not uncommon to have good analysis on bad quality samples, and unfortunately in these cases the results can be severely misleading.

One of the main tasks of the reservoir engineer is therefore to gain an understanding of the inherent reliability of the available fluid samples.

Towards this objective, the first check to perform is to locate the wells that have been sampled. Ideally, these wells should be newly drilled, have stabilised GOR, no water cut and high productivity, in order to assure the lowest possible drawdown.

The following step is to check whether the wells have been properly conditioned prior to sampling. The objective of conditioning is to remove all the non-representative fluids existing around the well bore and replace it with original reservoir fluid, flowing from the virgin part of the reservoir. To achieve this, the well is flowed for few hours at the lowest stabilised oil rate, in order to guarantee the highest possible bottom hole flowing pressure, while the surface GOR should remain constant.

Subsequent checks to be performed depend on the adopted sampling procedure. There are basically two procedures for sampling reservoir fluids: bottom hole sampling and surface recombination.

6.2.3.1 Bottom Hole Samples

In bottom hole, or subsurface sampling, a sampler is run in the borehole to the reservoir depth and a fluid sample is collected at the prevailing bottom hole pressure.

If the reservoir is initially undersaturated, the sample can be collected when the well is still flowing and this is possibly the ideal sampling case. However, if the reservoir is initially saturated or only slightly undersaturated, the well is usually shut-in after the conditioning, to allow for the re-dissolution of the gas that could be present in the vicinity of the well bore.

Saturated reservoirs actually present the biggest problems in terms of sampling, because the saturations around the wellbore never correspond to the original reservoir conditions, due to the liberation of some solution gas. In general, two situations can be encountered:

- If the gas saturation is lower than the critical gas saturation, then the observed GOR will be lower than the Rs_i, hence measured bubble pressure could be lower than the actual bubble pressure.
- If the gas saturation is higher than the critical saturation, then some free gas is present in the reservoir and the observed GOR will be higher than the Rs_i. In this case, the measured bubble pressure could be higher than the actual bubble pressure.

A typical clue that the reservoir fluid has not been properly sampled is when the measured saturation pressure is very close to the prevailing bottom hole pressure at the time of sampling.

6.2.3.2 Recombined Fluid Samples

Recombined fluid samples are created in the laboratory by recombination of separate volumes of oil and gas taken at separator conditions. Some corrections have to be applied to the measured GOR, because the gas sample is usually taken at the separator, while the GOR refers to stock tank conditions.

The selection and conditioning of the wells prior to sampling are not dissimilar from subsurface sampling. Again, in the case of saturated reservoirs, care must be taken in the evaluation of the GOR, since the presence of free gas may result in measured saturation pressures higher than the actual values.

This kind of procedure is usually preferred to bottom hole sampling in the case of volatile oils and gas condensate.

6.2.3.3 Reliability of the Fluid Samples

The quality of the sampling operations is usually assessed first at the well site, by measuring the bubble point on the various samples taken and checking the consistency of the results. This allows for the detection of tool malfunctioning or fluid contamination. However, this preliminary check does not guarantee in itself the reliability of the samples. In any case, it is always good practice to read the sampling reports carefully.

It should also be pointed out that a number of methodologies exists, that allow for an improvement of the quality of the data of an otherwise unreliable sample. Ref. [7] describes some of these techniques.

6.2.3.4 Vertical and Lateral Fluid Property Variations

It is commonly assumed that hydrocarbon reservoirs are uniformly saturated, with a constant amount of gas in solution, which implies that the same initial saturation pressure is present in any part of the field. Similarly, at any given pressure, the viscosity properties of the reservoir fluids is expected to be the same throughout the reservoir.

In fact, many reservoirs exhibit vertical and/or lateral variations in PVT properties. Typically, for example, steeply dipping or thick reservoirs show a vertical compositional gradient, with higher proportions of heavier hydrocarbon compounds towards the bottom of the accumulation. The process responsible for the existence of this gradient is commonly known as **gravitational segregation**.

In other cases, lateral hydrocarbon properties variations are observed, for example in reservoirs with great lateral extent or in the presence of large permeability barriers. These variations can be related to primary processes (migration) or secondary processes (biodegradation, alteration), which determine incomplete fluid composition equilibrium across the field.

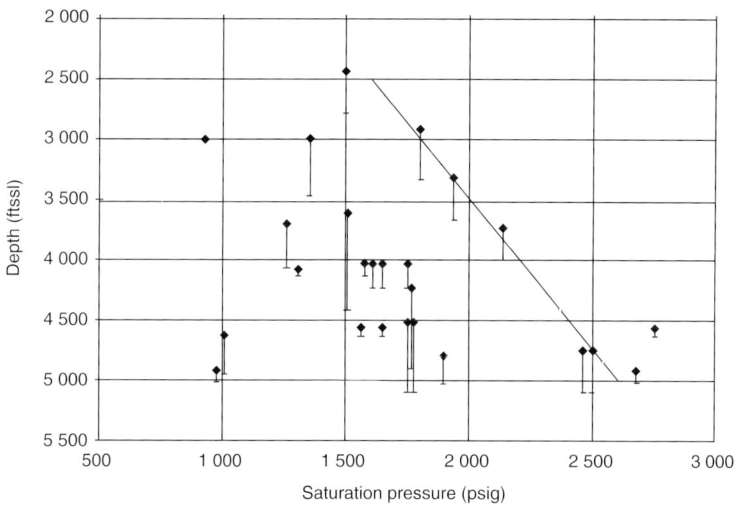

Figure 6.12 Measured saturation pressures vs. depth for a heavy oil field.

The problem is illustrated in Fig. 6.12, which shows the measured saturation pressures relevant to all the available oil samples collected in a giant, heavy oil field. For each sample, the interval depth is shown and the initial formation pressure gradient is also indicated.

There is a considerable dispersion of the measured values and to draw a single correlation is clearly impossible. The whole problem here consists in understanding how much of this dispersion is related to samples reliability and how much to actual variations in the fluid properties throughout the field.

A careful check of the available samples and the sampling procedure normally allows for the identification of the most reliable analyses. Additionally, as discussed in paragraph 6.2.5, the inspection of the production data (variation of GOR and oil and gas stock tank gravities), often provides useful information about the existence of spatial variations in the fluid properties.

6.2.4 PVT Laboratory Analysis

Laboratory analyses on reservoir fluids provide measurements of the main PVT properties of the liquid and gas hydrocarbon phases. For some of these characteristics, e.g., chemical compositions, laboratory tests represent the only source of data, while in other cases estimates of the fluid properties can be obtained from independent sources, e.g., generalised correlations or field production data.

A comprehensive treatment of the laboratory procedures and their use in reservoir engineering applications is beyond the scope of this work and can be found in Ref. [7]. Only the main tests will be dealt with briefly here, since they allow for the calculation of some of the most important fluid properties, which eventually should be compared and integrated with other sources.

There are three main PVT experiments that are routinely performed on reservoir fluid samples:

1. **Flash expansion**. In this experiment the fluid sample is charged to the PV cell and raised to the initial reservoir pressure and temperature. Data are collected through an isothermal expansion, i.e., lowering the cell pressure in a number of stages, while keeping the temperature constant. When the bubble point is reached, gas is liberated from the liquid phase, however no fluid is withdrawn from the cell during the experiment, therefore the overall hydrocarbon composition in the cell remains unchanged. For this reason, this test is also called **equilibrium** expansion (or vaporisation). The test is used to compute the bubble point pressure and the fluids relative volumes at different pressure steps. Data are usually normalised to the bubble point volume.
2. **Differential expansion**. This experiment is identical to the flash expansion until the bubble point pressure is reached. However, at each lower pressure, the total amount of gas liberated during the last depletion stage is removed from the cell and therefore the overall composition of the hydrocarbon in the cell changes at each stage of the experiment, the remaining phases becoming progressively richer in heavier hydrocarbon compounds. As in the flash experiments, liquid and gas relative volumes are measured at each stage. From these basic data, a number of PVT parameters of interest can be derived, the main ones being B_o, B_g and Rs.
3. **Flash separator tests**. These tests are performed by connecting the PV cell to a single or multi-stage separator system, and flashing the reservoir fluids through the separator

system to stock tank conditions. The resulting volumes of gas and residual oil are measured at the end of the experiment. Note that, in the case of a single separator, this test approximates a flash liberation under non-isothermal conditions, while in the case of a multi-stage separator it is closer to a differential test.

It should be pointed out that, in general, the three tests generate different volumes of residual fluids. In other words, the main PVT properties of the reservoir fluids (B_o, B_g and Rs at the bubble point) will be different when calculated on the basis of each of these experiments. When crude oils are concerned, the flash processes leaves in most cases (but not always) less residual liquid with respect to a differential liberation and show therefore a higher oil shrinkage factor. The reverse is often true for volatile oils.

Before applying the laboratory derived PVT values it is therefore important to understand the physical meaning of each experiment and to compare it with the particular reservoir under study. This will provide a way to derive a consistent set of PVT values.

6.2.4.1 Physical Meaning of the Laboratory Experiments

The three tests described above are used in conjunction to describe the phase behaviour of a hydrocarbons mixture through the different stages of expansion it undergoes, from the reservoir to the stock tank.

Flash expansion occurs when the gas liberated below the bubble point is allowed to remain in contact with the liquid phase from which it evolved. From this point of view, it can be considered representative of a reservoir whose pressure is only slightly below the saturation pressure and whose gas saturation is still below the critical value. Therefore, the flash process is rarely, if ever, the predominant gas liberation process at reservoir conditions.

A flash type process also occurs when the reservoir fluids enter the production string and travel to the surface. This flash, however, differs from the laboratory experiment, since it is not isothermal.

In the case of the differential vaporisation, on the contrary, the gas liberated is constantly removed from the solution and the overall composition of the system changes. Therefore, the differential liberation is representative of reservoirs where the liberated gas is separated from the liquid from which it evolved, as for examples in the case of solution gas reservoirs beyond the critical gas saturation (when the mobility of the gas is far in excess to that of the oil) or where gravitational segregation prevails. As in the case of the flash separation, the differential process is rarely the only gas liberation process acting at reservoir conditions, but it is considered to predominate in most reservoirs.

Separator tests, as the definition implies, are representative of the separation process working at the surface and provide the actual values of oil and gas volumes that will be observed at the surface facilities. As we will see in the next section, these values will be used to correct the results of the other tests to account for the actual production infrastructures.

6.2.4.2 Laboratory Data Conversion for Reservoir Engineering Applications

The results of the laboratory experiments provide values of the fluids volumes relative to the bubble point volume. However, the most common way of presenting such results is to normalise them with respect to the volume of the residual oil at stock tank conditions, which is obtained as the last step of the differential test by flashing the residual oil at standard temperature (60°).

This type of presentation has the advantage of providing a set of PVT parameters that could be used directly in reservoir engineering calculations, since, being normalised to stock tank conditions, they are expressed as rb/stb (in the case of B_o) or scf/stb (in the case of Rs). Table 6.2 shows a typical report of a differential liberation, as it is often provided by commercial laboratories.

Table 6.2 Typical report of a differential liberation.

Pressure (psia)	***Rs* (scf/stb)**	**Rel. oil vol.** $V/V_{\text{resid.}}$	**Viscosity (cP)**
2335 (BP)	0	1.360	0.851
2230	22	1.348	0.943
1718	126	1.302	1.084
1130	242	1.251	1.253
800	307	1.222	1.485
510	366	1.195	1.832
250	427	1.163	2.128
14.7	553	1.067	3.342
14.7 and 60°F		1.000	

The use of these data in reservoir engineering calculation, however, should be made with caution. Actually, the volume of the residual oil at standard conditions is dependent upon the number of pressure steps performed during the differential liberation process and therefore this type of normalisation does not provide an absolute set of PVT parameters. Additionally, the values of the differential test do not take into account the actual surface separator conditions, which may provide significantly different values for the stock tank oil and gas volumes, especially in the case of volatile oils.

The correction of the differential liberation values is done through the integration of the flash separation data, using the following relationships:

For the oil formation volume factor:

$$B_o = B_{o_D} \frac{B_{o_{bF}}}{B_{o_{bD}}}$$

with:

- B_o corrected oil formation volume factor, often called **composite** B_o
- B_{o_D} differential oil formation volume factor at any pressure stage
- $B_{o_{bF}}$ flash oil formation volume factor at the bubble point
- $B_{o_{bD}}$ differential oil formation volume factor at the bubble point

Similarly, for the solution gas oil ratios:

$$Rs = Rs_{i_f} - (Rs_{i_D} - Rs_D) \cdot \frac{B_{o_{bF}}}{B_{o_{bD}}}$$

with:

- Rs corrected solution gas oil ratio, often called **composite** *Rs*
- Rs_{i_f} flash initial solution gas oil ratio
- Rs_{i_D} differential initial solution gas oil ratio
- Rs_D differential solution gas oil ratio at any stage

The graphical behaviour of the differential and composite B_o and Rs is illustrated in Fig. 6.13, compared to the raw differential data given in Tab. 6.2. These data are relative to a medium gravity, low shrinkage oil. It should also be pointed out that these are not the only equations that can be used to correct the differential liberation data for the separator conditions, since alternative expressions exist [9].

It is also useful to note that, when the PVT characterisation is to be input to a numerical simulator, care must be taken in the verification of the required PVT format. In fact, some models can handle the distinct concepts of composite and differential tables, while others, on the contrary, do not offer this capability and need composite data.

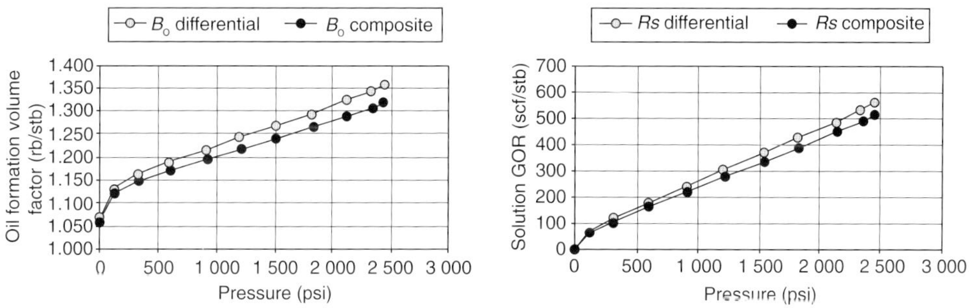

Figure 6.13 Composite and differential B_o and Rs.

To summarise, the representation of the fluid expansion process from the reservoir to the stock tank by means of laboratory data is performed by integrating the results of 2 basic tests, which simulate the behaviour in the reservoir (differential liberation) and the expansion to the surface and through the separation facilities (flash separator tests).

The sets of PVT parameters obtained can be used in any reservoir engineering application, from material balance to reservoir simulation. It is however important to check the consistency of the results through a comparison with similar PVT data, obtained from independent sources. This will be the topic of the next sections.

6.2.5 Field Production Data

Field production data provide a valuable source of information, as far as reservoir fluid properties are concerned.

Referring to the definitions provided in paragraph 1.5, these kind of measurements can be considered as **low precision data**, in the sense that the PVT properties cannot be estimated with precise values, as happens in the laboratory. However, they provide direct information about the actual behaviour of the reservoir, free from any sampling or analytical error and, from this point of view, field production data often represent a very accurate source of information.

Laboratory measurements should always be validated against the actual behaviour of the field, especially when uncertainties exist over the adopted sampling procedure or when the reservoir is suspected to be at, or close to, the saturation pressure.

At least three field production parameters can be utilised to verify the analytical results:

1. **Static pressure**. The measured pressure decline against time or cumulative production, offers a simple and reliable mean to estimate the actual saturation pressure of the reservoir. The liberation and the expansion of the gas phase below the bubble pressure provide a supplementary energy to the reservoir, which has the effect of reducing the pressure decline. Fig. 6.14 shows an example relative to an actual pressure dataset, where the saturation pressure can be identified with good confidence around 2 100 psi, at the interception of the saturated and undersaturated pressure gradients.

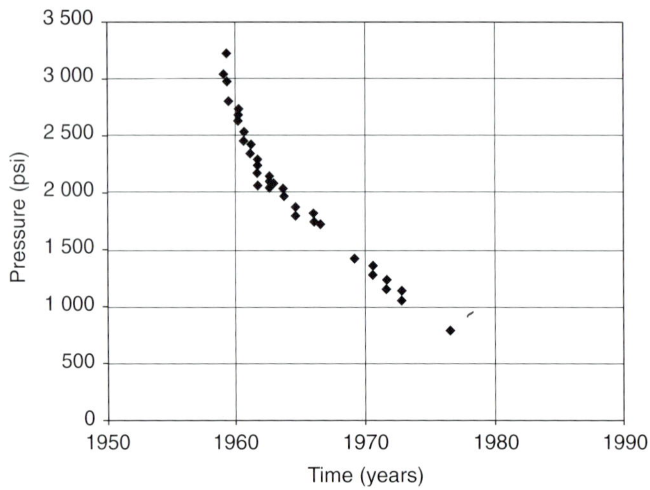

Figure 6.14 Identification of the bubble point by means of reservoir pressure data.

2. **Gas-oil ratio**. The behaviour of the production GOR also gives important information concerning the saturation pressure of the reservoir. If the GOR remains stable to a

value close to the measured or assumed Rs_i, it can be inferred that the reservoir fluid is undersaturated and that the bubble point has not yet been reached. When this happens, the GOR tends to increase rapidly. The pressure existing in the reservoir at the moment of this increase corresponds to the saturation pressure. It should also be noted that, when the producing GOR is observed to increase at different pressures in different areas of the field, this might suggest the existence of spatial variations in the PVT properties of the reservoir fluid. However, high quality data are needed in this case to be able to quantify such variations.

3. **API gravity**. The stock tank gravity of the produced oil is a routinely measured quantity in all producing wells. Therefore, this parameter has the noteworthy advantage of being available as a high density information, both in time and space. API gravity maps can therefore be generated, which offer a good insight into the delicate problem of PVT property variations areally. The example shown in Fig. 6.15 illustrates an API gravity map, where a clear trend of lower values is visible towards the South. Since this is also the structurally lowest part of the reservoir, it can be inferred that such behaviour could be related to gravitational segregation of the oil in the reservoir, i.e. to the existence of a vertical compositional gradient. In turn, this may also suggest the presence of a variation of saturation pressure with depth, which represents a highly valuable piece of information for reservoir engineering purposes.

In the next section, dedicated to PVT correlations, it will be discussed how field production data can be integrated into generalised analytical expression, to generate maps of fluid properties adjusted to the field under study.

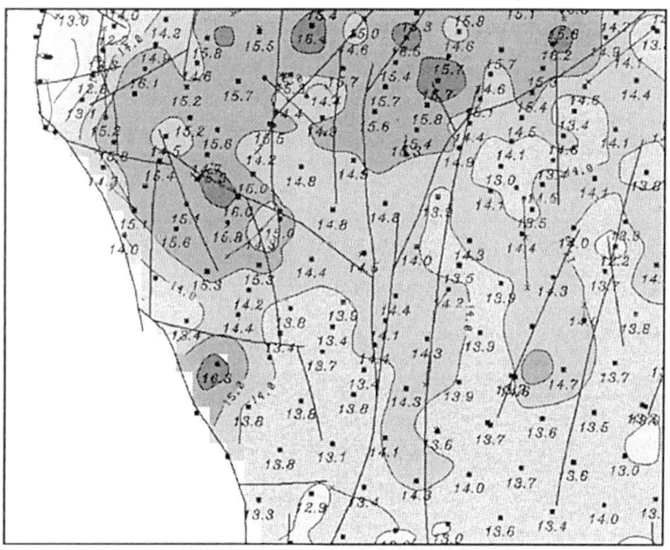

Figure 6.15 API gravity map showing gravitational segregation of the reservoir fluid.

6.2.6 Generalised PVT Correlations

Generalised PVT correlations have been used since the 1950's to obtain a simplified description of reservoir fluid properties based on surface measurements. Over the years, the literature has been growing constantly and nowadays several correlations are available, which often allow for surprisingly accurate estimations of the most important fluid properties.

Limiting the discussion to the main PVT parameters (saturation pressure, formation volume factors and solution GOR), correlations can be applied with the previous knowledge of some basic production parameters, i.e., the API gravity of the produced oil, the gas gravity of the associated gas, the producing GOR and the reservoir temperature.

The most common of these correlations have been derived by Standing, on the basis of crude oils and gases of the California area [10], however, other correlations are available that can prove to be more suitable depending on the particular reservoir under study. We also note the correlations developed by Lasater, Vasquez and Beggs and Glaso.

An interesting point, dealing with empirical correlations, is the possibility of using *local* correlations, i.e., correlations developed for the particular basin under study. These correlations are usually computed by statistically integrating a large number of reservoir fluid analyses carried out over the years and are normally available for all the major producing basins, for example the Gulf of Mexico [11]. These correlations can prove to be extremely robust and reliable and their existence should always be investigated.

Generalised correlations also represent one of the main sources for the determination of most of the other oil and gas PVT parameters, from viscosity, to isothermal compressibility, to gas deviation factor. Ref. [12] provides an interesting review of the main existing empirical correlations.

- **Extended Use of PVT Correlations**

Historically, generalised correlations were utilised when no laboratory measurements were available or when, for any reason, these are considered unrepresentative of the actual reservoir fluids.

In recent years, however, PVT correlations have found renewed interest among reservoir engineers. Recent PC-based PVT packages allow for a much expanded use of empirical correlations compared to the past, since they can be matched against the laboratory experiments through common regression procedures, by slightly modifying some of the constants that appears in the equations. The big advantage of this approach is that empirical correlations make use of surface production data as input, therefore they can be utilised to generate field-wide PVT models, in the forms of maps, that at the same time honour the available laboratory experiments.

These types of application, totally unpractical in the past, allow the reservoir engineer to better explore the existence and the impact of possible spatial variations in the reservoir fluid properties.

6.2.7 Integrating the PVT Information

Modelling the reservoir fluid properties is in most cases a difficult task. General rules do not exist, however some general guidelines can be identified. These guidelines refer as usual to the integration of all the available information and are summarised in the following points:

- Laboratory procedures provide the most precise source of information concerning the reservoir fluid properties, however such information may not be accurate. In addition, being based on individual samples, these methods provide a discrete piece of information that must be integrated with other sources, when spatial variations are to be investigated.
- The reliability of laboratory measurements depends largely on the representativeness of the sample. This should always be carefully checked and, when necessary, corrections can be applied to increase the reliability of the results.
- Differential data should always be corrected for actual separation conditions, by means of flash separator test results.
- In all cases, the results of the laboratory analyses must be compared with field production data, since the latter reflect the actual behaviour of the reservoir. When discrepancies are detected, laboratory data should be adjusted to be consistent with field production data.
- Generalised empirical correlations may provide adequate models of the PVT behaviour of the field. They are used when laboratory tests are not available or are not considered representative of the actual reservoir fluids.
- Generalised correlations can also be used to match the laboratory results and, being based on surface production parameters, they make an interesting integration tool, which allows for the spatial modelling of the reservoir fluid properties.

Finally, it should be mentioned that the modelling of an Equation of State, while not explicitly treated in this context, represent another powerful tool for integrating the available reservoir fluid information. This approach should be considered whenever a compositional simulation is to be performed.

6.2.8 Reservoir Water Properties

Reservoir water is always closely related to hydrocarbon accumulations and must be carefully considered in any integrated reservoir study.

On one hand, water properties are important in the petrophysical computation and consequently on the OOIP determination (see paragraph 4.1.4.2).

On the other hand, when considering the dynamic model of the reservoir, water properties have to be determined for a number of reasons, ranging from the evaluation of its expansion capability for material balance calculation, to compatibility issues related to waterflooding projects.

The properties of interest include the solubility of natural gas in water, the formation volume factor, compressibility, density and viscosity. All of these properties depend on the reservoir pressure and temperature and, of course, on the chemical composition of water.

6.2.8.1 Chemical Composition

The chemical composition of a reservoir water refers to the ion concentration that can be determined through standard laboratory procedures.

The most common ions encountered in reservoir waters are:

- among the cations, sodium (Na^+), potassium (K^+), calcium (Ca^{++}) and magnesium (Mg^{++}),
- among the anions, chlorides (Cl^-), sulphates (SO_4^{--}), and carbonates (CO_3^{--} and HCO_3^-).

The results of these analyses are normally represented through diagrams, which allow the main ions present in the solution to be visualised and different analyses quickly compared. The most popular of these diagrams is the so-called Stiff diagram [13], where the anions are shown at the right-hand side and the cations at the left-hand side of a set of four horizontal straight lines. Concentrations are expressed in milliequivalents per litre (Fig. 6.16). The use of the Stiff diagram has the advantage of showing different and typical patterns for different reservoir waters, which can quickly be recognised.

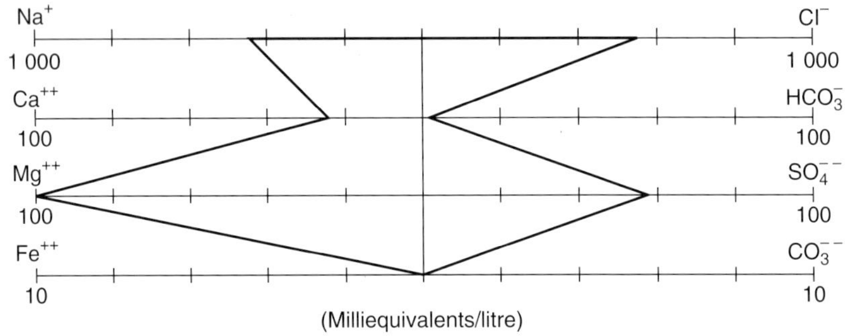

Figure 6.16 Stiff diagram.

The importance of a systematic and careful collection of all the available water analyses is in many cases of paramount importance in a reservoir study, especially in fields where there are active waterflooding projects or where unidentified water encroachments are observed.

In the case of waterflooding, the study of the available water samples may allow for the distinction between natural aquifer and injected waters, which in turn is an essential step in the characterisation of the displacement process and its optimisation.

Another area where water analyses may prove to be extremely useful is in the detection of suspicious water encroachments. For example, if the presence of an active hydrodynamism is suspected, the review of the water compositions should reveal the presence of high quantities of sulphates and carbonates, which are characteristic of meteoric waters. Likewise, well completion problems may be explained by proving that the produced water has a composition which is not compatible with the known formation water.

6.2.8.2 PVT and Other Properties

The characteristics of reservoir waters are seldom measured in the laboratory, since in most cases they can be safely determined through appropriate empirical correlations. These correlations make use of the basic reservoir parameters (pressure and temperature) and of the salinity of the water, usually expressed as total solid salinity.

Refs. [12] and [14] provide the most common correlations for the reservoir water properties, as well as their usual graphical representation.

6.3 ROCK-FLUID PROPERTIES

Rock-fluid properties are used in reservoir engineering and simulation to describe multiphase flow in the reservoir. As a consequence, the definition of correct sets of properties and their spatial distribution is of primary importance in the accuracy of the results.

Three properties will be discussed in this context: wettability, capillary pressure and relative permeability.

6.3.1 Wettability

Wettability is defined as the tendency for one fluid to adhere to a rock surface in the presence of other immiscible fluids.

Different wettability states can be defined for actual petroleum reservoirs, depending on the relative distribution of reservoir fluids with respect to the rock framework [2]:

- **Water wet**. The whole rock surface is coated with water, while oil or gas occupy the central position of the largest pores.
- **Oil wet**. The relative positions of oil and water are reversed with respect to the water wet state, the oil coating the rock surface and the water residing in the centre of the largest pores.
- **Intermediate wettability**. This term applies to reservoirs rocks where there is some tendency for both oil and water to adhere to the pore surface.

The importance of wettability is related to the fact that the relative distribution of fluids within the pore network is critical to the microscopic flow properties. In other words, water wet and oil wet reservoirs behave differently with respect to a displacement process. Fig. 6.17 illustrates the schematics of a waterflood process in a water wet (imbibition process) and in an oil wet (drainage process) reservoirs. As it can be appreciated, the saturation evolution of the reservoir fluids is completely different in the 2 cases and eventually will lead to different recovery factors.

For many years, wettability has been the object of extensive theoretical and experimental work. Ref. [15] through Ref. [20] summarise most of the knowledge that has been collected on this subject.

It is generally accepted that wettability results from the adsorption of molar compounds on the rock surface. However, several factors are believed to affect the preferential wettability of a reservoir rock:

- Oil and formation water compositions. In particular, oils with high content of asphaltenes are more likely to be related to oil wetting conditions.

2. Note that gas is always assumed to be a non-wetting phase.

- Rock mineralogy. Carbonates are more likely to be oil wet than siliciclastic rocks.
- Amount of connate water saturation. The lower the connate water, the higher the oil wetting character of the rock. This implies that wettability is also related to the height above the oil water contact.

Wettability is determined in the laboratory on core plugs. Several types of experiments have been described in the literature that will not be discussed here. The most common of these experiments involve spontaneous and/or forced imbibition using reservoir fluids and result in a **wettability index** which provide a semi-quantitative indication of the preferential wettability of the rock. In particular, a wettability index equal to 0 indicates a neutral rock, while values of + 1 and − 1 indicate strongly water wet and strongly oil wet characteristics, respectively.

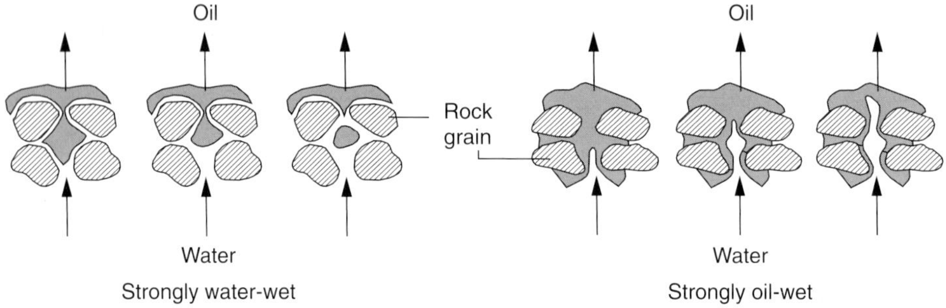

Figure 6.17 Oil displacement in water wet and oil wet reservoirs during waterflooding.

Note that wettability, as such, is not directly used in reservoir engineering calculations. However, it is of paramount importance in the evaluation of other rock-fluid properties, like capillary pressure and relative permeability. If the analysed plugs are not representative of the actual reservoir formation in terms of wettability, the results of subsequent special core analyses will provide totally misleading results.

In this respect, it is worth mentioning that in the early days of petroleum engineering, it was assumed that all formations were preferentially water wet. This was attributed to the fact that reservoir rocks were deposited in an aqueous environment, while oil migration happened only later. Note that this belief had a strong impact on the special core analyses conducted in the past, since core samples were extracted, dried and brine saturated prior to testing, with the specific aim of restoring a water wet condition. Therefore, whenever it is believed that conditions in the reservoir are not water wet, such analyses should be discarded.

More recent investigations seem to indicate that intermediate wetting reservoirs are probably the most common worldwide, while preferentially oil wet reservoirs are not uncommon either. In such cases, the preservation or the restoration of the original wettability conditions of the reservoir rock becomes an essential step of any experimental procedure.

6.3.2 Capillary Pressure

The discussion of capillary pressure measurements, as well as their integration with log data, has been presented in detail in paragraph 4.1.4.1.B, while the methodology to derive consistent drainage capillary pressure functions has been presented in paragraph 4.2.2.4.

In this context, it is useful to mention that, in addition to establishing the initial fluids equilibrium, capillary pressure also acts as a dynamic mechanism in the reservoir, together with viscous and gravitational forces. Both drainage and imbibition processes can be invoked. The drainage curves are used for establishing the initial saturation conditions (initialisation of the model), and the latter curves in the simulation phase.

Actually, imbibition processes, like water displacing oil in a water wet reservoir or water or oil invading a gas cap, require the definition of imbibition capillary pressure curves. These curves are usually determined in the laboratory on core samples. Fig. 6.18 compares 2 sets of drainage and imbibition capillary pressure functions, for a typical sandstone rock.

In the majority of reservoirs, the influence of the capillary forces in the global energetic balance are negligible, especially when viscous gradients are important. In some cases, however, capillary pressure may play an important role in the recovery. This is the case, for example, of fractured, water-wet reservoirs, where a significant contribution to the global recovery is obtained by capillary imbibition of the matrix. Other examples refer to heterogeneous reservoirs, where important amounts of oil are trapped in low permeability layers.

In such cases, the definition of proper capillary pressure curves, both in the drainage and imbibition cycles, is critical in the accuracy of the final results.

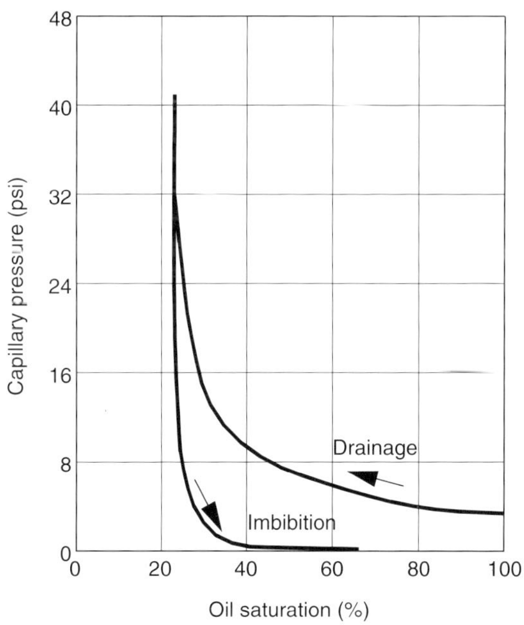

Figure 6.18 Drainage and imbibition capillary pressure curves.

6.3.3 Relative Permeability

The absolute, or specific permeability, is a property of the porous medium and it is independent of the saturating fluid, provided that there is no reaction between the rock and the fluid.

When more than one fluid phase is present in the pore space, as it is the case in petroleum reservoirs, the concept of permeability must be applied to each phase separately, because it depends upon the quantity and distribution of the particular fluid phase within the pore system. On this basis, we can define an **effective permeability** to a specified fluid, which, like absolute permeability, can still be determined from the application of Darcy's law (under the assumption that the fluids are immiscible, incompressible and that no gravity forces are affecting the steady flow of each phase).

An alternative way to define the permeability of a particular fluid phase is to normalise it to the value of the absolute permeability. This is the widely used concept of **relative permeability** (relative to the absolute), which can be expressed as:

$$kr_o = \frac{k_o}{k} \qquad kr_g = \frac{k_g}{k} \qquad kr_w = \frac{k_w}{k}$$

where k is the absolute permeability and k_o, k_g, k_w refer to the effective permeability to oil, gas and water, respectively.

The concept of relative permeability is fundamental in the simulation of the dynamic behaviour of a reservoir, since it expresses the relative contribution of each phase to the total multiphase flow. As any reservoir engineer has experienced, the correct definition of a set of relative permeability functions is one of the most difficult and, at the same time, one of the most important steps in the construction of a reliable simulation model and for this reason a great deal of attention must be paid to this phase of the study.

In the next sections, the available techniques to derive sets of relative permeability functions will be reviewed briefly. A short digression on residual oil saturation to water (Sor_w) is also presented, as this is one of the most important relative permeability end-points.

It should be noted that, since the treatment is basically methodological, only the general case of a water-oil system will be discussed. Reference to other systems will be made only for particular issues.

6.3.3.1 Laboratory Measurements

Relative permeability can be measured in the laboratory on core samples. A wide number of techniques have been described in the literature (see Ref. [21] for a comprehensive summary), but basically these can be divided into either steady-state or unsteady-state experiments.

- **Steady State**. In the steady state method, a fixed ratio of fluids is forced through the test sample, until pressure and saturation equilibrium is reached. The effective permeability of each fluid phase is calculated as a function of saturation by direct application of Darcy's law, by measuring the volumetric flow rate, the pressure drop and the saturation of each individual phase. A number of techniques have been developed and employed through the years, the best known being the Penn-State and the Hassler methods.

- **Unsteady State**. These types of experiments are performed by measuring the effluent from a core during an imposed displacement process, in terms of cumulative production, and back calculating the relative permeability ratio consistent with that outcome. The function which is commonly used to compute the relative permeability is some form of the Buckley-Leverett equation. Unsteady state methods are faster and cheaper than steady state methods.

As far as the accuracy of the results is concerned, it is generally assumed that steady-state methods provide more reliable results, since they are based on a direct measurement of the parameters that define effective permeability, i.e., the parameters that appear in Darcy's law. Unsteady-state methods, while easier and quicker to apply from an operational point of view, are more subject to interpretation problems.

However, as it has been observed [22], the degree of equivalence of the different methods for measuring relative permeability has not been established yet, therefore no *a priori* decision can be taken about the best or more accurate method.

In all cases, a significant degree of uncertainty is likely to be present in these kinds of measurements. A recently published comparative study [23] revealed that, under fixed laboratory procedures applied on homogeneous, water wet cores, measurements from four different core analysis contractors showed differences of 20 saturation units in residual oil values and a factor of 3 between the lowest and the highest endpoint water relative permeability. Even worse, when the laboratories applied their own standard procedures, these differences increased to respectively 35 saturation units and a factor of 10 in the endpoint water relative permeability.

Fig. 6.19 shows the end point oil and water relative permeabilities measured on similar samples by the 4 laboratories, using their preferred procedures. Of course, these conclusions cast some doubts on the actual usefulness of these types of measurements.

In fact, the intrinsic reliability of the laboratory measurements is only one aspect, and perhaps not the most important one, concerning the definition of sound relative permeability functions. Actually, at least 3 major factors must be considered, which may affect the meaning of laboratory-derived relative permeability curves.

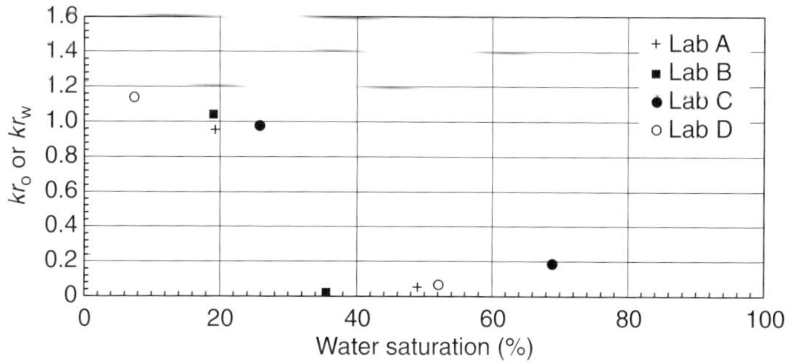

Figure 6.19 End point oil and water relative permeabilities from four laboratories [23].

A. Wettability

Wettability affects relative permeability because it is a major factor in controlling the location, flow and distribution of fluids in a porous system. The influence of wettability on relative permeability is thoroughly discussed in Ref. [19].

Typically, water-wet rocks exhibit a lower permeability to water at residual oil saturation than oil-wet rocks, as well as a higher permeability to oil at irreducible water saturation. Fig. 6.20 shows the characteristic shape of water-oil relative permeability curves for different wettability conditions. Note that in a gas-oil system, where gas is always the non-wetting phase, wettability-related problems are much less relevant.

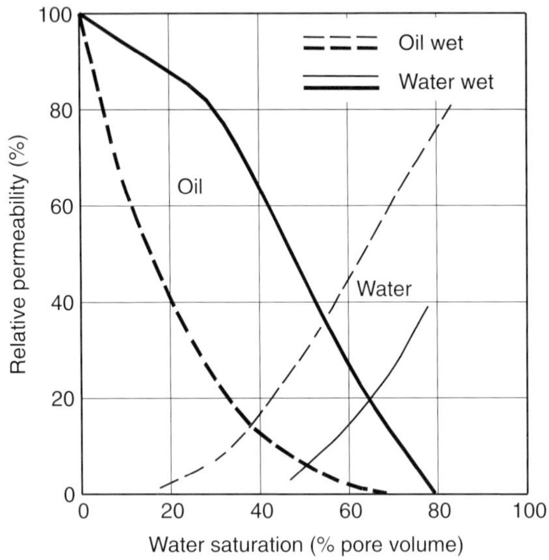

Figure 6.20 Typical relative permeability curves for water wet and oil wet reservoirs.

As far as relative permeability is concerned, the problem is that the original reservoir wettability of the tested core samples used has often been altered. Many factors are known to have an influence over the original wettability: among them, we have the process of core cutting and recovery, the invasion of coring fluids, weathering and contamination during preservation and storage, cleaning and preparation procedures and handling during the measurements [15]. The most reliable samples are probably the so-called preserved, or native state samples, taken from cores cut with low invasion techniques.

When preserved samples are not available, an alternative procedure is to age the extracted cores in reservoir oil for some weeks, with the objective of restoring the original wettability which may have been altered by the cleaning procedure.

Figure 6.21 shows the relative permeability curves of a high porosity and permeability sandstone sample, comparing a prior measurement performed over a cleaned and toluene-extracted plug and a later measurement, performed after an ageing process of 4 weeks in reservoir oil. The difference between the 2 sets of data, both regarding the shape and the end-points of the curves, is striking and would lead to completely different results in the simulation phase. Note also that the aged curves look quite similar to the native state sample.

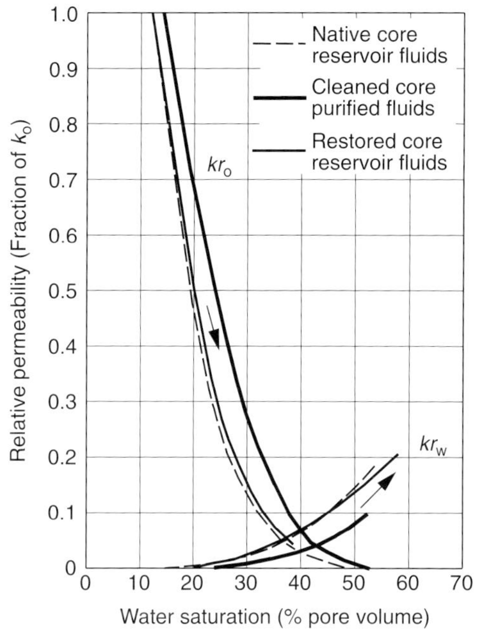

Figure 6.21 Cleaned, native and restored plug relative permeability.

B. Core Scale Heterogeneity

The presence of core-scale heterogeneity and its influence on fluid flow has been recognised for a long time [24]. The problem is illustrated in Fig. 6.22, which shows sets of relative permeability curves obtained on plugs cut with a different orientation than the prevalent heterogeneity at the core scale.

This problem is generally avoided by cutting plugs in the more homogeneous parts of the cores but in fact small scale laminations are always present in reservoir formations and they have an impact on fluid flow. The issue has been extensively documented in the literature [25, 26] and is illustrated in Fig. 6.23, where various types of core scale laminations are shown, compared to the typical core plug size. The implications for relative permeability measurements are also indicated.

C. Higher Scale Heterogeneity

The effect of large scale reservoir heterogeneity in fluid flow has already been discussed in Section 3.4. When relative permeability is concerned, it becomes particularly important to assess what is the nature of the internal heterogeneity represented within the core plug sam-

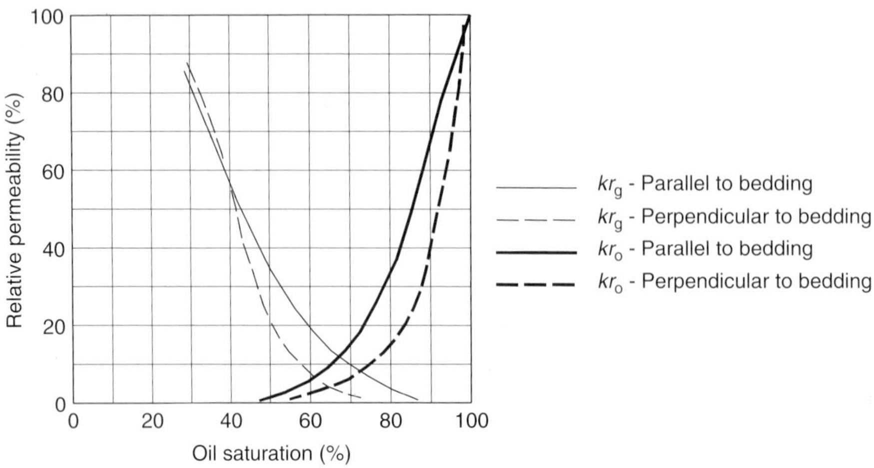

Figure 6.22 Effect of anisotropy on relative permeability measurements.

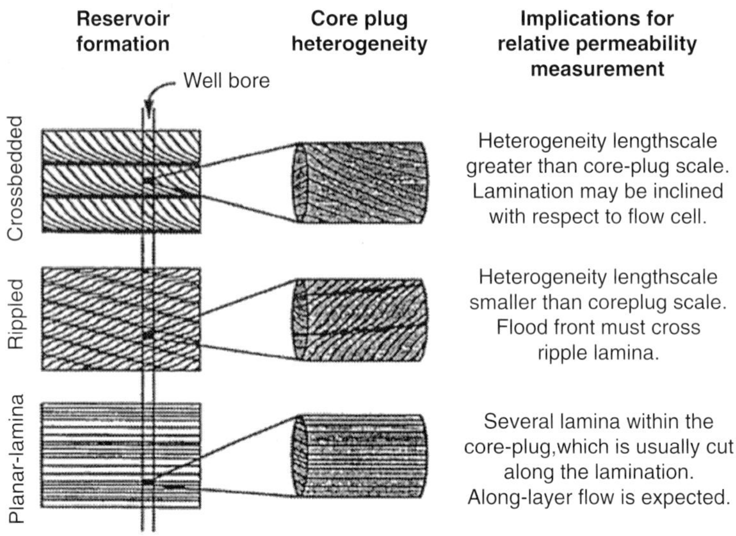

Figure 6.23 Core scale laminations and impact on relative permeability (from Ref. [25]).

ple and whether or not larger scale heterogeneities exist, whose effect cannot be captured in the plug. In such cases, the direct assignment of core-derived relative permeability curves to the simulator gridblocks will lead to an incorrect simulation of the actual fluid flow in the reservoir.

In conclusion, the quality of laboratory relative permeability data should be accurately verified, through a close inspection of the coring, cleaning and measurements procedures, as well as by considering the type of reservoir under study and the heterogeneities that will have an impact on the dynamic behaviour of the reservoir.

When doubts exist over the reliability or the representativeness of the available laboratory measurements, it becomes essential to compare the results with independent source data.

6.3.3.2 Empirical Correlations

Empirical models of relative permeability have been proposed by many authors since the beginning of the 1950's. These models allow for the generation of relative permeability curves compatible with the rock under study in the absence of experimental data and in most cases they have the advantage of providing reasonably reliable data in a quick and convenient way.

These models can be divided into 2 broad categories:

- **Capillary models**. These models are based on the assumption that the rock can be represented by a bundle of capillary tubes of different diameters, with a tortuosity parameter being introduced to take into account the actual geometry of the flow path. This approach allows for the application of the common Darcy and Poiseuille equations in their derivation. The equations proposed by Corey [27] are among the most popular of this type.
- **Empirical equations**. These equations are based on the modelling of sets of measured relative permeability curves, generally through regression analysis. Different equations have been derived depending on the rock type (sandstones vs. limestones vs. conglomerates), the rock fabric (consolidated vs. unconsolidated) and the wettability (water vs. oil wet).

A comprehensive presentation of the various models available in the literature can be found in Ref. [21]. It should be appreciated that the application of these types of equations should not be disregarded even when experimental measurements on core plugs are available, since they may provide data of an accuracy acceptable for reservoir simulation purposes.

6.3.3.3 Field Data

Under favourable circumstances, relative permeability data can be derived from field data. The procedure in this case relies on the construction of fractional flow curves, either for the gas-oil or the water-oil systems, that relate the observed producing GOR or water cut to the prevailing saturation conditions existing in the reservoir.

Figure 6.24 shows an example, for a field with active waterflooding. The values of the ratio k_o/k_w have been derived from a fractional flow calculation, based on the Buckley-Leverett frontal displacement theory, while water saturation is computed from material balance. The points in the figure refer to measured water cut values and may be used to calibrate the fractional flow curve and hence relative permeability.

This technique may provide a useful insight into the actual behaviour of the reservoir, even when only a few points are available. In turn, this information can be used to calibrate the laboratory curves when they are available.

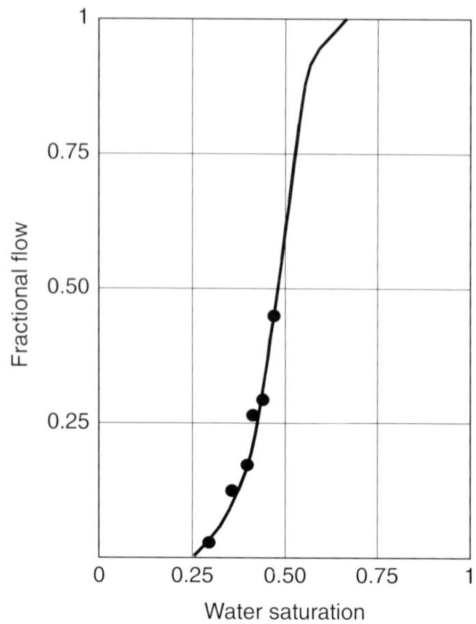

Figure 6.24 Relative permeability from field data.

Of course, the reliability of these types of calculations depends on a number of factors. In particular, it is mandatory to have good quality production data, and the limitations of the classical material balance approach still apply (see paragraph 6.6).

6.3.3.4 Relative Permeability from Numerical Simulation (Pseudofunctions)

One of the most serious drawbacks of core-derived relative permeability data is related to the typical size of the simulation gridblocks. The larger the gridblocks, the less representative will be, in principle, the core measurements. Note also that, from this point of view, the problem of allocating the correct relative permeability functions to the simulation gridblocks can be viewed as an upscaling problem.

Different approaches to the upscaling of relative permeability have been proposed since the beginning of numerical simulation. The most widely used technique relies on the construction of a fine scale simulation model, usually a vertical cross-section, to be used in the computation of average relative permeability (and capillary pressure) curves. These average curves are usually called **pseudofunctions**, in the reservoir engineering jargon. The idea is that pseudofunctions should reproduce, in the coarse model, the same initial fluid distribution, fluid movement and pressure behaviour observed in the fine scale model.

Two types of pseudofunctions have been commonly used in reservoir studies: VE (Vertical Equilibrium) pseudofunctions [28], that assume that gravity and capillarity control the vertical fluids distribution, and dynamic pseudofunctions [29], which represent the most general case, where the viscous forces are also taken into account. A complete discussion of the various types of pseudofunctions and their use can be found in Ref. [30].

Pseudofunctions have been widely used in the past, when the limitations imposed by available computer resources imposed on reservoir engineers the use of a reduced number of vertical layers. Now it is not uncommon to make use of fine grids even in full field simulation studies, therefore the need for pseudofunctions is less stringent than before.

However, for particular cases, the derivation and the application of pseudofunctions can still provide the reservoir engineer with a valuable tool for investigating the relative permeability behaviour at different scales.

6.3.3.5 Three-Phase Relative Permeability

All reservoirs potentially contain 3 fluid phases, i.e., oil, gas and water, therefore in most cases the simultaneous flow of the 3 phases can be envisaged. To cater for these complex flow conditions, 3-phase relative permeability curves must be specified in the simulation model [3].

In the laboratory, 3-phase relative permeability data are difficult and time-consuming to measure. For this reason, most simulators offer indirect techniques to derive this kind of data, like the equation proposed by Stone [31]. However, it should be appreciated that such techniques provide in general a very approximate description of the physics of the process and, from this viewpoint, the definition of reliable 3-phase relative permeability functions is (and will likely remain) a rather uncertain issue.

A comprehensive review of three-phase relative permeability models can be found in Ref. [32].

6.3.4 Residual Oil Saturation

Residual oil saturation to a water displacement (Sor_w) is a parameter of paramount importance in waterflooding, since it defines the microscopic sweep efficiency of the process and hence the final recovery. Additionally, when an EOR process is under consideration, the value of Sor_w is often critical in evaluating the economics of the project.

Residual oil saturation can be determined by means of a number of different techniques, both in the laboratory and in the field, which provide alternative estimations of such parameter. A complete review of the available techniques and their limitations can be found in Ref. [33]. In general, the following techniques may be utilised:

- **Core analysis**. Core analysis is a direct method for measuring the residual oil saturation in the laboratory. Two techniques are usually applied: end-point relative permeability and Dean-Stark extraction tests. In the former case, attention must be paid in the measurement procedures, since very different results may be expected [23]. In the latter case, the retrieval technique used in cutting the core is a very important factor, since oil expulsion during the recovery of the core barrel often results in an underestimation

3. It should be appreciated that simultaneous 3-phase flow is probably rare in nature. However, in the simulation approach, 3-phase flow happens as soon as the saturation condition of the gridblock (hence the 3-phase relative permeability) allows it. In other words, 3-phase flow is more likely to happen in the model than in nature.

of the residual oil saturation. Better results are obtained using pressure coring and more recent applications like *sponge coring*.

- **Estimation from conventional logs**. In the presence of effective sweeping, the water saturation value in a water swept zone corresponds to the residual (or remaining) oil saturation. Routine log analysis techniques can be applied to compute this value, which has the advantage of representing a large-scale estimation when compared to core analysis. In this respect, the value derived from log analysis is much more similar in scale to the value needed in the simulation model. Fig. 6.25 shows an example for a pair of twin wells, drilled with a time span of approximately 30 years. During this period, the aquifer advanced significantly and invaded the pay zone with the more recent well appearing completely swept.

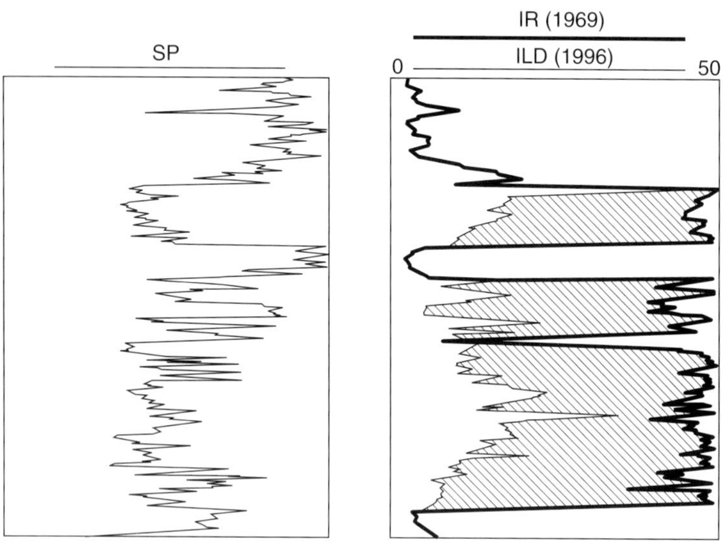

Figure 6.25 Residual oil saturation from well logs in twin wells.

- **Estimation from production logs**. As mentioned in paragraph 4.1.4.2, water saturation can be computed by means of production logging tools (pulsed neutron and carbon-oxygen tools). When run in cased hole in water swept wells, these production logs can often provide reliable estimates of the residual oil saturation. Special applications, known as *log-inject-log*, have also been successfully applied. This technique is based on multiple passes of the logging tool, before and after the injection of contrasting salinity waters. This approach has the advantage of eliminating some of the interpretative parameters, thus allowing for an accurate determination of Sor_w in swept intervals.
- **Single well tracer tests**. These tests consist of injecting a tracer into the well, which has the property of dissolving to form a second tracer when the well is shut-in. The concentration of the 2 tracers during the subsequent production phase can be used to estimate the residual oil saturation. These tests give accurate results and have also the

advantage of providing a Sor_w value averaged over a significant portion of the reservoir rock. On the other hand, they are expensive and they are usually run only in the pilot phase of EOR projects.

The residual oil saturation is an important parameter to determine. The utilisation of a single estimation, i.e., the end-point water relative permeability, can lead to substantial error in the final recovery calculations. However, the confidence level of the determination is greatly enhanced when several techniques are compared.

Figure 6.26 shows the results of the determination of residual oil saturation values, using different techniques [34]. Comparison of the results and a knowledge of the limitations of the various methods will allow a confident estimation of this parameter in such cases.

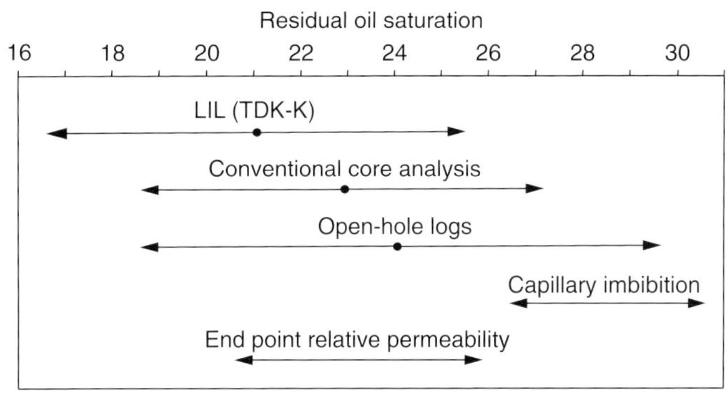

Figure 6.26 Comparison of residual oil saturation techniques [34, mod.].

6.4 PRESSURE ANALYSIS

The practice of using bottom hole pressures to define the dynamic behaviour of a field was started about 1930. Since then, the analysis of static and dynamic well pressures and their evolution with time is one of the typical and most important steps in a reservoir study.

Several types of pressures can be measured in a well and at least 4 types are of interest to the reservoir engineer:

1. Static Tubing Head Pressure (STHP), i.e. the pressure at the wellhead when the well is shut-in.
2. Flowing Tubing Head Pressure (FTHP), i.e. the pressure at the wellhead when the well is flowing.
3. Static Bottom Hole Pressure (SBHP), i.e. the pressure at the bottom hole when the well is shut-in.
4. Flowing Bottom Hole Pressure (FBHP), i.e. the pressure at the bottom hole when the well is flowing.

All these pressures are used in a variety of engineering applications, from wellbore friction loss calculations to the determination of the Productivity Index (PI) of the well.

As far as an integrated reservoir study is concerned, static reservoir pressure is by far the most important well pressure, since it directly reflects the dynamic behaviour of the field. In the remaining of this chapter, we will focus on this type of pressure.

6.4.1 Formation Pressure

Formation pressure is defined as the total fluid pressure in the pore space. As the reservoir is exploited, formation pressure tends to decline. This decline may or may not be important depending on the energy mechanisms of the reservoir and in fact, conversely, the dynamic behaviour of the reservoir can be understood through the analysis of the formation pressure decline. For this very reason, pressure data are an essential piece of information in order to evaluate past performance and predict the future behaviour of any producing field.

The existence of a pressure decline in the reservoir can be detected by monitoring the formation pressure in a number of key wells, where surveys are performed on a regular basis during the producing life of the field. The key information that can be derived from these surveys is the individual well decline rates, i.e., how fast the pressure is declining with time at each well location, and the existence of different rates of decline in different parts of the field, i.e. the existence of lateral pressure gradients in the reservoir.

This last point is of particular interest to the geoscientist: differences in the withdrawal rates, the presence of gas or water injection wells, lateral changes in the rock properties and mostly the existence of large scale heterogeneities in the reservoir formation (faults, stratigraphic changes like local pinch-outs ...), are the most likely factors responsible for the presence of pressure gradients in the reservoir.

In the following section, the sources of static pressure data and their characteristics will be reviewed. Later, it will be discussed how this information can be integrated to build a consistent model of pressure variation with time and space.

6.4.2 Reservoir Pressure Data Sources

Several types of sources of data can be used in pressure analysis. The basic types are static pressure measurements, well-testing derived pressures and WFT measurements. In the next paragraph we will briefly discuss the information brought by each of these data.

6.4.2.1 Static Pressure Measurements

Pressure data are collected on a regular basis in all reservoirs throughout the whole life of the field by running a pressure gauge in the shut-in well. These data represent the basic and frequently the only information as far as static pressure is concerned.

Pressures are usually measured during periodic surveys, which provide the main information to check the depletion stage of the reservoir. During these surveys, a number of wells are shut-in for a specified length of time, to allow the restoration of the static pressure level in the neighbourhood of the wells.

From a general viewpoint, any reservoir has a typical closing time, i.e., a value that has been established over the years and that results from the exploitation experience. This typical time period is normally a compromise between the contrasting needs of having a shut-in time as long as possible and reducing the production loss to the minimum.

The duration of the shut-in time is in most cases an issue of great importance. In general, the time needed for a complete restoration of the static pressure can vary from a few hours to months, mainly depending upon the mobility of the reservoir fluid. In many reservoirs, the pressure will not reach equilibrium within the specified shut-in time, especially in the case of heavy oils or low permeability rocks. In this case, methods can be applied to estimate the theoretical average static pressure by interpreting the build-up pressure profiles (see next section).

However, it should be noted that, even in the case of partially restored pressure values, the inferred rate of decline of the reservoir pressure should be reasonably accurate, provided that the measurement conditions in different surveys are the same. In fact, in reservoir engineering applications, consistency is at least as important as accuracy, therefore the amount of pressure depletion (ΔP) is often more relevant to the engineer than the precision of the absolute pressure values.

It is important therefore to guarantee the consistency of the available information by careful quality checking of the recorded pressure data. Suspicious measurements, related for example to mechanical failures and too short shut-in time, must be verified and possibly disregarded.

Another important issue related to pressure data is the depth extrapolation of the recorded measurement. In fact, the pressure gauge is usually run in the well to a position which is the deepest safe operational depth and the measured values have therefore to be extrapolated to the midpoint of the producing interval, using the static gradients of the fluids present in the borehole. This may be a source of error whenever the location of the fluid interface is not exactly known or when uncertainty exists over the fluid densities.

When the pressure at the perforation midpoint is computed, a further extrapolation has to be applied to the reservoir datum depth, applying the prevalent reservoir fluid gradients at the time of the survey. Note that, even though this is often considered a trivial operation, errors are possible in this extrapolation also, especially in the presence of fluid segregation in the reservoir.

6.4.2.2 Reservoir Pressure from Well Test Interpretation

Well tests are performed throughout the producing life of any reservoir, with the aim of quantifying the producing behaviour of the wells and assessing the average reservoir pressure in the region around the wellbore. The former objective is of particular interest in the case of appraisal well testing, while the latter is of major importance in producing fields.

Evaluating the average static pressure from a well test is a routine, basic calculation for any reservoir engineer, however it may not be a straightforward task. Actually, even leaving aside any issue related to the quality of the data and the inherent difficulty of some interpretations, it should be appreciated that different pressure values can be derived from a well test interpretation:

1. **Maximum recorded pressure**. This is simply the highest value recorded by the gauge. In high mobility systems, this value may be very close to the average pressure, however in low permeability formations or in the presence of high viscosity fluids this

value can seriously underestimate the actual average pressure. In all cases, it represents the lower limit for the actual formation pressure.

2. **Computed average pressure**. Methods for determining the average reservoir pressure within the well drainage area have been presented since the beginning of well testing [35]. These methods are based on the estimation of dimensionless Pressure/Time functions and require prior knowledge of the shape of the drainage area, as well as the position of the producing well within this area. Modern well-testing packages allow for a quick computation of the average pressure, but still this value depends upon the assumed geometry of the drainage area. When uncertainty exists over the actual geometrical configuration of the well-drainage area system, then the computed average pressure may be incorrect.
3. **Horner extrapolated *P****. This a familiar issue to more than a generation of engineers and refers to the extrapolated value to infinite shut-in time of the straight line obtained in a Pressure vs. Pseudotime plot. It represents the theoretical value of the reservoir pressure in a homogeneous, infinite-acting system for an infinite shut-in time. Despite the widespread use of this kind of interpretation, it should be stressed that a number of limitations apply to P^*, which derive from the physical assumptions that are behind Horner theoretical development. In particular, the assumption of linearity of the final portion of the straight line is not justified in many cases and may lead to erroneous results. The presence of a boundary, e.g., a fault, would deviate the late time pressure trend from a linear behaviour. Fig. 6.27 shows a typical misinterpretation of a Horner plot. A complete review of the applicability and limitation of this kind of interpretation is obviously beyond the scope of this work however an interesting digression on the meaning of P^* can be found in Ref. [36].

When the 3 types of pressure values described above are in reasonable agreement, then the definition of an average value is straightforward. However, when large discrepancies exist, as is sometimes the case, then there is no general rule for how to derive a representative average value. From a general viewpoint, the choice of utilising one of the three types of pressure indicated above will depend upon the particular case under study, as well as the individual reservoir engineer's judgement. In all cases, whenever possible, we should be looking for consistency with other data coming from independent sources.

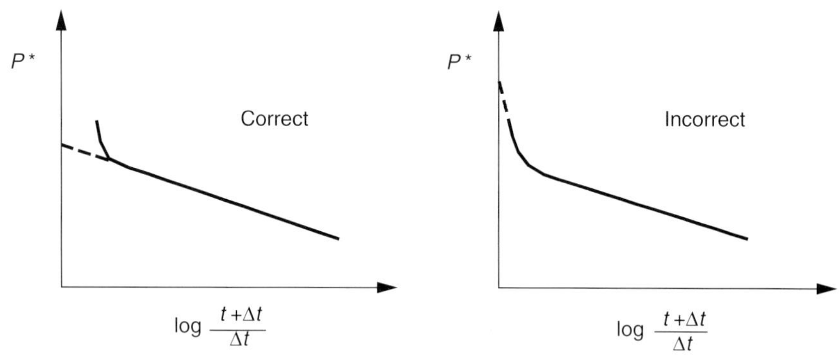

Figure 6.27 Correct (left) and incorrect (right) Horner P^* interpretation.

6.4.2.3 WFT Pressure Data

Wireline Formation Tester Tools provide an alternative means of measuring the formation pressure [4]. The tool is run in newly drilled wells before running the final completion and, in the case of undeveloped reservoirs, it generally provides accurate and precise measurements of the original formation pressure.

However, the most interesting application of the tool is probably in developed reservoirs, since in this case it provides useful information concerning the depletion of the reservoir at the newly drilled location.

Compared to the traditional sources of pressure information (static surveys and well test interpretation), the WFT tool has the following differing characteristics:

- It provides a vertical pressure profile along the reservoir section.
- It allows for a much higher resolution and accuracy in the pressure data determination.

These characteristics allow for the utilisation of these pressure data for some extremely interesting applications, like the determination of the reservoir fluid gradients and the definition of the position of the fluid contacts (Fig. 3.28). However, the most important utilisation of WFT data is perhaps the possibility of defining the level of differential depletion of distinct reservoir units in heterogeneous reservoirs. Fig. 6.28 (left) shows an example of a reservoir where the production has created a differential depletion of the various sand bodies that make up the reservoir. By comparison, the pressure behaviour of a homogeneous reservoir is shown to the right.

This type of information is extremely useful to the geologist as well as to the reservoir engineer, since it allows for the evaluation of the actual reservoir connectivity through the identification of the main flow units. It also gives useful information on the impact of vertical heterogeneities on fluid flow, as well as on the existence of unsuspected crossflows among the different geological units. Finally, this type of data provides a means to validate the pressure information coming from existing well test interpretations, since when large pressure differentials exist among individual flow units, as in the above example, the pressure value interpreted by a single well test performed in the whole reservoir section may have little meaning.

6.4.3 Pressure Modelling

Modelling the pressure behaviour of a field means to study the formation pressure variation in time and space, in order to infer some information concerning the main driving mechanisms and the reservoir performance.

To achieve this aim, the quality checked pressure information coming from static surveys, well test interpretation and WFT measurements must be consolidated into some kind of database. This database, in turn, will represent the basis for subsequent engineering applications.

4. Principles and characteristics of the tool are briefly described in paragraph 4.1.5.3.A.

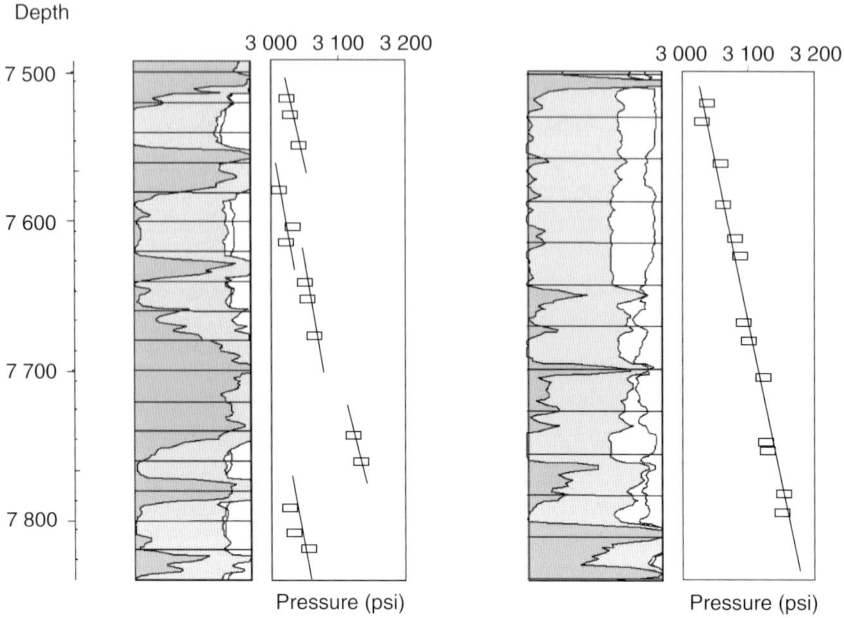

Figure 6.28 WFT surveys in heterogeneous (left) and homogeneous (right) reservoirs.

Many tools are commonly utilised to this end in the reservoir engineering practice. A complete review of all these methods is certainly beyond the scope of this work, however in general they fall in one of the following categories:

- **Pressure maps**. The available pressure data can be utilised to draw pressure, or isobaric maps. Pressure data relative to a limited time period are plotted in a base map and contoured, in the same way as is commonly done for any geological parameter. Maps are built for a number of selected periods of the life of the field, depending on the production history of the reservoir under study. It is common for example to draw maps before and after the implementation of a secondary recovery project like waterflooding, as well as at the end of the historical production. These isobaric maps allow for the direct identification of different pressure regions and hence the main pressure gradients acting in the field in different periods of the exploitation. Fig. 6.29 shows one set of such maps, where a pressure increase is clearly recognisable in the whole reservoir, as a consequence of a water influx coming from the western and northern flanks of the field (light grey). Attention must be paid in the construction of these maps. In fact, since pressure is time dependent, only the measurements relevant to a short time window must be selected in order to insure the chronological consistency of each map. Unfortunately, this often conflicts with the need for a large enough pressure data set to make a proper job of mapping.
- **Pressure profiles**. Pressure profiles are another common way of representing the pressure information. Pressure can be plotted as a function of time or cumulative produc-

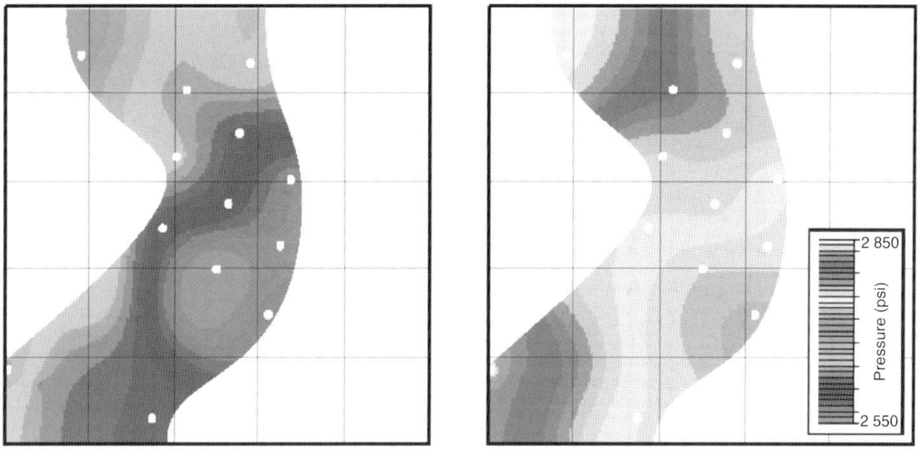

Figure 6.29 Isobaric maps at different time periods.

tion, for a single well, group of wells, a fault block, a lease or the entire reservoir. These kind of plots allow for the study of pressure evolution with time and are extremely important in defining the main driving mechanisms, as well as in assessing the possible compartmentalisation of the reservoir. Fig. 6.30 shows an example relative to a solution-gas drive reservoir, with a saturation pressure of about 4 800 psi. In a pressure-cumulative oil plot (P vs. N_p), the slight change in slope observed after a cumulative oil production of 10 MMbbl, testifies to the increased energy provided by the liberation of gas in the reservoir. In the same plot, another change in the pressure decline trend can be observed at $N_p = 32$ MMbbl, that in this case can be related to the start of a gas injection project in the field.

Volumetric Pressure

In some engineering applications, in particular material balance, it is important to estimate the average pressure of the reservoir under study at selected time periods. The objective in this case is to build a pressure decline curve that represents the average depletion behaviour of the reservoir. To build such a curve, a method to average individual well pressures at each time period is needed. Ideally, individual wells pressures should be averaged by weighting on the drainage volume. However, since this is rarely known with any precision, the common averaging procedure is based on a hydrocarbon pore volume scheme:

$$P_a = \frac{\Sigma_i\, p_i V_i}{\Sigma_i V_i}$$

Where p_i are the individual static pressures and V_i are the relevant reservoir pore volumes, as computed for example in a hydrocarbon thickness map. Fig. 6.31 shows an example relative to a reservoir with a non-homogeneous pressure decline, due to lateral permeability variations in the field. Note that applying the averaging technique described

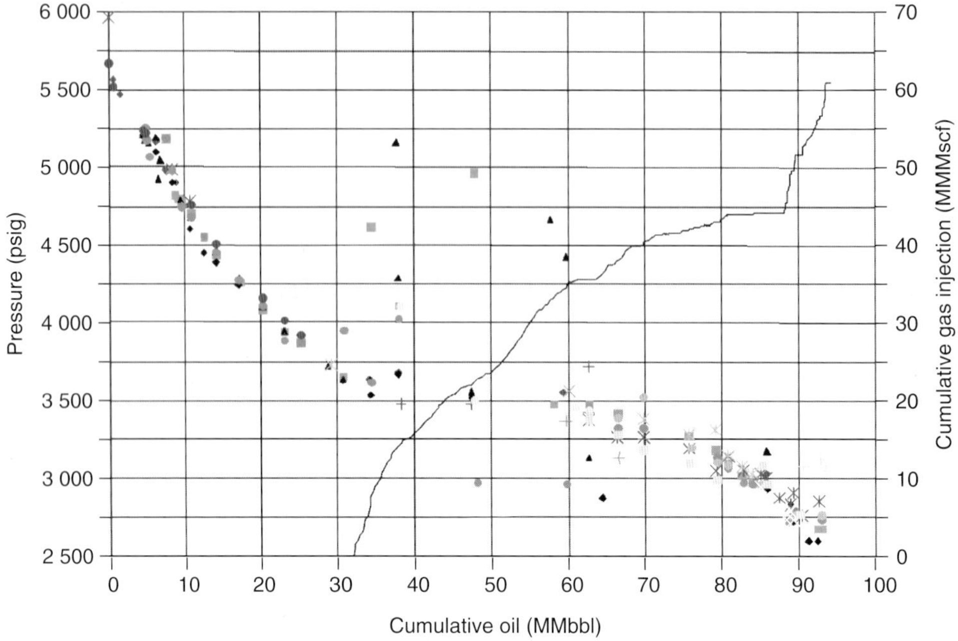

Figure 6.30 Pressure vs. cumulative oil production for a solution gas drive reservoir.

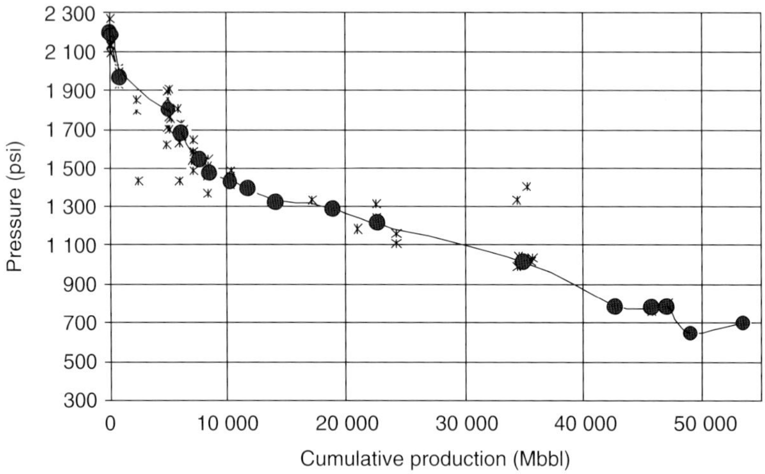

Figure 6.31 Individual well pressure data and average profile.

above, it is possible to obtain a relatively homogeneous pressure trend decline, which could safely be used for computing a material balance for the reservoir.

Other techniques to obtain average pressure profiles can be found in Ref. [36].

6.5 RESERVOIR FLUIDS DISTRIBUTION AND MONITORING

At the time of discovery, hydrocarbon reservoirs are at dynamic equilibrium. In fact, while processes like hydrocarbon migration, diffusion or convection are continuously acting throughout the reservoir from the start of the generation, these mechanisms are very slow and they have no impact on fluid distributions at the reservoir exploitation timescale.

However, when production is started, major changes are induced to the fluid distribution in the reservoir. The extraction and the injection of fluids generates viscous pressure gradients that propagate into the reservoir and ultimately result in fluid movement and replacement.

The understanding of how fluids move within the reservoir is one of the most important tasks of a reservoir study. This work allows for the understanding and the characterisation of the main displacement processes in the reservoir and it provides a reference framework for matching the results of the simulation phase.

Furthermore, this study provides the location of unswept areas in the reservoir and it therefore represents the basis for any development study. For these reasons, proper modelling of the fluid saturations as a function of time is a necessary condition both to a proper simulation phase and to a sound reservoir exploitation.

After a short digression on production reallocation, we will discuss how reservoir saturation changes can be detected and modelled, when water and/or gas enter the reservoir as a consequence of natural encroachment and/or injection. As usual, the attention will be focussed on the different kind of data that can be used to address the issue properly.

Finally, the basics of 4D seismic will also be briefly reviewed, since this technique has a strong potential for imaging and monitoring the fluid saturation changes in the reservoir as a function of time.

6.5.1 Production and Injection Reallocation

Reservoir studies typically involve a revision of the geological correlation scheme. Since the fluid distribution study basically relies on dynamic (production and injection) data, it is therefore imperative to make sure that the existing production database is consistent with the new geological model.

As a matter of fact, when different vertical units (reservoirs or pools) are defined, modifications in the geological correlation scheme imply a different allocation of the produced/injected fluids with respect to the original database.

Such redistribution of fluids is generically called **production reallocation** and it often represents a major issue in a reservoir study. Actually, the procedure requires the simultaneous handling of 3 types of data, i.e., the geological markers, the production/injection data and the completion data (Fig. 6.32).

Whenever a modification is generated in the geological model, this should be propagated throughout the production database, through some reallocation procedure. These procedures are typically based on a net pay or a producing thickness criteria.

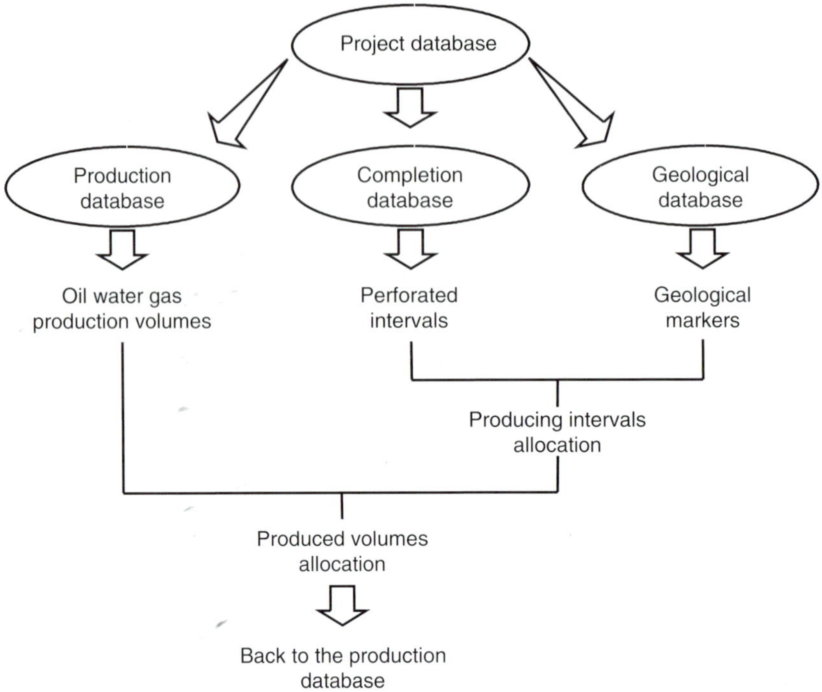

Figure 6.32 Simplified procedure for production and injection reallocation.

Ideally, in a tight integrated database system, such a loop should be transparent to the user. Once the reallocation procedure is defined, each modification should automatically propagate through the whole set of data, both of static and dynamic nature. Unfortunately, in the majority of cases, such tight integration does not exist and the geoscientist is left with the problem of using some sort of existing interface or, even worse, of generating an *ad-hoc* application.

In the workflow of a reservoir study, the need for a production reallocation phase and its related problems in terms of available applications is something to be anticipated. When this is not taken into account, the project manager will experience unexpected problems, which could in turn generate severe delays to the study.

6.5.2 Water Advance with Time

As exploitation starts, a pressure disturbance propagates from the producing wells, eventually reaching the oil-water interface. From this moment, water will enter the reservoir. The quantity will in general be dependent upon the global pressure drawdown, the compressibility of the water phase and the volume and the permeability of the interconnected aquifer. A number of models have been developed in the reservoir engineering literature, which describe the natural encroachment of aquifers into the reservoir (Schilthuis, Hurst-Van Everdingen, Fetkovitch ...) and they will not be dealt with here. What should be noted is that, when the reservoir is connected to a natural aquifer, a water front enters the reservoir sometime after the beginning of the exploitation, starting a process of progressive invasion of the producing wells.

The geometrical configuration of the water front as a function of time is related to the displacement process taking place in the reservoir. In simply structured, homogeneous reservoirs, with a favourable mobility ratio, the displacement is generally dominated by gravity forces and generates rather a stable aquifer invasion. Under these conditions, water breaks through in a regular way, from the structurally lowest to the highest wells.

In the more general case of heterogeneous reservoirs and/or in the presence of unfavourable fluid mobility ratios, the displacement process is dominated by viscous forces and usually results in highly irregular water fronts. Water may intrude in highly conductive paths, like faults or high permeability streaks (*water fingering*) generating early breakthrough in some of the producing wells. At the same time, large scale geological heterogeneities (sealing faults or stratigraphic variations) may prevent water advance in some areas of the reservoir.

In the presence of injection water, the displacement process is basically a function of the injection pattern, however in general the resulting water advance will be even less regular than a natural encroachment, since injection rates tend to result in higher velocities in the water phase and hence in unstable fronts.

The study of the water saturation changes in the reservoir can be usefully tackled through the analysis of the available static and dynamic data as a function of time. This approach provides a means to understand the geometry of the water front, the impact of the structural and sedimentological features and the importance of the viscous gradients.

The study should be performed separately for each identified flow unit and for a number of selected time periods. The choice of the number of periods to consider obviously depends on the duration of the production history, the available data and ultimately the time allocated to this phase of the study, since this work can be very time consuming, especially for large reservoirs with many wells and a long production history. From a very general point of view, the whole historical production profile should be divided into perhaps 3-6 time phases, dominated by characteristic production/injection behaviour. In any case, the choice of the time periods should be made to be consistent with the pressure analysis work.

For each of these periods, the position of the water front should be identified and mapped, trying to distinguish, whenever possible, between the aquifer and the injected fronts.

The available information for the definition of the water front position is diverse. In most cases, for each of the selected periods and each reservoir layer under study, we can make use of the following sources:

- **Production data**. An observed water breakthrough in a particular well gives a useful indication of the position of the water front at a given time. Of course, care must be taken in order to eliminate all the wells whose water cut can be related to well completion problems.
- **Well logs**. All the wells drilled and logged in a particular period, bring some information about the position of the water front. The only **exact** information can be obtained in those wells that encounter a Oil Water Contact (OWC) in the reservoir layer under study. However, in all other cases, we can extract what we could refer to as an **interval** data, i.e., an information that bounds in time the position of the OWC. For instance, if

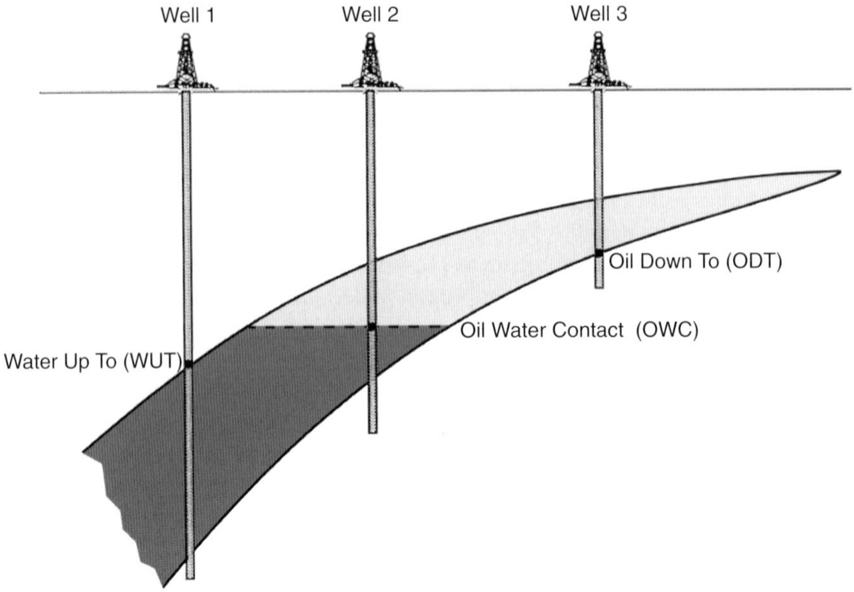

Figure 6.33 Aquifer advance and OWC, ODT and WUT.

a well encounters a reservoir section completely filled with hydrocarbons (**Oil Down To**, ODT), the water front has not yet reached the position of the well and has to be located somewhere down in the structure. Additionally, this information is also valid for the previous time periods, since if the well had been drilled anytime before the actual drilling date, it would have encountered an ODT. In the opposite way, any well that tags a water section (**Water Up To**, WUT) will prove that the water front has already passed through that location and provide us with an information that can be used also in the following time periods. To summarise, while the OWC gives us the exact information that can be used only in the time period under study, the ODT and the WUT provide us with an interval information that can be extended to, respectively, the previous and the following periods (see Fig. 6.33).

- **Production logs**. Production logging measurements (PLT) and cased hole pulsed neutron logging (Thermal Decay Time or Carbon Oxygen tools) are sometimes run in particular key wells, usually when some workover is performed or for reservoir monitoring purposes. These tools provide usable information about the saturation condition of the reservoir at the time of logging, which can be integrated with the available open hole logging data. Ref. [37] provides an example of a large scale, full field application of these tools for continuous fluids monitoring.

The deliverables of a water advance study are in general a set of structural maps, where the progressive position of the water front is illustrated as a function of time. Fig. 6.34 shows an example relative to a water drive field, where the time lag selected for the fluid distribution study was approximately 10 years.

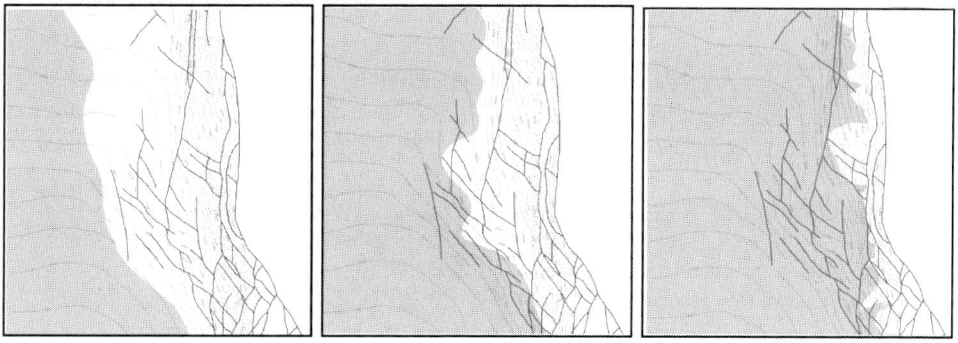

Figure 6.34 Study of the water advance as a function of time.

The most interesting of these maps is of course the last one, since it illustrates the present position of the OWC and therefore the remaining hydrocarbon zones. This map can be used to quantify the unswept oil through a conventional volumetric estimate and therefore it represents the basis for any further development program for the field.

As a conclusion, it should be stressed that proper modelling of the water advance is a major step in the workflow of an integrated study, since it allows for the identification of the major fluid paths in the reservoir, as well as the evaluation of large scale heterogeneities. From this point of view, this work may also help in clarifying some of the unresolved issues of the reservoir characterisation phase.

6.5.3 Gas Advance with Time

As the reservoir pressure falls below the bubble point, gas starts being liberated in the reservoir. When the total liberated gas reaches a given critical saturation, the gas is free to flow independently from the liquid phase and, under favourable conditions of gravity segregation, it migrates towards the higher part of the reservoir structure. Here it may join an existing, primary gas-cap or develop a new, secondary gas-cap.

The presence of a gas-cap, either primary or secondary, is rather a common feature in petroleum reservoirs. As depletion continues, the gas-cap increases its volume and may eventually reach the producing wells, an unwanted situation in most cases.

Monitoring the gas front position with time is therefore as important a task as the study of the water front advance.

The procedure for identifying the position and the evolution of the gas front as a function of time is similar to that which has been previously discussed for the water. The data used for this study are basically the same:

- **Production data.** An observed clear gas breakthrough in a particular well gives a useful indication about the position of the gas front at a given time. Since the measured GOR's may have some uncertainty, care must be taken to make sure that the observed increase in GOR is related to an actual gas-cap breakthrough and not, for example, to a local increase in the gas saturation around the well due to producing conditions.

- **Well logs**. As in the case of water, wells drilled and logged over a particular period, bring some information about the position of the gas front. Again, the only exact information can be obtained in those wells that encounter a clear Gas Oil Contact (GOC), nevertheless in the other cases we can derive interval data for the GOC, i.e., information that bounds in time the position of the contact. Of course, this can only be done when well logs that permit the distinction between oil and gas, typically the Density and/or the Neutron log, are available.
- **Production logs**. When available, production logging measurements (PLT) and cased hole pulsed neutron logging (Thermal Decay Time or Carbon Oxygen tools), as well as more traditional cased hole neutron logs are other useful sources of information concerning the evolution of the position of the GOC with time.

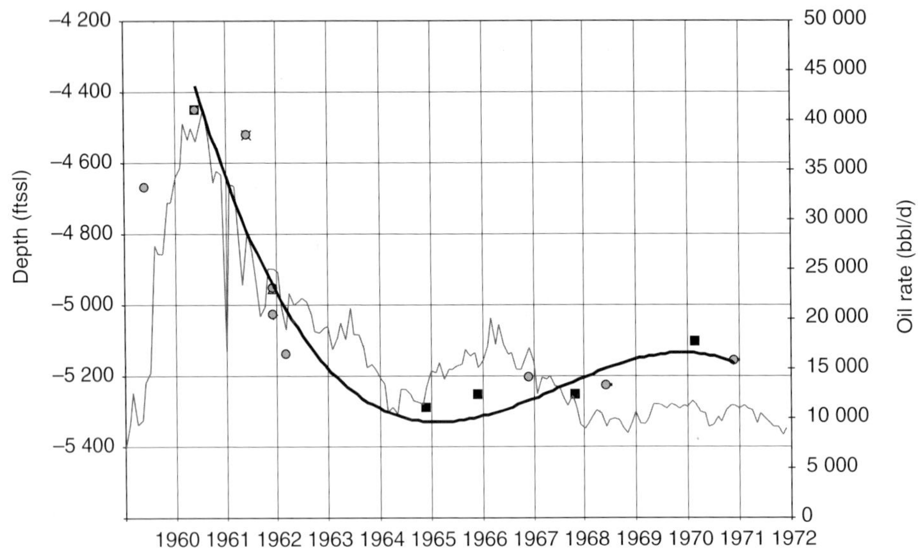

Figure 6.35 Evolution of the position of the GOC with time.

The procedure to draw maps showing the position of the gas contact as a function of time can be performed in the same way as previously described for the water-oil system. Again, the last map of the set will display the current position of the GOC and it will inherently define the remaining oil zone.

An alternative and possibly faster way to represent the process is to build a cross-plot of depth vs. time, illustrating the evolution of the position of the contact (Fig. 6.35). All the information described above can be input in this kind of plots, and a line can be drawn by interpolation through the available points. Production data (circles in the figure) provide exact data points and are described by the depth of the top of the perforated interval on the Y axis and the date of the gas breakthrough on the X axis. Well logs that show a distinct GOC are also exact information and are described by the depth of the contact on the Y axis and the time of logging on the X axis (squares in the graph).

The kind of representation of Fig. 6.35 provides an average position of the GOC as a function of time, and it should be used when the displacement process is reasonably stable (gravity-dominated). Of course, when these conditions are met, the method can also be applied to a water-oil system, in substitution to the more laborious mapping approach illustrated in the previous section.

6.5.4 4D Seismic Monitoring

4D seismic refers to the process of repeating 3D seismic surveys in a given field in a time-lapse mode. It is one of the most interesting and promising techniques as far as fluid monitoring and reservoir management are concerned, since it has the potential to image variation in the saturation and pressure conditions of the reservoir with time.

The basic principles behind this technique are simple. Seismic waves respond to variations in static properties of the reservoir formation (e.g., lithology, porosity), as well as dynamic properties, i.e., properties that change with time (fluids saturation, pressure). The relative contribution of these 2 components is not usually known and therefore when interpreting a traditional 3D seismic survey we are not able to make an unambiguous distinction between static and dynamic contributions.

However, the availability of repeated 3D seismic surveys in the same area, allows in principle to eliminate by subtraction the static component of the signal, since this does not change with time. The result of such operation is a direct image of the time-dependent dynamic components of the signal, related to field exploitation.

The potential of this technique is significant. In fact, the availability of a number of time-lapse images of the dynamic properties of the reservoir has several possible applications of the utmost importance, including the definition of the efficiency of the displacement process, the identification of bypassed reserves or undrained fault blocks, the tracking of the fluid interfaces with time and the monitoring of the injected fluids fronts.

The technique is relatively recent and a number of technical and operational issues must be carefully evaluated before its implementation [38]. Relatively few large scale applications are known in the literature, the most well-known being possibly relevant to the Duri Field, Indonesia, the world's largest active steamflooding project. In this field, 4D seismic monitoring has been successfully applied to monitor the injected steam fronts and the relevant sweep efficiency, thus allowing proper reservoir management [39].

However, most of the major oil companies are testing or developing the methodology and there is little doubt that in few years geoscientists will be dealing with 4D seismic much more than today.

6.6 MATERIAL BALANCE

For many years in the past, the material balance equation has been the most important tool available to reservoir engineers in the study of the production performance of oil and gas fields.

The situation changed, however, starting in the 1960's, when the availability of increasing computer resources sidelined the role of material balance in favour of emerging numerical methods. For a long period, especially during the 1980's and early 1990's, sophisticated numerical simulation techniques seemed to have completely superseded the material balance approach, which was considered as a kind of relic of the old times. In a period where new technologies were rapidly emerging and changing the traditional way of working at all levels, there is little doubt that there was also a fashion component in this attitude.

Nevertheless, an inversion of tendency has perhaps started in recent years, which again could be related to the prevailing computer evolution. The widespread availability of personal computers has led to the development of user-friendly, windows-based material balance software, that allows the engineer to apply material balance-based techniques in a much less tedious way than had to be done in the past.

From a reservoir engineering point of view, there is little doubt that such a comeback will have a positive effect. Material balance, in fact, is still a powerful tool for analysing the performances of hydrocarbon fields and, as will be shown later, its objectives and possibilities are not the same as numerical simulation, but rather are complementary. For this reason, whenever the basic conditions exist, as it is very often the case, material balance should always be applied as a pre-requisite to the numerical simulation phase.

6.6.1 Why Run a Material Balance?

Why apply the material balance equations, when a 3D simulation model is eventually to be built? This question reflects a common attitude, which somehow deprives material balance of some of its many merits. In fact, possibly more than numerical simulation, material balance is a useful technique to investigate the dynamic behaviour of a hydrocarbon field. In most cases, it allows for the identification of the main drive mechanisms acting in the reservoir and the relevant impact of each. It also provides good estimates of the hydrocarbon originally in place, to be compared with the available volumetric figures. Furthermore, it provides an indication of the global consistency of the available dynamic data. As it will be discussed more in detail in later sections, these applications are not within the typical domain of numerical simulation.

Finally, under favourable circumstances, it can be applied to compute the advance of the water or gas front and also to evaluate the efficiency of the displacement process.

Undoubtedly, the most important feature of the material balance method is its simplicity and rapidity. As it has been implemented in recent PC-based software, material balance can be used to quickly investigate various reservoir drive hypotheses and the impact of each energy component. A reservoir model can be built in a matter of hours, while the computer runs themselves last no more than seconds. Different PVT, geometric and production configurations can be tested for screening purposes. No other reservoir engineering technique pays so well.

The general formulation of the material balance equation and its characteristics are reviewed in some detail in the Appendix. In the following sections, some typical applications to reservoir studies will be reviewed.

6.6.2 Material Balance Application to Reservoir Studies

The material balance equation, in its essence, expresses a very simple concept. It states that, for a given pressure drop, the volume of fluids produced must equal the total expansion of the reservoir system plus any natural water influx. The qualitative formulation of the material balance principle, expressed in reservoir conditions, is therefore:

$$\textit{underground withdrawal} = \textit{reservoir system expansion} + \textit{fluid influx}$$

In this definition, underground withdrawal represents the total quantity of produced fluids (oil, gas and water), while the system expansion represents the (virtual) increase in total reservoir volume related to the expansion of reservoir fluids (oil, gas and connate water) and the formation itself.

In practice, once a pressure decline has been identified and the basic reservoir and fluid parameters have been introduced, the material balance equation can be solved simultaneously for several reservoirs unknowns and a defined number of time steps, thus allowing for the history match of reservoir performance.

In a classical application, pressure data and reservoir and PVT properties are utilised to calculate the expected reservoir performance in terms of fluid withdrawal. This computed production profile is compared to the actual production data of the field, while the input parameters (typically the aquifer constants) are varied until the match is performed.

Figure 6.36 shows an example relative to a water drive reservoir, comparing the actual and computed production profiles. In this case, the match has been obtained by varying the aquifer type and permeability. As a way to crosscheck, most software also offers the possibility to reverse the computation and derive a pressure trend starting from an imposed production profile. Note that this is the way a typical numerical simulator would work.

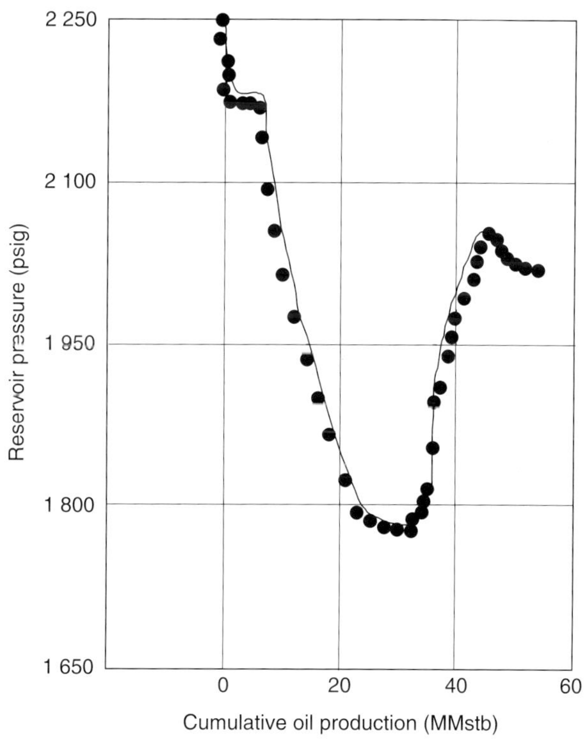

Figure 6.36 Material balance: match of the observed production.

Another interesting output of a typical material balance study is what could be referred to as an *energy graph*. This kind of plot, shown in Fig. 6.37, summarises the energy supplied by each drive mechanism in the various phases of the development, according to the material balance model. This allows for a significant insight into the reservoir mechanics as a function of time. In the example, it is clear that water drive was the main producing mechanism in the initial part of the field life, while fluids and rock expansion only played a minor role. Later, around 1975, a water injection project started, which in the last years provided up to 30% of the total energy of the system.

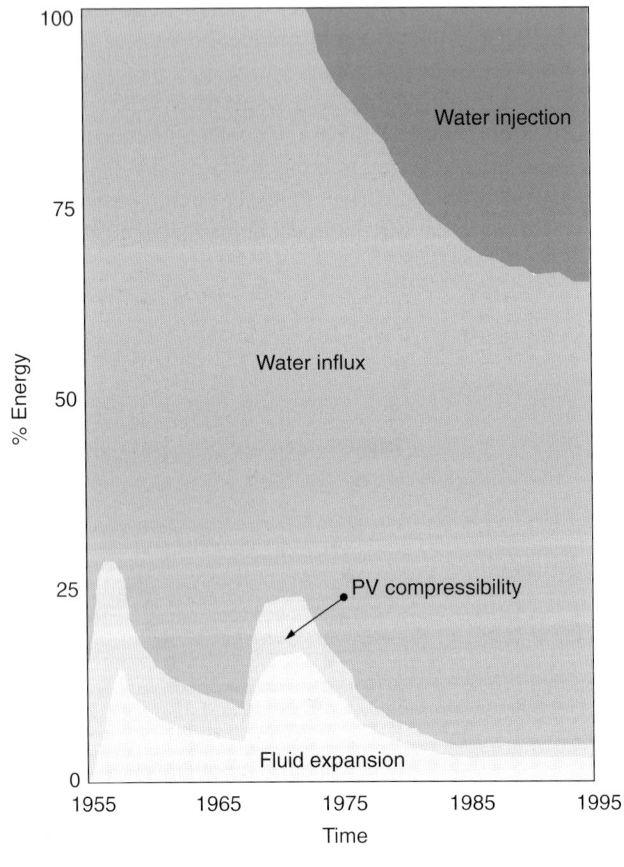

Figure 6.37 Material balance, energy graph.

When the distribution of fluids in the reservoir as a function of time is known, the material balance technique can also be used for more sophisticated applications, which allows for history matching the observed position of the contacts and, ultimately, for the estimation of a global sweep efficiency. Fig. 6.38 shows one of such applications. The example is for a solution gas drive field, where a simultaneous gas and water injection project have been active for over 30 years. A hydrocarbon pore volume vs. depth relationship is provided as an input, while the position of the gas and water contacts are computed by assuming some parameters of displacement efficiency (see paragraph 6.1.3). In other words, if the actual position of the contacts is known, the global sweep efficiency can be estimated. In this example, the match of the water contact has been achieved by imposing an areal sweep efficiency (E_a) equal to 40%, thus implying that over 60% of the reservoir had not been contacted by the injected water. On the other hand, the efficiency of the injected gas proved to be much better.

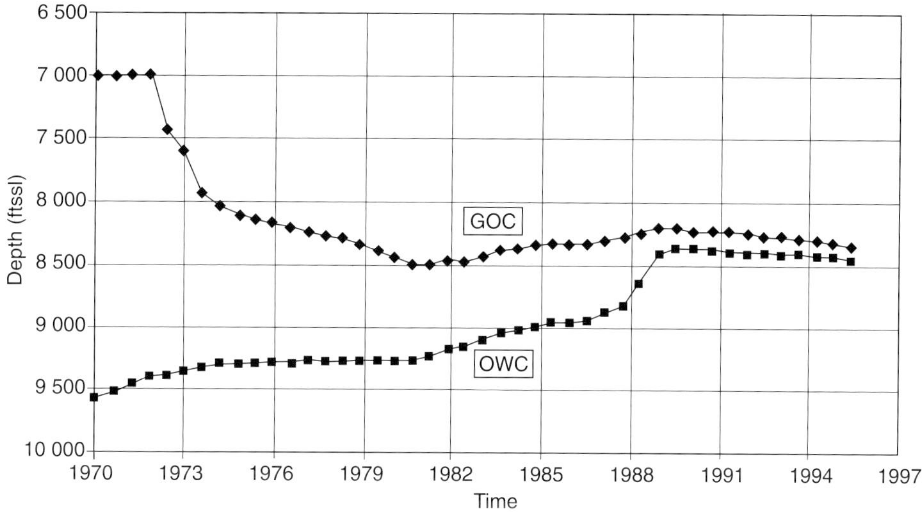

Figure 6.38 Material balance: match of the position of the contacts.

6.6.3 Material Balance vs. Numerical Simulation

It is common belief that numerical simulation can be considered a more sophisticated, 3D version of the traditional material balance technique. In fact, this is far from reality. Even though material balance calculation is an essential part of the simulation routines, the global approach of the simulator is completely different to the material balance method.

The differences between the 2 techniques are illustrated in the diagram of Fig. 6.39. The numerical simulator takes the geometry and the petrophysics as input to the calculations, together with the PVT properties and the measured field production. Therefore all the volumetric parameters have to be evaluated beforehand and imposed on the model. This implicitly means that the OOIP is an input, rather than an output of the model. The same holds for the aquifer and/or gas cap volume. The numerical simulator provides solutions for the pressure and saturation at each time step, which have to be matched against the measured data.

In the material balance approach, on the contrary, pressures are given as an input to the equation. No assumption is made as far as the geometry of the system is concerned and in fact the volumetric parameters can be computed as an output, thus providing an independent assessment to the classical geological estimates. In addition to that, material balance provides an insight into the reservoir drive mechanisms, while providing in most cases a reliable quantification of the reservoir energy sources.

In conclusion, material balance should be considered as a preliminary step to the more complex phase of reservoir simulation. In fact, most of the assumptions that need to be made in the numerical simulation can be investigated and clarified first through the application of this simple technique.

For these reasons, over 60 years after its first publication by Schilthuis [40], material balance still represents an essential step in reservoir studies.

MATERIAL BALANCE		***NUMERICAL SIMULATION***
Production Pressure PVT	**Input data**	Production OOIP PVT Scal Water influx
OOIP Water influx	**Output data**	Pressure Saturations

Figure 6.39 Material balance (left) and numerical simulation (right) approach to reservoir modelling.

6.7 STREAMLINES SIMULATION

Streamlines and streamtubes methods have been in use for almost 50 years for modelling injection patterns and sweep efficiency.

As with material balance, streamlines simulation has found renewed interest in recent years. Such a comeback is related to the development of more sophisticated geostatistical techniques and to the need for simulating large heterogeneous geological models. Refs. [41] and [42] provide the state of the art of the methodology, as well as the most recent theoretical developments.

As a modelling tool, streamlines simulation cannot be regarded merely as a simplified version of the traditional finite-difference (FD) models. The main feature of streamlines simulation is that the transport of components is decoupled from the pressure solution. Streamline-based flow simulation relies on two support grids: a traditional simulation grid, where the initial petrophysical properties are defined and pressure is computed, and a dynamic streamline grid which is used to compute the fluid transport. Fluid is transported along each streamline independently of the underlying grid used to define the petrophysical properties and to solve for pressure. The streamline grid is updated (i.e., streamlines are recomputed) whenever the boundary conditions change (well rate variations, new or shut-in wells …), or the total mobility has changed significantly.

As a consequence, timestepping also differs from that in a traditional numerical simulator. In streamlines simulation, there is a global timestepping, related to how often the 3D pressure field is solved and the streamlines are computed, and a lower order 1D timestep for the saturation calculation along the streamlines. The former is the most expensive component of the whole simulation. As a result run times for streamline-based simulators are practically proportional to the number of gridblocks in a model.

Streamlines simulation finds its typical application in modelling incompressible, heterogeneous systems, where the flow is convection-dominated. The technique is not suited for complex fluid flow physics: while recent theoretical developments allow for the application of streamlines to compressible and gravity dominated systems, the main advantage of the technique, i.e., rapidity, is lost in such complex configurations. Fig. 6.40 shows the typical application domains of streamlines and FD simulation.

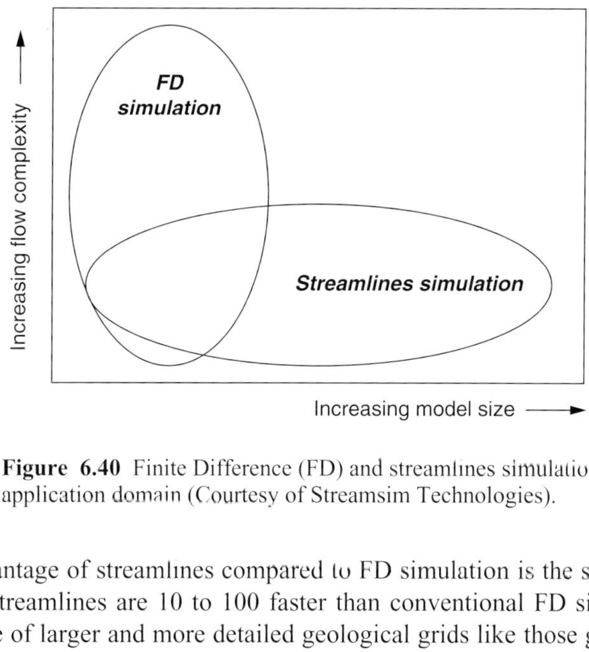

Figure 6.40 Finite Difference (FD) and streamlines simulation application domain (Courtesy of Streamsim Technologies).

The main advantage of streamlines compared to FD simulation is the speed of computation. Typically, streamlines are 10 to 100 faster than conventional FD simulations, which allows for the use of larger and more detailed geological grids like those generated through geostatistical modelling, while limiting or even eliminating the need for upscaling.

More generally, streamlines simulation has a number of interesting applications, which are listed below:

- **Pattern studies**. This is probably the most typical application of the technique. Streamlines provide an instantaneous visualisation of the flow pattern as a function of reservoir heterogeneity. Additionally, they also provide an estimation of the well allocation factors, i.e., how much of an individual well production/injection rate is related to other production/injection wells. Such information is particularly useful in optimising and balancing well patterns.

- **Simulation of heterogeneous systems**. Streamlines simulation represents an alternative to FD models for the simulation of heterogeneous systems dominated by convection. In such reservoirs, the streamlines approach proves to be faster and more efficient. Even when a FD model is eventually to be built, streamlines represent a powerful tool for screening purposes and provide a means to accelerate the history match phase. Fig. 6.41 shows the streamline pattern in an oilfield dominated by a strong aquifer action. Such images offer a significant insight into the actual sweep process in the reservoir.
- **Ranking geostatistical models**. Streamlines represent a fast and efficient way to simulate and rank different geostatistical realisations. This allows for the selection of a limited number of significant realisations, to be eventually simulated through FD models.
- **Sensitivity studies**. Streamlines simulation is a convenient tool to perform sensitivity studies to evaluate flow mechanics and displacement efficiency.

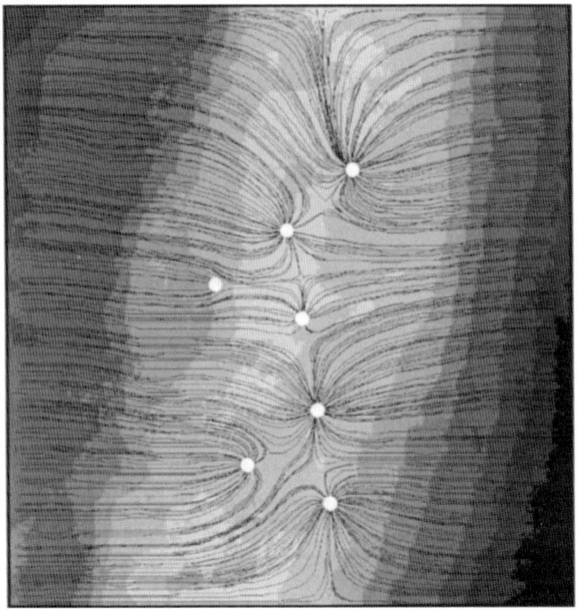

Figure 6.41 Streamline pattern during natural depletion of a water drive reservoir.

In the framework of an integrated reservoir study, streamlines simulation finds its typical application as a complementary modelling tool, prior to the full-field FD simulation model. Results can be used to better understand the displacement processes and to derive useful guidelines to be used in the subsequent simulation phase.

Under favourable circumstances, however, streamlines may represent an efficient alternative to FD modelling. The project manager should always evaluate such possibility, since it may result in a faster and easier simulation study.

References

1 Pietraru V (1996) New analytical water/gas-coning solution for vertical and horizontal wells. World Oil, June.

2 Dietz DN (1953) A theoretical approach to the problem of encroaching and by-passing edge water. Proc. Konikl. Ned.-Akad. Wetenshap, Series B, 56, 83.

3 Merle HA, Kentie CJP, van Opstal GHC, Schneider GMG (1976) The Bachaquero study: A composite analysis of the behaviour of a compaction solution gas drive reservoir. JPT, Sept.

4 Cook CC, Jewell S (1996) Simulation of a North Sea Field experiencing significant compaction drive. SPE-RE, Febr.

5 Sulak RM (1991) Ekofisk Field, the first 20 years, JPT, Oct.

6 Meisingset KK (1999) Uncertainties in reservoir fluid description for reservoir modelling. SPE-REE, Oct.

7 Amyx JW, Bass Jr DM, Whiting RL (1960) Petroleum Reservoir Engineering – Physical Properties. McGraw-Hill.

8 McCain Jr WD (1990) The Properties of Reservoir Fluids. PennWell Books.

9 Moses PL (1986) Engineering applications of phase behaviour of crude oil and condensate systems. JPT, July.

10 Standing MB, Volumetric and phase behaviour of oil field hydrocarbon systems. SPE, Richardson, 124.

11 Petrosky Jr GE, Farshad FF, Pressure-volume-temperature correlations for gulf of Mexico crude oils. SPE paper 26644.

12 McCain Jr WD (1991) Reservoir-fluid property correlations – State of the art. SPE-RE, May.

13 Stiff HA (1951) The interpretation of chemical water analysis by means of patterns. Trans. AIME, 192, 376.

14 Gravier JF (1986) Propriétés des Fluides de Gisements. Ed. Technip, Paris.

15 Anderson WG (1986) Wettability literature survey – Part 1: Rock/oil/brine interactions and the effect of core handling on wettability. JPT, Oct.

16 Anderson WG (1986) Wettability literature survey – Part 2: Wettability measurements. JPT, Nov.

17 Anderson WG (1986) Wettability literature survey – Part 3: The effect of wettability on the electrical properties of porous media. JPT, Dec.

18 Anderson WG (1987) Wettability literature survey – Part 4: Effect of wettability on capillary pressure. JPT, Oct.

19 Anderson WG (1987) Wettability literature survey – Part 5: The effect of wettability on relative permeability. JPT, Nov.

20 Anderson WG (1987) Wettability literature survey – Part 6: The effect of wettability on waterflooding. JPT, Dec.

21 Honarpour M, Koederitz L, Harvey AH, Relative Permeability of Petroleum Reservoirs. CRC Press.

22 Rose W (1989) Relative permeability. *In:* Bradley's Petroleum Engineering Handbook, Chapter 28.

23 McPhee CA, Arthur KG, Relative permeability measurements: An inter-laboratory comparison. Paper SPE 28826.

24 Corey AT, Rathjens CH (1956) Effect of stratification on relative permeability. JPT, Dec.

25 Ringrose PS, Jensen JL, Sorbie KS, The use of geology in the interpretation of core-scale relative permeability data. SPE paper 28448.

26 Honarpour MM, Cullick AS, Saad N, Humpreys NV (1995) Effect of rock heterogeneity on relative permeability: Implications for scaleup. JPT, Nov.

27 Corey AT (1954) The interrelation between gas and oil relative permeabilities. Prod. Mon. 19, 38.

28 Coats KH, Nielsen RL, Terhune MH, Weber AG, Simulation of three-dimensional, two-phase flow in oil and gas reservoir. SPE Reprint Series 11.

29 Kyte JR, Berry DW (1975) New pseudo-functions to control numerical dispersion. SPEJ, Aug.

30 Barker JW, Thibeau S (1997) A critical review of the use of pseudorelative permeabilities for upscaling. SPE-RE, May.

31 Stone HL (1973) Estimation of three phase relative permeability and residual oil data. J Cdn Pet Tech, Oct.-Dec., 53-61.

32 Blunt MJ, An empirical model for three-phase relative permeability. SPE paper 56474.

33 Chang MM, Maerefat NL, Tomutsa L, Honarpour MM (1988) Evaluation and comparison of residual oil saturation determination techniques. SPE-FE, March.

34 Kidwell CM, Guillory AJ (1980) A recipe for residual oil saturation determination. JPT, Nov.

35 Matthews CS, Brons F, Hazebroek P (1954) A method for the determination of average pressure in a bounded reservoir. Trans. AIME.

36 Dake LP (1994) The Practice of Reservoir Engineering. Elsevier, Amsterdam.

37 Harness P, Shotts N, Hemingway J, Rose D, van der Sluis R, Accurate oil saturation determination and monitoring in a heavy oil reservoir. SPE paper 46245.

38 Lumley DE, Behrens A, Practical engineering issues of 4D seismic reservoir monitoring. SPE paper 38696.

39 Jenkins SD, Waite MW, Bee MF (1997) Time-lapse monitoring of the Duri steamflood: A pilot and case study. The Leading Edge, Sept.

40 Schilthuis RJ (1936) Active oil and reservoir energy. Trans. AIME.

41 Thiele MR, Batycky RP, Blunt MJ, Orr Jr FM (1996) Simulating flow in heterogeneous media using streamtubes and streamlines. SPE-RE, Febr.

42 Batycky RP, Blunt M, Thiele MR (1997) 3D field-scale streamline-based reservoir simulator. SPE-RE, Nov.

CHAPTER 7

Numerical Reservoir Simulation

Numerical Reservoir Simulation has been practised since the beginning of the 1960's, as a way of estimating the future behaviour of petroleum fields. Beforehand, reservoir engineering calculations were largely based on analytical methods like material balance and displacement theories like Buckley-Leverett.

The birth of reservoir simulation, in its modern definition, is closely related to the availability of fast digital computing machines and the parallel evolution of numerical techniques that allowed for the solution of large sets of finite difference equations, describing 2D and 3D multiphase flow in heterogeneous media.

The potential of the application of such techniques in the context of petroleum engineering soon appeared evident and in less than a decade most of the major oil companies already had their own numerical models and applied reservoir simulation to their most important producing fields.

Thirty years later, reservoir simulation is an everyday practice in oil and service companies and is handled by most reservoir engineers. Applications are varied, from conventional field production forecasting under different exploitation schemes, to more specialised tasks like phenomenological models. At the same time, new developments are ongoing, especially in the parallel-computer hardware and software domain, and large scale simulations (what have been referred to as *megacell reservoir simulations*) are becoming increasingly common. Ref. [1] provides the state of the art of the technology in this respect.

The current widespread use of this tool in the reservoir engineering community is in fact related to many factors, not all of them strictly technical:

- **Applicability**. The applicability of the tool has no competition from any other conventional technique. It could argued there is not a single problem, among those encountered in the daily routine of the reservoir engineer, that cannot potentially be tackled through numerical simulation.
- **Ease of use**. Modern simulation packages are provided with interactive pre and post-processors, which tremendously facilitate the use of the model. The availability of default options and different expert levels allow even the most inexperienced engineer to end up with some kind of result for the problem at hand.

- **Acceptance**. Management has been trained through the years, often by the same reservoir engineers, to accept reservoir simulation as the standard technique. Currently, in most companies, high level management requires simulation-supported studies.

In spite of these points, reservoir simulation is not a panacea and its application may prove to be dangerously misleading, or in many cases simply unnecessary. In the next section, some of the issues related to the applicability of this technique will be reviewed.

7.1 WHEN TO RUN A SIMULATION MODEL?

The widespread use and acceptance of reservoir simulation in petroleum applications is not free from danger, of course. The limitations of the technique and its possible misuse have been highlighted many times through the years, in some cases by those very experts who are considered among the pioneers of the technique [2, 3].

As a matter of fact, a number of points should be considered before applying numerical simulation, especially as far as the specific objectives of the study are concerned. In petroleum engineering, as in any other scientific application, a well posed problem is the first step of the solution and some preliminary analysis work is always necessary, in order to evaluate the real need for a simulation study.

This preliminary analysis will demonstrate the applicability of reservoir simulation, it will provide an indication of the expected results and finally it will indicate the required degree of complexity of the model itself. In particular, the following points should be considered:

- **Accuracy of the expected results**. Leaving aside the numerical errors (the simulator provides a reproducible but approximate solution), the accuracy of the results is related to a correct problem statement and to the quantity and quality of the available input data (*garbage in, garbage out*). The experience and knowledge of the engineers involved in the study represent another important factors.
- **Inherent uncertainty**. Simulation work is subject to a degree of uncertainty, which derives from the incomplete knowledge of the geological model, from the representativeness of the input data and finally from the numerical solution approach (numerical dispersion, grid orientation effects …).
- **Non-uniqueness of the results**. The reliability of the model predictions depends on the quality of the history match achieved. However, as has been noted for many years [4], the history match procedure always results in a non-unique combination of variables, since in a typical simulation study we have many more unknown than known parameters. This means that different reservoir descriptions can produce the same history match, and in general they will provide different forecast profiles. In this respect, simulation results should best be regarded in a relative, rather than in an absolute sense.
- **Technical overkill**. In many instances, reservoir engineers are faced with relatively simple problems, which can safely be tackled with simpler, analytical techniques. In such cases, the use of a complex tool like numerical simulation may result in technical

overkill. Reservoir engineers should always evaluate the degree of complexity of the problem and use the right degree of technology accordingly.

- **Available resources**. Numerical reservoir simulation is more expensive than other techniques, because it requires the allocation of significant human and technical expertise. The decision to run a simulation model and the relevant level of complexity should be compared with the resources available.

In conclusion, before starting any study, the project manager must evaluate all the aspects involved in the decision of running a reservoir simulation model. The basic question is always: is it really worth? From a general perspective, problems should always be solved by the simplest and least expensive method that will provide an adequate answer. In some cases, this preliminary analysis may show that conventional reservoir engineering techniques represent a simpler, faster and less expensive approach than reservoir simulation and should therefore be preferred. For example, when the short-terms production profiles are to be evaluated, decline curve analysis (Fig. 7.1) represents a reliable and cost-effective tool, while simulation would prove to be a long and expensive alternative.

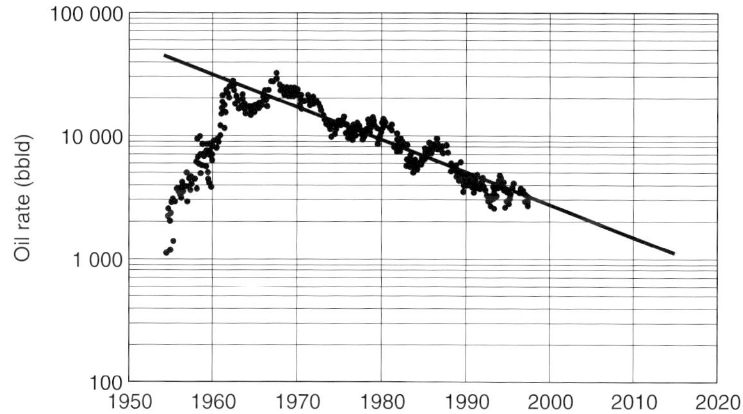

Figure 7.1 Decline curve analysis for short term production forecasting.

In an old but evergreen paper about the use and misuse of reservoir simulation [2], Coats states that valid applications should fulfil the following three features:

1. A well posed question of economic importance. A typical question would challenge for example the choice of a waterflooding project over a natural depletion scheme.
2. Adequate accuracy of reservoir description and other required input data.
3. Strong dependence of the answer upon non-equilibrium, time-dependent spatial distributions of pressure and fluid saturations. This dependence will rule out traditional analytical techniques like material balance.

7.2 WHY RUN A SIMULATION MODEL?

There are many reasons to perform a simulation study. Perhaps the most important, from a commercial perspective, is the ability to generate oil production profiles and hence cash flow predictions.

In the framework of a reservoir study, the main objectives of numerical simulation are generally related to the computation of hydrocarbon production profiles under different exploitation options. In this context, there is little doubt that reservoir simulation is the only qualified technique that allows for the achievement of such objectives. Simpler techniques like material balance are particularly useful for evaluating the reservoir mechanisms, but are not suited for reservoir forecasting.

Reservoir simulation, on the other hand, offers the required flexibility to study the performance of the field under defined production conditions. All commercial simulators are provided with sophisticated well-management routines that allow the engineer to specify the operating conditions at the levels of producing interval, well, well group, reservoir and field. In its simplest definition, these well-management routines assign specified rates and pressures to the wells, but they can also perform much more complex tasks, like shut-in or workover a well according to some GOR or WOR criteria, optimise individual well production to match facilities capacity, control gas or injection rates and so on. This is why reservoir simulation is considered the best technique for reservoir management. No other engineering tool offers such capabilities.

7.3 DESIGNING THE SIMULATION MODEL

Once the decision to run a simulation study is taken, the following stage is to design the simulation model. This phase implies the selection of the type of geometry to utilise, whether 1D, 2D or 3D and the choice of the simulator, whether a black-oil, a compositional, miscible, thermal or chemical.

In this respect, a number of factors have to be taken into consideration, some of which are listed below and described briefly.

- **The recovery process of the reservoir**. This is the most important parameter, since the model must be able to correctly reproduce the main reservoir drive mechanisms. This influences the type of model to use and also the degree of detail to attain. For example, when a water-oil displacement process is the main driving mechanism, a black-oil simulation will be adequate, but on the other hand the model must be sufficiently refined both areally and vertically to properly reproduce the complex geometry of the displacement fronts.
- **Quality and type of the available information**. These influence the level of detail to use in the model. Complex reservoir and fluid descriptions based on few and/or poor quality data may be seriously misleading and generate unrealistic solutions.
- **Type of answer required**. In most studies, relatively simple outputs are required, typically oil, gas and water production profiles. In such cases, a black-oil simulator may

be sufficient even when complex hydrocarbon interactions happen in the reservoir. However, if for the same reservoir the composition of the produced phases is required, then a compositional model must be run. The desired accuracy of the expected results will also influence the design of the simulation model.

- **Available resources**. The study must be measured against the human, economic and technological resources available. It is dangerous to start complex studies, without assessing the global effort required, in terms of expert level, software, hardware and the budget limits.

This preliminary analysis will help in defining the degree of complexity required for the particular study. The bottomline is that the model design phase should always lead to the construction of the simplest model able to meet the objective of the study.

7.3.1 Selecting the Model Geometry

The first step of the design phase is to define the geometry of the model. Several types of geometry can be utilised, the most common being the following [5]:

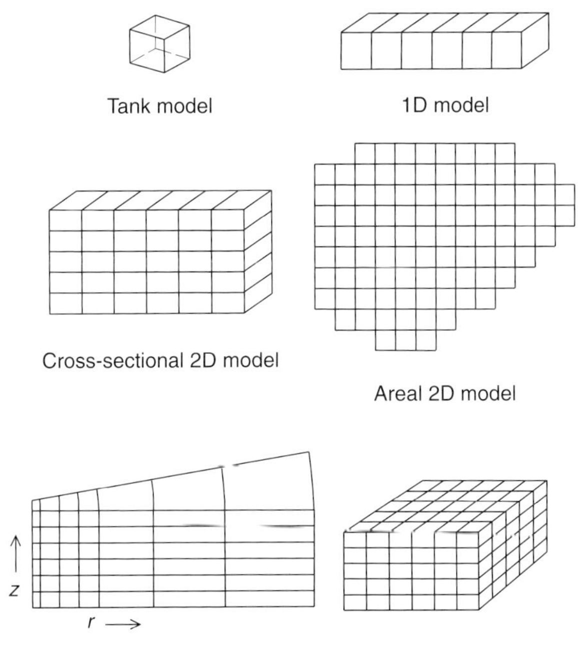

Figure 7.2 Basic reservoir simulation models [5].

- **1D models**. These types of models are never used for field study, since they do not represent the actual reservoir geometry and cannot simulate the displacement process.

They can be used, however, for sensitivity purposes in variations of reservoir parameters or to realise dynamic upscaling of petrophysical properties.

- **Cross sectional 2D models**. They are used when vertical displacement processes are to be studied, for example in the cases of flank water injection or crestal gas injection. Fig. 7.3 shows an example relative to the study of the unstable water-oil displacement in a reservoir with unfavourable mobility ratio. These types of models can also be used to define pseudofunctions, when a vertically coarse 3D model is to be built eventually (paragraph 6.3.3.4).
- **Areal 2D models**. They are used when areal flow patterns dominate reservoir performance and when vertical heterogeneities are not relevant to fluid flow. The typical application of these models concerns therefore pattern injection studies. They can also be used in the case of solution gas drive reservoirs, where gravity effects are negligible. In most cases these models require pseudofunctions to account for vertical flow.
- **Radial models**. These models are limited to the region surrounding a well and are usually built to evaluate the well production behaviour in the presence of large vertical gradients. The typical application is related to the study of water and gas coning or cusping in vertical and horizontal wells.
- **3D models**. These are the most commonly utilised models. They can account for the actual distribution of geological and petrophysical properties within the reservoir and therefore they must be used in the presence of large scale vertical and/or horizontal heterogeneities and in general whenever geology is too complex for a 2D representation. Theoretically, these models can be used to represent any recovery process in the

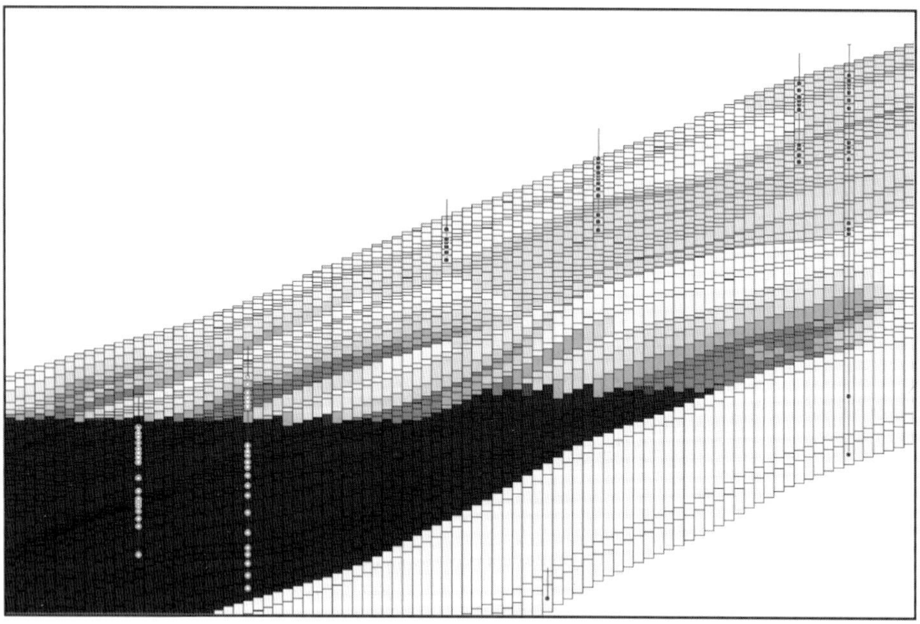

Figure 7.3 Cross-section model for water fingering study [6].

reservoir, the only limitation being the number of total cells, which in turn limits the degree of detail of the description.

The full field, 3D model is the most obvious choice for an integrated study, since the entire reservoir can be effectively modelled. Moreover, this approach allows for the integration of all the available static and dynamic information.

However, it is not always necessary or desirable to build full field studies, especially in the case of large, old fields, when the amount of information to handle and its quality makes the building of a reliable model a prohibitive or questionable task. In these cases, a combination of small scale phenomenological studies and conventional reservoir engineering techniques may represent a wiser alternative.

In other studies, a combination of models may prove to be a sound approach: for example, the results of a detailed, phenomenological 2D cross-section may be integrated in a later stage in a coarser, full-field 3D model. Streamline models can also be used in such context. Ref. [7] presents an excellent example of use of different types of models to design a miscible CO_2 flood.

7.3.2 Selecting the Simulator Type

Different types of simulators are used to represent the mechanisms related to different types of reservoirs. The selection basically depends on the nature of the original reservoir fluids and the predominant recovery process. Leaving aside the chemical models, which are seldom used, the basic types of simulators are the so-called black-oil, compositional and thermal. Their features are briefly described in the following points, while for a more comprehensive treatments we refer to the available textbooks [7, 9].

- **Black-oil models**. This type of isothermal model applies to reservoirs containing immiscible oil, gas and water phases. The black oil model treats hydrocarbons as if they had 2 components, i.e., oil and gas, with a simple, pressure-dependent solubility law of the gas in the liquid phase. No variations are allowed for gas and oil compositions as a function of pressure or time. These models can be used to reproduce most reservoir mechanisms, including solution gas-drive, gas-cap drive, water drive, water injection, and immiscible gas injection. They can deal with vertical variations of the PVT properties, by defining a saturation pressure/depth relationship. They can also deal with lateral PVT variations, through the definition of different equilibrium zones.
- **Compositional models**. In an isothermal compositional model, the hydrocarbon phases are represented by N components, whose interaction is a function of pressure and composition and is described by some Equation Of State (EOS). The number of hydrocarbon components N is usually related to the desired detail of the results but is often limited by practical computational time and is normally between 3 and 7. Compositional models are used whenever the hydrocarbon phase compositions and properties vary significantly with pressure below the bubble point or the dew point. Typical applications of these models are the depletion of volatile and gas-condensate reservoirs, gas-cycling projects or injection of CO_2.

- **Thermal models**. When the temperature varies in the reservoir, a thermal model must be used. In a thermal model, the usual components are H_2O, in water or steam phase, a light (volatile) and a heavy (non-volatile) hydrocarbon phases. The fluid and rock-fluid properties are characterised as a function of pressure and temperature. These models are used to simulate cyclic steam injection, continuous steamflooding or more complex processes like in-situ combustion.

A last type of model to be mentioned is the **dual media** (fracture-matrix) model, where the reservoir rock is considered to be composed of 2 interconnected networks, the fracture and the matrix, each characterised by its own properties. These models are run under both black-oil or compositional formulation, using different configurations usually called dual-porosity or dual permeability, depending on whether or not the fluid flow in the matrix is explicitly permitted. These models are typically applied to the study of naturally fractured reservoirs.

7.4 BUILDING THE SIMULATION GRID

Grid selection and building is an essential part of the simulation work, especially when geologically complex reservoirs are concerned. The choice of a correct grid representation or, conversely, of an incorrect grid, may have a considerable impact on the accuracy of the results, as well as the time and resources required by the simulation exercise.

In the context of integrated reservoir studies, grid building is also particularly relevant because it represents the phase where the geological architecture of the reservoir, both in terms of external and internal geometry, is transferred into the simulation model. It is here that we eventually decide which degree of simplification can be applied to the geological description without jeopardising the quality of the final results.

Several types of grids are normally available to the geoscientists, from the conventional Cartesian grids, to 3D Corner Points grids and the more sophisticated hybrid grids. These types of geometry will be briefly discussed in paragraph 7.4.4. In any case, the choice of the geometrical representation to be used must account for a number of geological, dynamic and numerical issues. These will be reviewed in the next sections.

7.4.1 Geological Issues

Naturally occurring hydrocarbon reservoirs often exhibit a high degree of geological complexity, both in terms of external architecture and internal heterogeneity. In order to obtain reliable results, the simulation grid must adequately reproduce such geometrical features.

The external boundary of the reservoir is the first and most obvious geometric element that has to be represented. The grid must encompass the whole hydrocarbon zone and also a sufficiently large part of the water zone, when an active aquifer exists. In the absence of a distinct permeability anisotropy within the reservoir, the external boundaries of the reservoir also define the main grid orientation.

The other important point to consider is the presence and complexity of internal reservoir heterogeneity. Structural features such as faults, either sealing or highly conductive, must be represented in the model and their throw should be carefully reproduced, if a correct distribution of the reservoir fluids is sought. In the presence of inverse faulting, the use of sophisticated gridding techniques should be considered, especially when the inclination of the fault plane is significant.

Likewise, in the vertical direction, the presence of laterally extensive barriers to fluid flow must be properly modelled and the grid geometry should be conformed to represent such heterogeneities. Extensive reservoir boundaries, e.g., flooding surfaces or shale breaks, define individual flow units and therefore should be individually modelled as simulation layers. Stratigraphic complexity such as layer pinch-outs should also be represented.

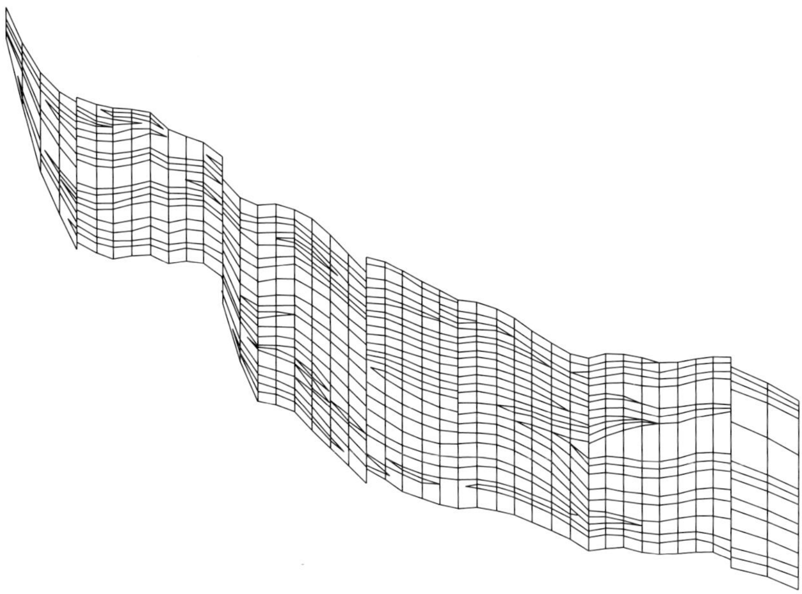

Figure 7.4 Vertical layering in a stratigraphically complex reservoir [10].

An interesting example of a geologically oriented simulation grid is shown in Fig. 7.4 (from [10]). The figure illustrates a complex stratigraphic model, with presence of truncation and onlap sequences whose impact on fluid flow is expected to be crucial. To properly take into account such complexity, a fine scale vertical grid was built, with more than 40 layers.

The correct integration of the geological model in the numerical simulation can only be performed through an adequate representation of the reservoir external and internal geometry in the simulation grid. Excessive simplifications would fail in capturing the flow characteristics related to the complexity of the architecture of the reservoir units.

7.4.2 Dynamic Issues

The analysis of the geological complexity of the reservoir provides some basic requirements for grid building, as far as the degree of detail is concerned. However, to correctly reproduce the observed field performance, dynamic issues should also be considered, which may indicate the need for some modification of the level of detail of the simulation grid. Actually, even in the presence of relatively homogeneous reservoirs, the simulation grid has sometimes to be refined, to cater for a correct representation of the fluid flow in the reservoir, especially in a multiphase flow case.

In the XY plane, the degree of refinement of the simulation grid largely depends on the number of producing wells. It is generally assumed that at least 2 or 3 cells should exist between neighbouring wells (possibly more between injector-producer pairs), in order to correctly reproduce the displacement process, while minimising numerical dispersion problems. In some cases, especially when the wells are not located in regular patterns, local grid refinements (LGR) can be applied to the regions around the producing wells, in order to improve the calculation in these zones.

Vertically, the grid layering must be fine enough to be able to reproduce viscous and gravity-related processes, like for example water fingering and gas overrun. In the latter case, when a vertically coarse grid is used, the segregation of gas and its migration to the structurally highest parts of the reservoir cannot be correctly modelled, thus resulting in a misrepresentation of the observed GOR's of the producing wells.

7.4.3 Numerical Issues

The selection of the grid size and orientation also depends on some numerical issues, the most important being related to numerical dispersion and grid orientation effects.

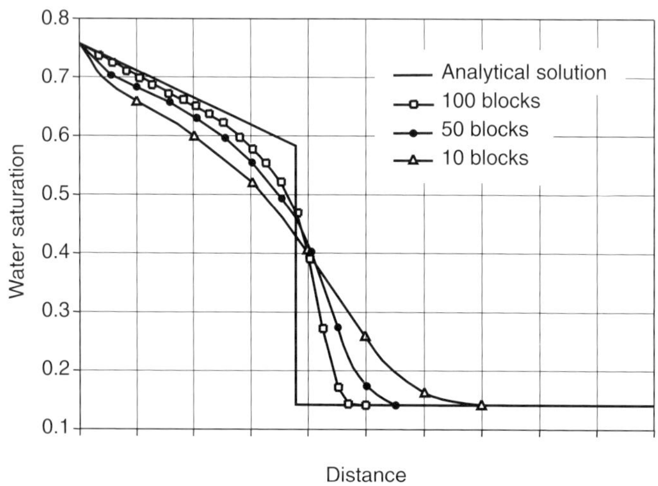

Figure 7.5 Effect of numerical dispersion.

Numerical dispersion is an artefact of current numerical techniques, which occurs in the simulation of processes related to rapid changes in saturation, e.g., a water-oil imbibition process. Numerical dispersion introduces a bias in the results, which is related to the dimension of the cells that are used in the simulation model.

Fig. 7.5 illustrates the problem. When the same displacement process is simulated using a different number of cells, the saturation profile changes. As it can be noted, the effect of numerical dispersion is to smear the saturation profile when few grid cells are used and therefore to decrease the displacement efficiency at breakthrough. In addition, this effect is more pronounced in the presence of favourable mobility ratio. More details on the subject can be found in Ref. [5].

Numerical dispersion is always present in any simulation model to some degree, even though reducing the cell size minimises the impact on the calculated results. A satisfactory balance should be sought in the selection of the grid dimensions, in order to limit the effect of numerical dispersion and to work with a practical number of cells.

Grid orientation effects, on the other hand, refer to a bias in the calculated results, which stems from the orientation of the grid with respect to the location of the injection and production wells. The problem is illustrated in Fig. 7.6, where 2 injection and 1 producing wells are shown, named A, B and C, respectively. Using the grid in the figure, the path from the injector B to the producer C is much larger than the path from the injector A, therefore the breakthrough time will be artificially longer. Note also that rotating the grid of 45 degrees, would reverse the situation.

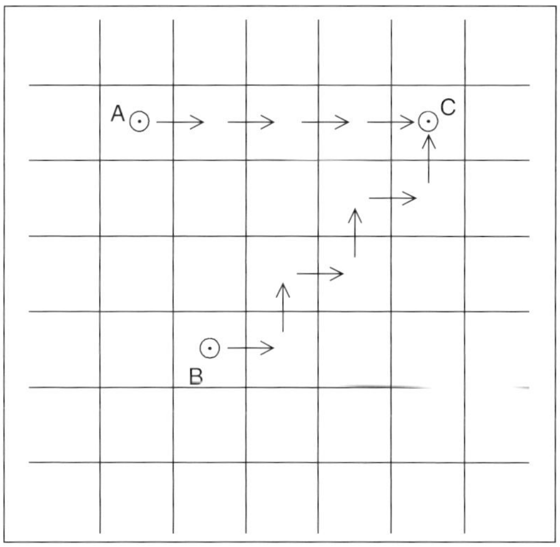

Figure 7.6 Grid orientation effects.

Grid orientation effects are particularly important in the presence of very high mobility ratios, as in the case of gas-oil or steam-oil displacement processes, but its impact should never be overlooked, even in a more conventional water-oil system. The most effective solution for limiting grid orientation effects is possibly the use of nine-points formulation of flow equations, where the diagonal connections are also taken into account (Fig. 7.7). This approach is more computer-intensive, however when computation time is not an issue, the nine-points scheme should be the preferred option.

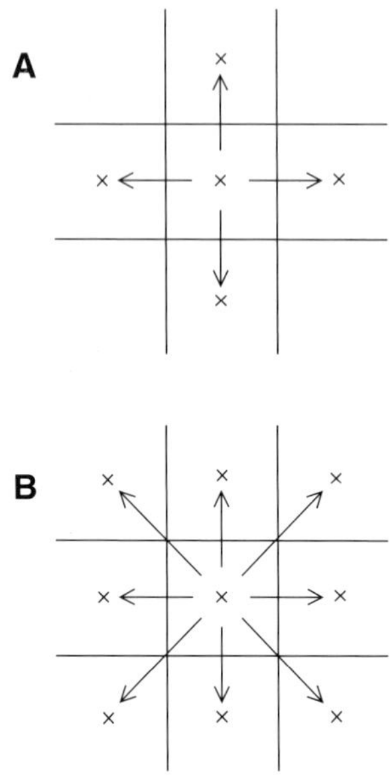

Figure 7.7 Five points (**A**) vs. nine points (**B**) computation schemes (from [5]).

7.4.4 Choice of the Simulation Grid

Once geological, dynamic and numerical issues have been properly considered, the decision can be taken about the type of grid geometry to use and its dimensions.

As far as the type of grid is concerned, two basic geometry are used in the current practice, the Cartesian and the Corner Point.

The block-centred Cartesian grid is obtained by aligning the grid blocks along the Cartesian coordinates in the 3 directions of space, thus resulting in a globally orthogonal grid. This kind of grid is the oldest and possibly still the most frequently used, because of its simplicity and ease of construction.

In the corner point geometry the coordinates of each grid block corner are specified, instead of the block centre coordinates. The resulting grid is not necessarily orthogonal and this allows for a more accurate representation of the actual geology of the reservoir. On the other hand, these types of grids would theoretically require the specification of all the components of the potential gradients at each block face. Since most of the simulators do not cater for these extra components, it is considered good practice to limit the local distortion of the grid, otherwise severe errors are introduced in the calculation.

Figure 7.8 compares the 2 types of grids. It should be noted that, even though the corner point geometry is much more appealing for the increased sense of geological realism, this does not necessarily implies more accurate result. In fact, while the choice of the geometry type remains a rather subjective issue, some engineers still recommend, as much as possible, the use of the old, familiar Cartesian grids.

Another point of interest concerning simulation grids is the possibility of using different block dimensions in different parts of the fields. This feature is of particular interest in most studies, since there are often some zones of the reservoir where it is desirable to have calculations as accurate as possible and other zones, usually the peripheral areas or the aquifer, where accurate calculations are not deemed necessary. The advantage in this case is a considerable reduction in the total number of active cells and therefore in the computation time.

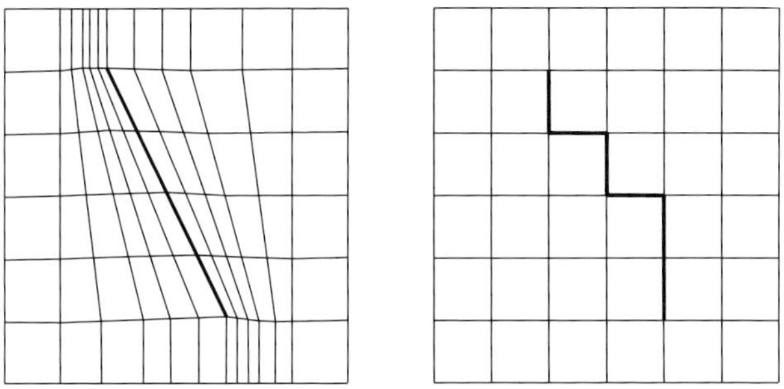

Figure 7.8 Cartesian vs. corner point (left) geometry grids.

All of the existing grid building software allows for the specification of different row and column sizes, thus leading to the so called *tartan grids* [1]. In more complex cases, an existing grid can be locally refined in regions of interest, usually where rapid changes of saturation and pressure occurs. These locally refined grids (LGR) are more demanding from a computational point of view, but allow for a much greater flexibility of use.

Figure 7.9 shows two example grids, relative to a variable size tartan grid and a locally refined grid. Note that in the latter case, a single block face in the coarse grid can be connected to two or more block faces in the fine grid.

Finally, it should be mentioned that in recent years a great deal of attention has been devoted to the development and application of more complex grid geometries, like Voronoi and PEBI grids [11, 12]. The driving forces behind such interest are related to recent technological advances in different petroleum disciplines:

- **Complex well geometries**. Complex well geometries are becoming increasingly common in operational practice (horizontal, multidrains, 3D wells). The correct description of the fluid flow around such complex configurations requires adapted gridding techniques.

1. Attention should be paid in limiting the deformation of the peripheral cells. It is commonly accepted that a maximum value of 3 should be accepted for the cells shape factor ($\Delta X/\Delta Y$).

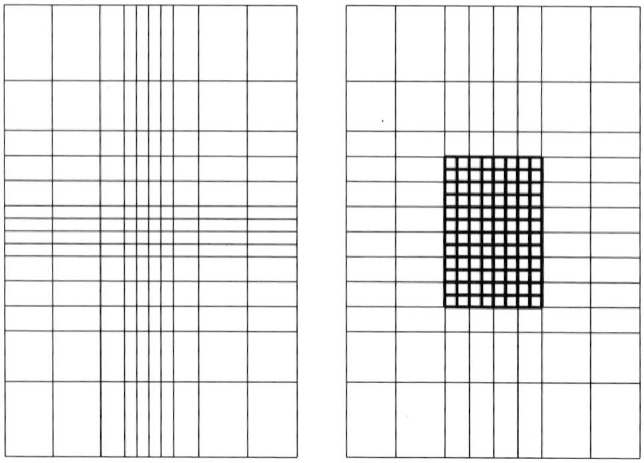

Figure 7.9 Variable size cartesian grid (left) and locally refined grid.

- **Complex geological models**. Recent advances in geological modelling allow for a 3D detailed description of complex geometrical reservoirs, including reverse faulting. In order to properly transfer such geological architectures to the simulation model, sophisticated gridding techniques are needed.

Reference 13 provides a review of the various standard simulation grids that could be utilised in a reservoir simulation exercise. Currently, the future trend in this field seems oriented towards the application of **hybrid grids**, i.e. grids which combine different geometries in different zones of the reservoir, while optimising the fluid flow solution and the practical running time. Fig. 7.10 provides an example of utilisation of hybrid grids for reservoir simulation.

7.4.5 Building the Simulation Grid: Conclusions

Building the numerical simulation grid is one of the important steps of an integrated study. It is here that the geological representation of the reservoir, both in terms of external shape and internal, large scale heterogeneities, is modelled using a simplified geometrical grid.

From a general viewpoint, it is important to represent the geological description as much as possible and to preserve the detail that has been attained in the geological study, especially the internal discontinuities. Typically, for example, it is important to conform the main grid boundaries in the vertical direction with the flooding surfaces or the sequence boundaries identified in the sequence stratigraphy study. In addition, the grid should represent, as far as possible, the structural configuration of the reservoir, including all the faults that are thought to have an impact on fluid flow.

It should also be appreciated that, in some cases, these needs may conflict on one hand with the need to use a practical number of grid cells and, on the other hand, with other dynamic and numerical issues.

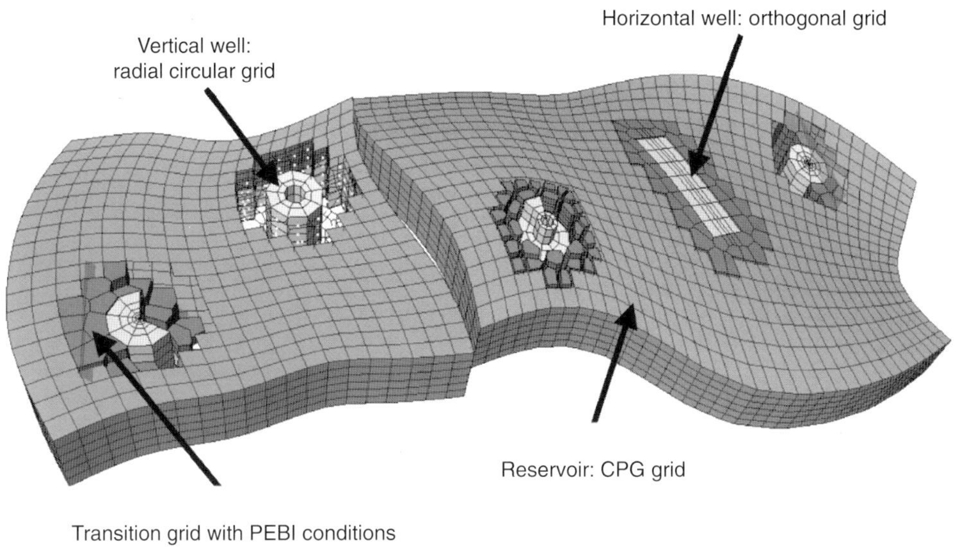

Figure 7.10 Hybrid grids for reservoir simulation (Courtesy of Institut Français du Pétrole).

Although the choice of the type of simulation grid and its geometry is dependent upon the particular reservoir under study, here are some points that should always be considered at this stage of the study:

- The building of a simulation grid is an important integration stage in the framework of a reservoir study, where geological, dynamic and numerical issues must be taken into account.
- The grid block dimensions and the total number of cells should be adjusted to the problem at hand, considering in particular the complexity of the geological setting, the prevailing recovery process and the number of active wells. The number of simulation cells should satisfy the above requirements, while keeping in mind that increasing the number of cells does not necessarily guarantee increased accuracy of the results. Results from a recently published paper [14] show that when only a few wells are concerned, relatively coarse grid representations provide accurate results, when compared to a fine scale simulation. The need for detailed grids becomes more stringent when a large number of wells are involved in the model.
- The choice of the type of grid geometry to use, block-centred Cartesian, corner point or hybrid, depends on the reservoir under study, as well as on the technical resources available. From a general viewpoint, corner point geometry should be preferred when the geological complexity is such that the use of a Cartesian grid would result in an oversimplification of the actual reservoir layer connectivity (i.e., in the presence of reverse faulting). In a simple geological setting, on the other hand, the Cartesian option may prove to be a better alternative.

- The use of special features like local grid refinement allows for a greater flexibility and accuracy of calculations in regions of interest, but their use should be limited because of increased computing effort.
- More sophisticated gridding options (hybrid grids) should be considered in the presence of complicated reservoir geometry and/or complex wells.

Finally, it should be noted that the availability of increasing computing power will help in the near future to build more and more accurate numerical grids, while at the same time limiting problems related to numerical dispersion and upscaling.

Figure 7.11 shows the evolution of the simulation model dimension with time. The extrapolation of the trend indicates that in the next few years it will be possible to simulate reservoirs with more than a million cells in the context of routine operational studies.

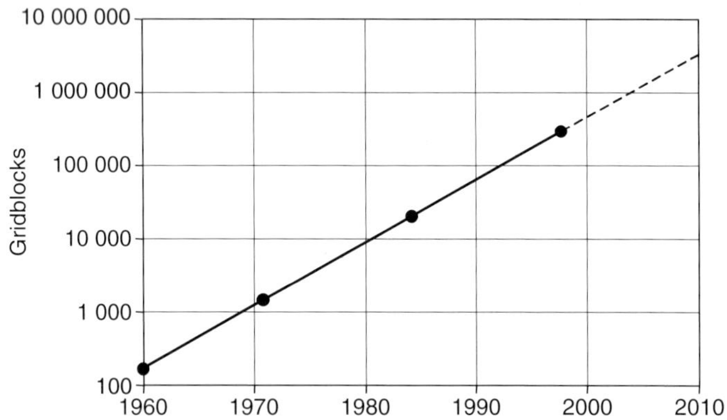

Figure 7.11 Evolution of the number of simulation gridblocks with time.

7.5 ASSIGNING THE INPUT PARAMETERS

This chapter deals with the assignment of the reservoir properties to the selected simulation grid. In particular, the discussion focuses on the assignment of rock and fluid properties for the general case of a 3-dimensional, 3-phases reservoir system.

All the rock and fluid parameters required in such a general case are summarised in Table 7.1. In the following sections each group of input data will briefly discussed.

7.5.1 Reservoir Geometry

Reservoir geometry is loaded into the simulator by introducing the reservoir structural top map, plus the gross and the net thickness maps of each reservoir layer. Alternatively to the net thickness, the Net to Gross ratio can be entered. Using these parameters the simulator is able to generate internally the reservoir geometrical configuration.

Table 7.1 Rock and fluid input parameters for a 3D-3 phase simulation model.

Geometry	
D	Structural top map
h_g	Gross formation thickness, for each reservoir layer
h_n	Net formation thickness, for each reservoir layer
N/G	Net to Gross ratio (alternative to h_n)
Rock properties	
ϕ	Porosity
K	Absolute permeability (3 directions)
c	Rock compressibility vs. pressure
Fluid properties	
B_o	Oil FVF vs. pressure
B_w	Water FVF vs. pressure
B_g	Gas FVF vs. pressure
ρ_σ	Oil density at standard conditions
ρ_w	Water density at standard conditions
ρ_g	Gas density at standard conditions
Rs	Gas in solution vs. pressure
μ_o	Oil viscosity vs. pressure
μ_g	Gas viscosity vs. pressure
μ_w	Water viscosity vs. pressure
c_o	Oil compressibility
c_w	Water compressibility
Saturation functions	
Pc_{wo} vs. Sw	Water-oil capillary pressure (drainage and imbibition)
Pc_{go} vs. Sg	Gas-oil capillary pressure (drainage and imbibition)
Kr_o, Kr_w vs. Sw	Oil and water relative permeability functions (drainage and imbibition)
Kr_o, Kr_g vs. So	Oil and gas relative permeability functions (drainage and imbibition)
Kr_o, Kr_g, Kr_w	3 phase oil, gas and water relative permeability functions

The assignment of these properties to the reservoir grid blocks is straightforward. The data are usually available in matrix format as output from some gridding package or geocellular modelling software, while all commercial simulators are currently suited for reading these large files in ASCII format or in some standard binary format.

It is usually good practice to utilise the visualisation options of the model to verify that the reservoir geometry has been correctly introduced into the simulator. In particular, the depth of the wells should be carefully checked, especially when the lateral dimensions of the grid cells are large. When discrepancies are found, the cell depth should be modified to honour the measured well depth.

7.5.2 Rock Properties

Leaving aside compressibility, which is usually introduced as a single value or a table as a function of pressure, the petrophysical parameters needed by the model are porosity and permeability (see Table 7.1).

As in the case of geometry, the assignment of the petrophysical properties to the grid blocks is, in itself, a simple operation, both in the cases of 2D input data (porosity and permeability maps) and in the case of 3D input data, derived from some geocellular modelling package. In many simulators, there is also the option to perform the input of all the geometrical and petrophysical parameters with a single operation, when both types of parameters are available within the same geological modelling software.

However, before executing this simple operation, attention must be paid to an important and sometimes overlooked issue, the upscaling problem.

7.5.2.1 The Upscaling Problem

In a typical 3D geological model, the reservoir is often described using a very fine support grid, whose dimensions involve a total of 10^6–10^8 cells. However, these models are not suitable for dynamic simulation, since typical numerical models are based on much coarser support grids, whose total number of cells is in the order of 10^4–10^5. A factor of 100 or more therefore exists between the two modelling approaches and a proper procedure must be applied in order to transfer the detailed geological description to the coarse simulation grid, while limiting the loss in geological detail. This is the well-known problem of **upscaling**.

The problem of upscaling of petrophysical parameters has been one of the main subjects of research in the last decade and many interesting papers have been published that properly discuss the matter [15]. We refer to those publications for a thorough treatment of this subject. In this context, we will briefly comment on some of the most popular techniques.

A. Porosity

The upscaling of porosity to a different support volume does not pose special problems. Actually, as porosity is an additive variable, the correct upscaling operator is the simple linear weighted average of the small scale values. One interesting feature to note is that, at higher scale, the effect of the larger support volume is to actually reduce the dispersion around the mean value.

B. Permeability

In the pioneer work of Cardwell and Parsons [16], the effective permeability of a heterogeneous system is defined as *the permeability of a homogeneous segment of the same dimensions that would pass the same flux under the same pressure drop.*

Therefore, the problem of permeability upscaling depends on the distribution of the heterogeneities and the boundary conditions applied, which in turn depend on the volume considered. When these conditions change, the resulting effective permeability will have a different value.

Various techniques have been developed in the last years, which provide satisfactory results in most cases. They can be grouped in 2 main categories: analytical and numerical techniques.

Analytical Techniques

Effective permeability values for different geometrical configurations can be derived by means of simple analytical techniques, based on different types of averaging:

- **Arithmetic average**. This represents the correct operator in the case of homogeneous layers of contrasting permeability, when the flow is parallel to the bedding.
- **Harmonic average**. This provides the average permeability value of a layered system, when the flow is perpendicular to the bedding.
- **Geometric average**. This provides the average permeability of a random heterogeneous system.

It is particularly interesting to note that the arithmetic and harmonic averages bound effective permeability on the high and low side, respectively [16]. In particular, when a heterogeneous system is concerned, the upper and lower boundaries can be found by taking, respectively:

- the arithmetic average over each plane perpendicular to the direction of interest, followed by the harmonic average of those arithmetic averages;
- the harmonic average over each column in the flow direction, followed by the arithmetic average of those averages.

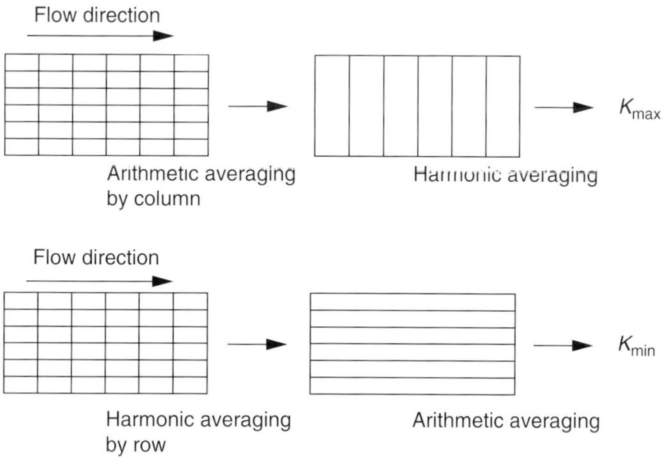

Figure 7.12 Minimum and maximum effective permeability calculation.

The procedure is illustrated in Fig. 7.12. For each coarse grid cell, the value of K_{min} and K_{max} can be computed, and can be used for approximating vertical and, respectively, horizontal effective permeability.

A general formulation of the analytical (or algebraic) methods can be provided by the so-called power-law averaging [17]:

$$K_e = \Sigma_i \left(\frac{k_i^{\omega}}{n}\right)^{1/\omega}$$

Actually, it can be appreciated that the lower bound harmonic average can be seen as a power average with an exponent ω equal to -1, while the upper bound arithmetic average can be seen as a power average with exponent ω equal to $+1$. The geometric average corresponds to the limit of the function when the exponent ω tends to 0.

In the general case of a heterogeneous 3D distribution of small scale permeability values, the power average will provide the general solution for the large scale effective permeability. Ref. [18] provides a method to estimate the power average exponent ω as a function of a global anisotropy factor, related to the vertical to horizontal permeability ratio and the correlation length.

Numerical Techniques

Effective permeability for a heterogeneous system can also be obtained through numerical techniques. One widely applied method is the so-called Pressure-Solver method [19]. Here, the heterogeneous medium is described on a fine scale grid, while the total flux is computed under known pressures at the inlet and outlet faces and fixed boundary conditions. This flux is then imposed to a homogeneous medium and its effective permeability is back-calculated. The procedure is displayed in Fig. 7.13. This technique should be applied to each reservoir facies or rock type, in order to derive the correspondent upscaled effective permeability.

The pressure solver method represents an accurate procedure for upscaling, however the results depend on the selected support volume and the boundary conditions. In particular, the volume to be simulated should be selected bearing in mind the type and dimension of the heterogeneities under study. The method is also more time-consuming than analytical techniques.

Analytical and numerical techniques are not the only available methods for permeability, or single phase upscaling. Other approaches have been proposed, one interesting example being the application of renormalisation techniques [20].

The choice of the upscaling method to apply is a difficult problem. As a general rule, when permeability variance is moderate, analytical methods provide a rapid and efficient approach to the problem. On the contrary, when permeability contrasts are important, numerical methods should be preferred. The presence of zeros, in particular, may heavily affect the algebraic methods, since effective null permeabilities can be obtained for particular arrangements of the fine scale values. These situations must be identified and solved, otherwise too low effective permeability distributions will be generated.

In general, each method will provide different results and no grounds actually exist to unequivocally determine which is the best approach or whether a specific upscaling procedure provides a good or bad approximation of the actual (and unknown) effective permeability values.

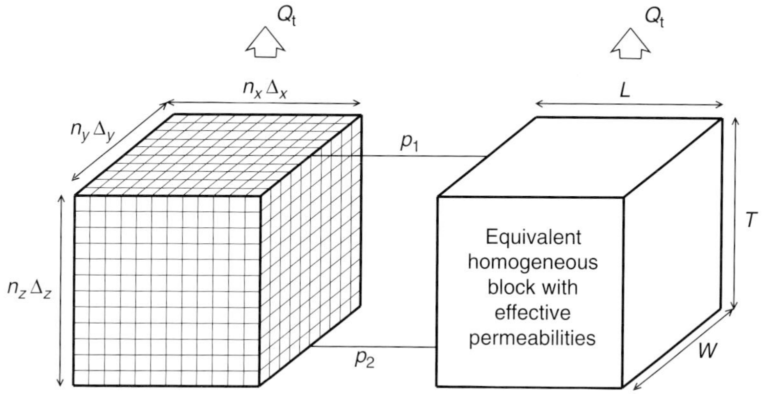

Figure 7.13 Pressure solver method for effective permeability calculation [19].

A quantitative assessment of an upscaling operation can be obtained by comparing fine scale and coarse grid results, when the formers are available. A recently published study [14], compared the effectiveness of a specific upscaling procedure with a reference fine scale simulation performed with a parallel simulator. Results showed that, in this case, coarse grid simulation provided accurate predictions on global basis, while the individual well performance was less satisfactory. Additionally, the relative ranking of the geostatistical realisations was preserved.

In all cases, adequate attention must be devoted to the upscaling problem, since the final results of the dynamic simulation are often heavily influenced by the chosen method. Whenever applicable, sensitivity studies provide a useful mean to evaluate the impact of the upscaling procedure and to select the proper procedure.

7.5.3 Fluid Properties

Fluid properties have been discussed in some detail in paragraph 6.2. Once the reservoir fluids have been satisfactorily characterised, the assignment to the grid blocks do not pose particular problems. Fluid properties (see Table 7.1) are normally introduced in the simulator as tables as a function of pressure.

One point of attention is the way the model treats the PVT tables. Some simulators require PVT values already corrected for the actual separator conditions, while others can compute the corrections internally, when all the relevant differential and composite data are input.

Separate sets of tables may be needed when spatial properties variations have to be managed, either vertically or areally. In this case, distinct PVT zones are identified that represent regions of different thermodynamic equilibrium. It is important to understand the behaviour of the simulator when fluids pass from one region to another, in order to identify possible anomalies.

7.5.4 Saturation Functions

Together with absolute permeability, saturation functions (capillary pressure and relative permeability) are among the most influencing factors as far as the final results of the simulation are concerned.

The definition of representative saturation functions has been discussed in paragraph 6.3. In this context, after a short digression on the hysteresis problem, the procedure of assignment to the numerical simulator will be discussed.

7.5.4.1 Hysteresis

Multiphase fluid flow is in general an irreversible process and, therefore, is path-dependent. One consequence is that the distribution of the fluid phases in the porous network depends not only on the level of saturation but also on the direction of saturation change. When the saturation of the wetting phase increases, we refer to an **imbibition** cycle, otherwise to a **drainage** cycle. These 2 cycles, in general, are different and this phenomenon is called hysteresis of the saturation functions.

Both capillary pressure and relative permeability curves are subject to a drainage or an imbibition cycle and it is therefore important to assess which is the predominant direction of saturation change in the reservoir under study and to observe whether or not a saturation reversal happens.

If this is the case, both cycles must be taken into account and explicitly input in the model. From this point of view, the previous knowledge of the production history of the field and of the predominant reservoir mechanisms normally provide useful indications.

In general, the difference between drainage and imbibition curves may be important and this means that the choice of the functions to use should be made while keeping in mind the main reservoir mechanisms. Using for example drainage relative permeability functions to describe an imbibition process, such as the waterflooding of a water wet reservoir, should be avoided. When it is known that saturation reversal occurs in the reservoir, the options of using both types of data (drainage and imbibition) should be considered. Ref. [21] provides a reference discussion on the treatment of saturation functions hysteresis in numerical simulators.

7.5.4.2 Assigning Saturation Functions to the Simulation Grid

In a typical study several models of saturation functions are usually defined. These apply to different zones of the reservoir, as a function of the prevalent lithological and/or petrophysical rock type.

In particular, most simulators allow for the definition of different curve shapes and endpoints sets, which can be combined in a flexible manner. Typically for example, when few experimental data are available, one set of normalised relative permeability and/or capillary pressure curves can be combined with different sets of end points.

This definition of different zones of the reservoir where different saturation functions apply is usually referred to as **reservoir zonation** and from a practical standpoint it comes down to the definition of subzones of the simulation grid, where different **rock types** can be identified.

The concept of rock type is actually one of the most important points of integration between the static and the dynamic characterisation of the field. Too often, the reservoir engineer creates and modifies the rock type zonation solely as a function of history match criteria, when in fact the rock type should be used to identify areas in the reservoir where a different dynamic behaviour is expected *as a consequence* of different lithological and/or petrophysical properties of the formation.

The relative permeability study illustrated in Fig. 1.6, where more than 200 curves were defined for a single reservoir, is just one extreme example of the possible misuse of the concept of rock-type. It is clear that no credible reservoir heterogeneity zonation is behind such modelling, while any extrapolation based on this set of curves is questionable.

For this reason, the generation of the rock-type zonation of the reservoir, both vertically and laterally, should be performed jointly by the geologist and the reservoir engineer to make sure that consistency is not lost during this phase.

Rock type zones can be created in a variety of ways. Whenever applicable, the concept of facies should be used as the driving tool in the definition of the distribution of rock type (paragraph 3.3.2.3). Maps of facies can be directly translated into rock type zones, especially when the different facies can be individually characterised from a petrophysical viewpoint. It is also important that any subsequent adjustment to the rock type zonation, which may be dictated by history matching purposes, be compared against the static characterisation of the reservoir, if a useful predictive model is to be built.

As an alternative to the facies-based approach, the distribution of saturation functions can be performed by means of analytical correlations. A typical procedure would involve the use of a correlating parameter (porosity, permeability or the square root of the ratio permeability/porosity) for the end points of the relative permeability functions measured in the laboratory.

Fig. 7.14 shows one example of such correlations and illustrates how the residual oil saturation, Sor_w, correlates with the square root of K/ϕ. As the values of porosity and permeability are known for each gridblock, the derivation of the corresponding value of Sor_w is straightforward. These types of graphs have been in use for many years to identify rock types within the reservoir. For capillary pressure, the assignment to the simulator can be done via a *J*-function [22], as the porosity and permeability distributions are known.

In conclusion, the definition and the assignment of saturation functions to the modelling grid is an important phase of the study. The following points illustrate some of the general guidelines that should always be considered.

- **Validate the experimental data**. Very often, especially in the case of old reservoirs, the geoscientist is faced with a fairly large amount of experimental data, that may be based on different operational practices and come from different laboratories. Validating the available information, for the reliability of the measurements and their representativeness, is a basic step in the definition of a useful set of saturation functions.
- **Integrate the information**. Whenever possible, the experimental data should be compared with independent source data, especially for the saturation end-points. This may reveal discrepancies that could be related to the existence of problems of scale.
- **Create a simple model**. It is always good practice to generate simple saturation function models, especially when data are insufficient to go into a greater detail or when

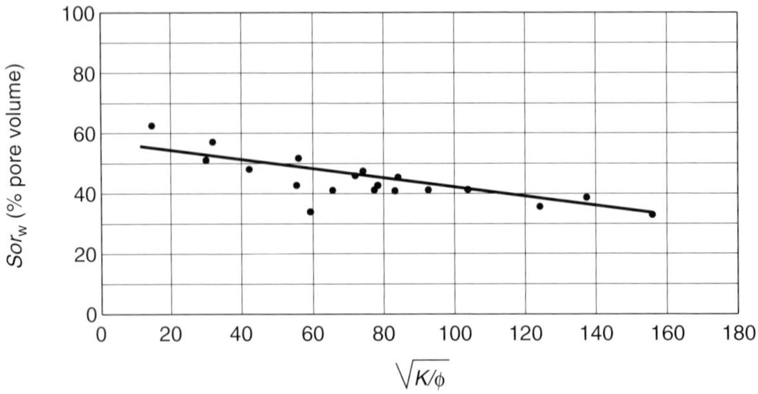

Figure 7.14 Plot of K/ϕ vs. residual oil saturation.

large uncertainties are thought to exist. Complex models, when not substantiated by good quality data, are likely to generate artefacts and bias in the predictions.

- **Respect the geology**. When generating a spatial model of saturation function, it is important to guarantee that the geological characterisation is respected. The same holds when corrections are applied to the initial model during the history match phase. The geological facies distribution, when available, provides a useful framework for defining the rock-type model to be used in the numerical simulation.
- **Concentrate on critical parameters**. Modelling the saturation functions for a given reservoir can be a lengthy and tedious work. In this respect, it is important to concentrate on the critical parameters (e.g., water end point permeability), while average or standard values may be used for the less important ones.

Whenever the available information is not sufficient for a safe modelling exercise, the project manager should ask for new additional data. The costs involved can be safely justified by the importance of this data on the final results.

7.5.5 Production and Completion Data

In a numerical simulation study production data (oil rates) are input to the model, together with water and gas injection profiles [2]. Gas and water production rates (or water cut and GOR), on the other hand, represent the output of the model, calculated using the simulated pressures and saturations.

The oil production history in the model is usually expressed as a monthly profile. However, most model pre-processors allow the smoothing and re-arranging of the input oil pro-

2. In the initial phases of the modelling, total fluid rates are often imposed, rather than oil rates. This allows for a correct computation of the total underground withdrawal and material balance. Total fluid rates are also used in the simulation of reservoir with very high water or gas production.

duction profile to different configurations, while still honouring the value of cumulative production. These features are useful in the case of long exploitation history, since they allow the production profile to be redefined, especially in the early exploitation periods (e.g., on a 6-months or yearly basis). In turn, this allows for a much faster and more stable numerical computation.

One point of attention concerning production data is the quality of the available information. Of course, good quality production data are essential for a reliable simulation study, both in terms of direct input data (oil rates) and in terms of reference data to evaluate the accuracy of the history match (water cut and GOR). It is always good practice to gain an understanding of the inherent quality of the available production information, by getting as close as possible to the source of data.

A recently published paper on the Greater Burgan Field [23], demonstrates how flaws in the production data can lead to unreliable simulation results. In this work, a systematic and critical review of all the historical information revealed that a significant part of the wells suffered from wellbore communication (crossflow) and/or production allocation problems.

For strictly commercial reasons, total field or separation plant oil production rates are probably one of the most reliable information in a reservoir study. Individual well rates are usually less accurate, since the allocation procedure is normally based on periodic production tests performed on each well.

Gas production data are, in general, even less reliable, especially when the produced gas is flared, when no contract sale exists and no other commercial use is envisaged for the produced gas.

Water production data may also carry a degree of uncertainty, especially in the case of old fields. Suspicious data should be carefully checked in the available field operational reports, where useful information can often be obtained about the possible origin of the produced water.

Water and gas injection data are, on the other hand, reasonably accurate measurements in most cases.

Completion data represent a further typical input for the simulator. The completion history of each well has to be specified in the model, in order to allocate the total well production to individual gridblocks. This phase is very important for a correct calculation of pressure and saturation in the model, therefore care must be exercised in the evaluation of the reliability of this information. It is not uncommon to discover that apparently anomalous model responses are related to flaws in the completion history of the wells. When doubts exist over the available information, it is always advisable to check the data, if possible, against the original operational documentation.

7.5.6 Model Initialisation

A further set of data is required by the model in order to establish the initial pressure and saturation (equilibrium) conditions. In the general case of an oil reservoir with a primary gas cap and underlain by water, the required parameters are the following:

- Oil reference pressure at a given datum depth.

- Water-oil contact (OWC) depth. This reference depth is used in conjunction with the drainage water-oil capillary function and the water and oil density to compute the initial pressure and saturation at each gridblock centre.
- Gas-oil contact (GOC) depth. As with water, the GOC reference depth is used in conjunction with the drainage gas-oil capillary function and the gas density to compute the initial gas pressure and saturation at each gridblock.

Particular initialisation procedures may be required when complicating conditions exist, as in the case of tilted contacts or variable PVT properties.

The initialisation phase allows for the calculation of the OOIP in the model, which is then compared with the available volumetric figures.

These two types of estimates never agree exactly. The differences may span from negligible fractions to significant percentages and are related to a number of factors, like the different support grids, the capillary functions used, the fault description and so on. One of the problem of this phase is that it is difficult, *a priori*, to understand to which particular parameter the observed difference is related. Even when the global estimate of the OOIP agrees with the volumetric figure, there is no certainty that the geometric, petrophysical and saturation estimates are equally correct. The apparent agreement may arise, for example, from an underestimation of reservoir gross volume and a compensating underestimation of water saturation.

One way to reconcile the volumetric and model OOIP estimates is to perform the initialisation in a stepwise fashion. From a practical viewpoint, this amounts to perform the following operations:

- Input the geometrical surfaces or grids (top and thickness), initialise the model and calculate the Gross Rock Volume. Compare with the geologic figure and, if it is the case, apply the necessary corrections.
- Input the Net to Gross grids and calculate the Net Rock Volume. Compare with the geologic estimate and if necessary adjust the model data.
- Input the porosity grids and calculate the Net Pore Volume. Compare with the geologic estimate and again, if it is the case, apply the necessary corrections.
- Initialise the model with the capillary pressure functions and calculate the OOIP (reservoir conditions). Compare with the relevant geologic figure.

This procedure has the advantage of showing exactly where the observed discrepancies lie, whether in the geometrical or in the petrophysical description of the reservoir. Furthermore, its application guarantees that consistency with the geological model is maintained in the simulation, for each step of the model building.

7.6 HISTORY MATCHING

History matching is the most important part of the simulation study. Basically, history matching is a model validation procedure, which consists in simulating the past performance of the reservoir and comparing the results with actual historical data. When differences are found, modifications are made to the input data in order to improve the match. History matching is therefore an iterative process, whose final objective is to reconcile all the different static and dynamic data into a coherent framework, representative to a specified degree of the actual reservoir behaviour.

More generally, history matching is a way of checking sensitivity to variations in the input parameters and eventually of understanding the representativeness of the model. From this point of view, the history matching process can be considered to be a valuable technique to improve the overall reliability of the simulation model which, if it is properly performed, will highlight flaws and inconsistencies in the existing reservoir description. Of course, only controlled and justified corrections should be applied to this end.

History matching is a complex procedure, which depends on the quality and amount of available data, the particular reservoir being studied, the resource allocated to the project and eventually the experience and personal attitude of the engineers working on the model. From this point of view, it is difficult to provide precise indications about the correct or best way to perform a match, since each reservoir (as well as each engineer) is different from the others.

This section is not intended to provide a systematic approach to history matching, since this has been usefully described in reference texts [5]. Rather, we will try to concentrate on some of the most relevant matching issues, that should always be considered when performing an integrated reservoir study.

7.6.1 Important Aspects of the History Match Process

There are a number of critical issues that need to be taken into account when performing the history match phase of a simulation study.

The most important aspect is the non-uniqueness of the results, i.e., the fact that an equally satisfactory history match can be obtained by means of different reservoir descriptions. This non-unique character of the history match process has been mentioned since the beginning of reservoir simulation [4] and it stems from the fact that the numerical simulator is a highly over-defined mathematical system, typically with only a few known variables (fluid properties, productions ...) and several thousands of unknown variables (e.g., porosity and permeability values for all the gridblocks). From a mathematical viewpoint, this generates an infinite number of solutions.

A recently published paper [24] provide a typical example of the problem of non-uniqueness. In this case, a simulation study conducted over a significant number of geostatistical realisations of the structural reservoir map, revealed that virtually identical production curves, all matching the field performance, could be obtained using different maps and different related OOIP.

The problem of non-uniqueness is somewhat disturbing, because it means that history matching is a validating technique which in fact cannot be used to state that the current reservoir description is *the right one*. At best, we could say that such a description is one among the many possible that *do not contradict* the few available input data. Furthermore, the non-uniqueness of the history match phase suggests an even more disturbing non-uniqueness of the prediction phase.

When we consider the problem of non-uniqueness from the point of view of stochastic modelling, one interesting issue becomes obvious. Many attempts have been made in the past to derive a methodology for the selection of the correct realisation of the stochastic process, among the infinite possibilities. Some of these attempts concerned the validation of a particular realisation on the basis of history matching. In fact, it should be acknowledged that the non-unique aspect of this process precludes its use as a screening tool for choosing among alternative geologic models (or realisations of the stochastic process).

Another critical aspect of history matching is the iterative nature of the process. There are several parameters that can be modified in a typical reservoir study, some of them belonging to the so-called static group (geological and petrophysical data) and some to the dynamic group (fluid data, productions). Before original data configurations are modified in the model, this should be consulted with the other professionals of the group, to make sure that consistency is not lost[3]. History match must not be achieved through uncontrolled adjustments, otherwise the efforts of a whole working team can be wasted in minutes.

7.6.2 Matching Parameters

The objective of history matching is to reproduce, as correctly as possible, the historical field performance, in terms of measured rates and pressure. The check should be always done both on a field and well basis.

Figure 7.15 shows a typical example, for a satisfactory history match performed in a well with a long production history. The matching parameters are in this case static pressure, water cut and GOR, while oil rates, being an input to the model, are exactly reproduced.

In the following paragraphs, the main parameters to be history matched in a typical study will be quickly reviewed.

7.6.2.1 Pressure

Static Bottom Hole Pressure values (SBHP) are practically always available for any reservoir. These values must be compared with the results of the model, keeping in mind that the measured pressures will not correspond directly to calculated pressures, since in general the 2 types of data represent different reservoir volumes (the well drainage radius and the gridblock volume, respectively). Most simulators allow for some corrections to be applied to the computed pressure in order to be comparable with the actual measurement.

Whenever possible, it is also useful to check the global pressure behaviour of the field through the generation of pressure maps at the reference datum depth, to be compared with the isobaric maps generated using the individual wells static pressures (see paragraph 6.4.3). This comparison is useful, in that it provides a global picture of the pressure distribution within the field, as well as the existence of lateral gradients, that may not be easy to pick in a well by well analysis. Fig. 7.16 shows one example of such comparison.

---

3. One typical example concerns the modification of permeability values for selected gridblocks. In such cases, to guarantee the consistency, the correlated parameters should also be modified (typically porosity).

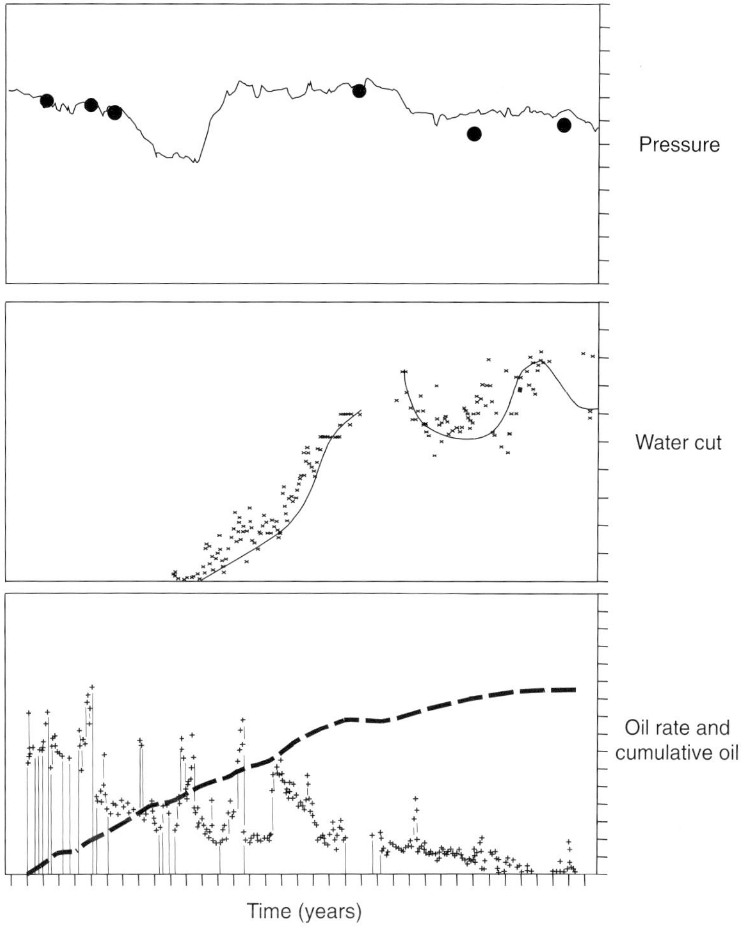

Figure 7.15 Example of a satisfactory history matched well.

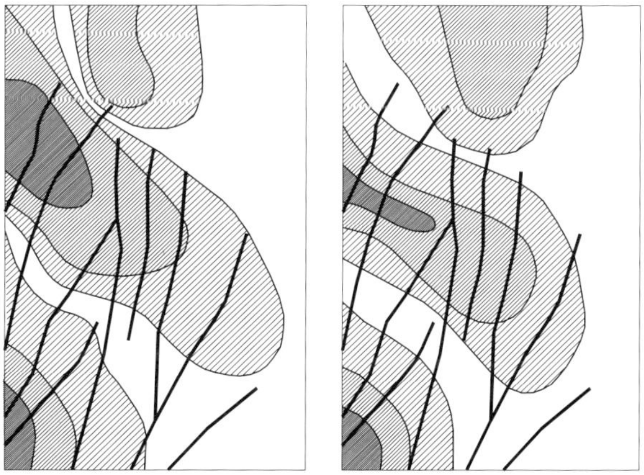

Figure 7.16 Isobaric (left) and model simulated (right) pressure maps comparison.

In addition to the SBHP, other types of pressures can sometimes be used to check the performance of the model, like static or flowing tubing head pressures (STHP and FTHP). These measurements have the advantage of always being available in large quantity, but are more difficult to handle, since they require the knowledge of the static and dynamic fluids gradients in the well completion.

7.6.2.2 Water Production

The simulated water production in the model should reproduce the observed field values, both in terms of breakthrough time and water cut evolution. The check should be done on a well by well basis, but it is always good practice to plot water saturation maps and to compare them with any available water advance maps (paragraph 6.5.2). This comparison provides a more complete image of the global displacement process and it also helps in identifying the critical or key wells, where a good match must be sought (typically, the wells located close to the water front). The total field water cut profile must also be checked and adjusted, in order to have a correct balance of the produced and injected fluids.

Figure 7.17 shows an example, relative to a field where a previous fluid monitoring study had led to the identification of the current position of the water front. On the same figure, the results of the simulation model, in terms of water saturation, are shown.

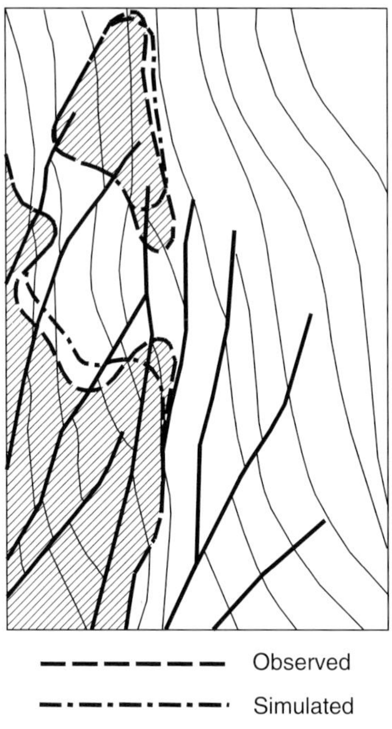

Figure 7.17 Actual water advance and simulated water saturation maps comparison.

7.6.2.3 Gas Production

The correct reproduction of the gas production profile, when the pressure falls below the bubble point, is critical to the performance of any reservoir model. Due to the high compressibility of gas, the quantity of produced gas will dictate, to a substantial degree, the level of energy of most reservoirs.

Deviation from the field observed profile may suggest problems in the PVT characterisation or in the relative permeability curves. Again, the check should be performed on a well by well basis and also for the total field production.

When a primary or secondary gas cap exists and it has been located in a previous phase of the study (paragraph 6.5.3), this should be compared with a gas saturation map. The coherence of these images will guarantee that the segregation process is correctly reproduced in the model.

Figure 7.18 shows one example, where the actual position of the secondary gas cap has been correctly reproduced by the model.

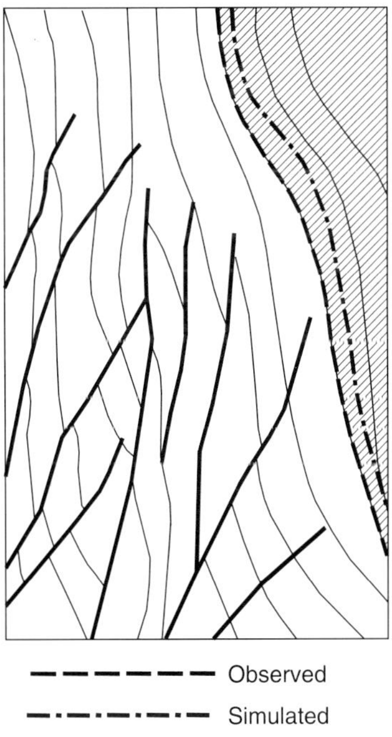

Figure 7.18 Actual gas distribution and simulated gas saturation maps comparison.

7.6.3 Matching Procedure

There is no standard procedure for history matching. Each field is different from any other, in terms of geological configuration, reservoir mechanisms, number of wells, production history, exploitation strategy and so on.

Therefore, each study has to deal with its own problems, which are generally tackled through unique solution procedures.

Nevertheless, few general steps can be identified that to a greater or lesser extent can be applied in most simulation studies.

The first stage in any simulation exercise is to define the critical parameters to be adjusted and the key wells.

Critical parameters are considered those that carry a high degree of uncertainty (which justifies some modification) and that have a significant impact on the final results. The identification of the critical parameters is usually related to the prevailing energy mechanism in the reservoir.

In water drive reservoirs, for example, typical critical parameters are the aquifer transmissibility and storage, while in the case of solution gas drive reservoirs the characteristic critical parameter is the gas-oil relative permeability.

Another critical parameter is of course permeability, which plays an important role in virtually all types of reservoirs.

Key wells, on the other hand, are considered to be those wells whose production behaviour is typical and must be correctly reproduced by the model. In studies that include a limited amount of wells, possibly less than 20, all the wells can virtually be considered as key wells and the matching effort can be addressed with the objective of correctly reproducing all these wells.

However, when the number of wells is considerable, as it is often the case in old fields, matching the observed behaviour of the totality of wells becomes impractical, and the effort would not necessarily lead to more accurate results.

The definition of key wells is based on a number of factors. They are usually characterised by long historical production with typical trends of water cut and GOR, they should possess fairly complete suites of logs, cores and pressure data and should be located in representative areas of the field. In addition, wells that are still on stream should be considered, whenever practical, as key wells.

The identification of the critical reservoir parameters and the definition of the key wells provide the reservoir engineer with a simplified reference framework for starting the history match phase. The following steps involve the matching of the pressure history and the subsequent matching of the saturations.

- **Pressure match**. This phase consists in the adjustment of the global energy balance in the reservoir. A simplified scheme for pressure match is shown in Fig. 7.19 (from [25]). It should be appreciated that the process first concentrates on establishing the global pressure level and the main gradients existing in the reservoir and, later, it focuses on the match of individual well behaviour. From a general viewpoint, permeability is the principal reservoir variable to modify, in order to improve the pressure match.

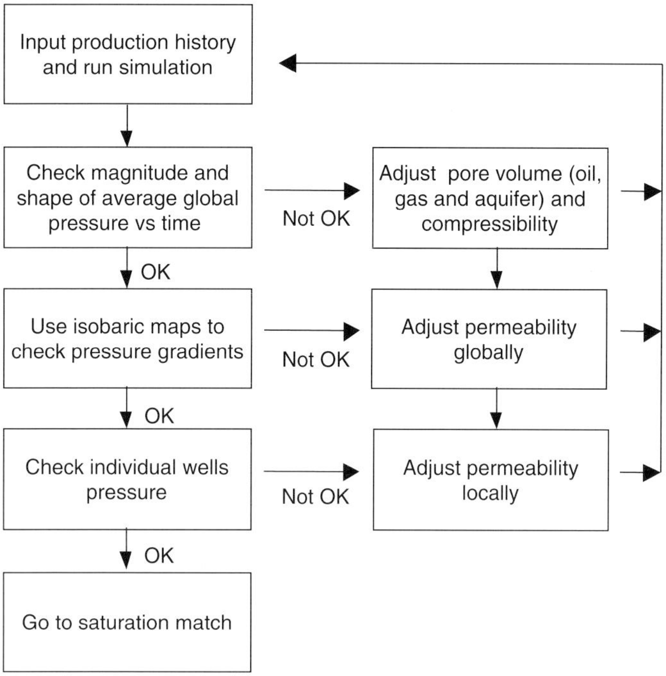

Figure 7.19 Simplified procedure for pressure history matching (from [25]).

- **Saturation match**. In this phase, the reservoir fluids distribution is matched, both in terms of arrival time of water and/or gas and evolution of the relevant production profiles after breakthrough.
  The reference scheme is shown in Fig. 7.20 (from [25]).
  Also in this case, the process should start from the adjustment of the total field performance and then concentrate on the behaviour of the key wells. Again, permeability is the main controlling factor concerning the arrival time, while the evolution of the water cut and GOR profiles after breakthrough is mainly governed by relative permeability curves. During this phase, control should be maintained over possible changes that may occur in the pressure match.

The process of history matching the historical performance of a reservoir is often the most time consuming part of a reservoir study and sometimes it proves to be a frustrating experience. In fact, a perfect match never exists and the reservoir engineer is therefore faced with the problem of deciding whether the history match can be considered satisfactory and, consequently, when it can be considered concluded. In the next section, we will try to define these points.

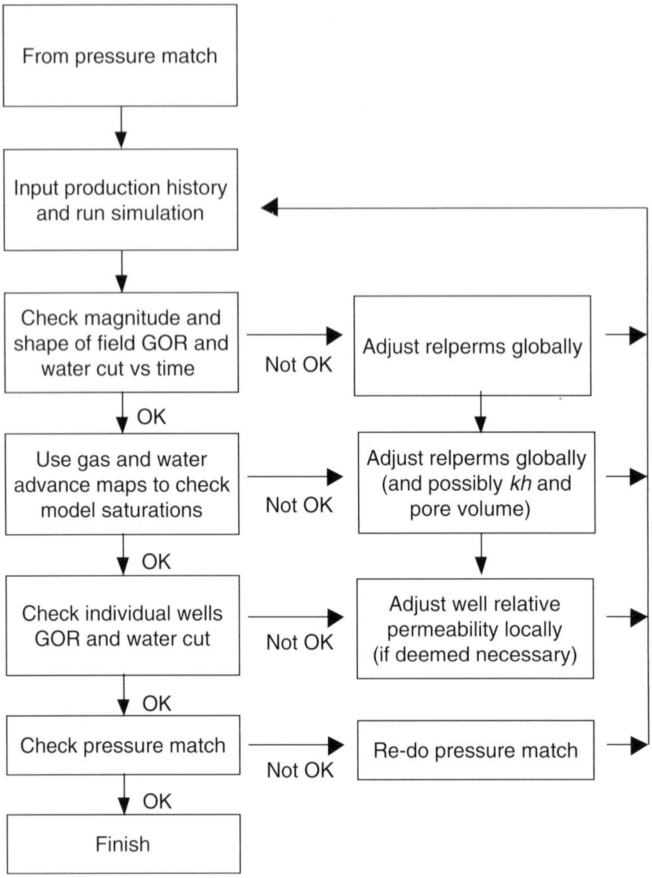

Figure 7.20 Simplified procedure for saturation history matching (from [25]).

7.6.4 Quality of the Match

Many reservoir engineers tend to consider the history match exercise as a sort of stand-alone phase of the study, whose results need to be *inherently* good. In other words, too often the model is modified purely for matching sake.

The problem is well known. To obtain a convincing match to show to management is virtually always possible, but often these results are obtained at the expenses of the geological integrity of the reservoir or the physics of the recovery process.

Modern numerical simulators are flexible tools that offer the user the possibility of varying a large number of parameters, but of course it is the engineer's responsibility to carry out the correct modifications. Local changes to improve the match of particular wells are no longer valid in the prediction phase, therefore their introduction in the model does not make any engineering sense.

In fact, the simulation model must be able to capture the main mechanisms that govern the field production and will never be able to foresee all the possible exceptions to the general depletion and displacement rules. From this point of view, as has been highlighted in an illuminating paper on reservoir simulation [26], the numerical model should be better considered as a probabilistic tool.

When the history match is viewed as a preliminary stage to the real objective of the model, which is the prediction phase, the definition of the quality of the match is straightforward: the model can be safely considered history-matched when the major controlling mechanisms of the reservoir have been correctly simulated, even though some wells are not matched (and possibly will never be [4]). Under these conditions, the model is likely to provide, from a probabilistic point of view, a reliable estimation of the future field performance.

7.7 PRODUCTION FORECASTS

Running production forecasts is usually the concluding phase of an integrated reservoir study. In its essence, the objective of this work is to visualise the future performance of the field under different operating strategies and to generate the production profiles needed for the economic evaluation of the project.

All the efforts of the integrated team, in terms of reservoir characterisation and simulation, converge in this phase of the study, where the most promising field exploitation strategies must be analysed and proposed to management for the short, medium and, more typically, long term periods.

As far as the inherent technical complexity is concerned, the production forecast phase of a simulation model can be substantially different from case to case. In simple studies, prediction runs can be performed in a matter of days, but in more complex cases they may take several months, depending on the size and complexity of the model, the implemented well-management routine and the number of predictions to be run.

Because of the general approach of this text, no detailed discussion of the possible procedures will be undertaken here. The focus will be kept on the general guidelines of this phase and on the integration aspects that should be taken into account. Excellent discussions on the process of running predictions can be found in reference textbooks [5].

7.7.1 Input Data for Predictions

This section discusses the usual input parameters that must be defined in the simulation model before running production forecasts.

The first step is always the definition of the cases to be run. Prediction cases are usually designed at the start of the forecast phase, but it is also worth noting that more prediction cases can be defined as the study proceeds, on the basis of the results of the previous runs.

---

4. Individual well behaviours are often related to near borehole heterogeneities that cannot be accounted for in the simulation grid.

The number and type of cases obviously depends on the particular study and the available time, however it is common practice to define a **base case**, which corresponds to the continuation of the field exploitation under the prevailing operating conditions. All the following prediction results are compared to this base case, which can therefore be considered as a benchmark for alternative development options.

The definition of the subsequent cases should be done with the objective of obtaining improved production and injection profiles, in terms of higher final recovery and/or accelerated production of the existing reserves. With this aim, a number of possible alternative development scenario should be tested (infill drilling, implementation of secondary recovery projects …), which in turn must be compatible with the existing (or foreseeable) infrastructures, the availability of fluids to inject (water, gas, CO_2 …), various financial constraints and so on. The knowledge of the reservoir acquired during the previous phases of the study should form a consistent base for a preliminary screening of the possible proposals.

In all cases, the definition of the prediction cases to be run must be made through a close co-operation with management, the economists and the production and facility departments. This will guarantee that no resources are wasted in simulating unrealistic production scenarios.

7.7.2 Setting Guidelines and Constraints

In order to simulate the future production performance of a field, we must specify a set of working rules in the model, which apply to both the surface facilities and to the individual wells. These rules are called guidelines or constraints, and express the expected operating conditions for the field under study. Typical surface constraints are maximum oil, gas and water production rates, maximum water and/or injection rates and pressure, minimum THP and maximum GOR. On a well basis, typical constraints are maximum WOR, GOR or total liquid rate and minimum oil rate.

Modern simulators allow the reservoir engineer to define the reservoir working conditions very flexibly. Table 7.2 provides a list of the most typical constraints that can be applied to a simulator, both on a field or group of wells and on a individual well basis. In addition to these simple constraints, fairly complete well management schemes can also be defined, that allow the modelling of complex field operations by automatically implementing a logical sequence of workovers, recompletions or drilling of new wells and starting artificial lift, according to some specified criteria.

Again, it should be stressed that these constraints must first be discussed with the production engineers, to make sure that a realistic exploitation strategy is being designed.

Another important point to consider is that these constraints do have a significant impact on the resulting production profiles, therefore care must be taken in applying the same set of rules to all the prediction cases, if comparable results are sought.

7.7.3 Inflow and Outflow Well Performance

The easiest way of obtaining production forecasts is to impose a constant total fluid rate (oil + water) to all the wells. This total rate is usually set equal to the average rate in recent years.

Table 7.2 Typical constraints for reservoir predictions.

Field/group production and injection constraints
Max oil production
Max water production
Max GOR
Max water injection rate
Max water injection pressure
Min average reservoir pressure
Separator pressure
Well production and injection constraints
Max GOR
Max WOR
Max total liquid rate
Min and max oil production rate
Min and max water injection rate
Min bottom hole pressure
Max water injection pressure
Well head flowing pressure

While such an approach has some merit in particular reservoirs (e.g., field under water injection and voidage replacement), a more typical approach to reservoir forecasting involves the definition of some surface constraints.

In such a case, the way the simulator works in the history match and forecast phases is basically different. During history match, field performance is known and the model translates the imposed oil rates into gridblock pressures, through a well management routine.

In the forecast phase, on the other hand, rates are unknown and must be calculated by putting a set of operating rules in the model, which from a practical standpoint are often represented by the flowing pressure conditions at the wellhead. This boundary pressure depends upon the surface infrastructures, the well producing conditions (natural flow, gas lift), the completion in use, the kind of flow (monophase vs. multiphase) and the pressure in the reservoir. To represent such conditions properly, a wellbore hydraulic model is therefore needed.

This model is aimed at defining the so-called VFP (Vertical Flow Performance) tables, which describe the outflow conditions of the wells. Most simulators have the facility to compute internally the VFP curves through empirical correlations and to use them in the calculation. Alternatively, these curves can be calculated using appropriate external friction loss software and then introduced into the simulator.

Outflow performance has a strong impact on well deliverability. In order to obtain realistic results, it is therefore important to define the input parameters carefully, through close co-operation with production engineers, and to check the results of the VFP calculations using field data. Typically, for example, the VFP tables must be calibrated against the results of those wells where both bottomhole and tubing head flowing pressures measurements are available.

Inflow well performance is equally important. The available well tests often provide a valuable database of well deliverability information and give useful estimates of the actual productivity index (PI) of the wells and the degree of damage (skin). These values can be used to correct the well PI's computed by the model, which in general are different, as the block pressure is not the same as the well drainage boundary pressure.

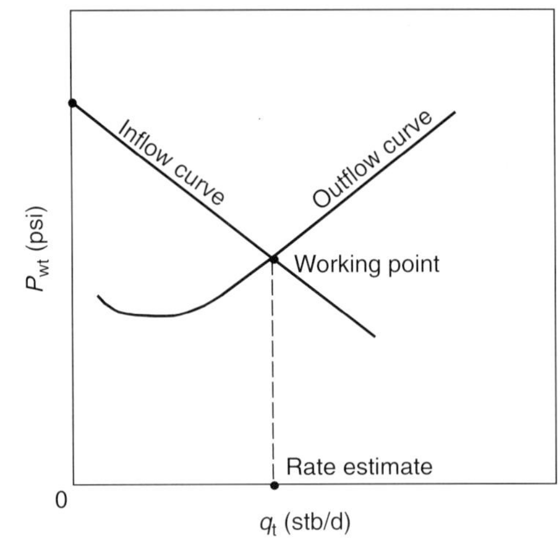

Figure 7.21 Inflow-outflow well performance.

The importance of the definition of the inflow-outflow performance of the wells in the model is related to the fact that well productivities in the forecast phase are calculated from such relationships. Fig. 7.21 shows an example of inflow-outflow graph, where the instantaneous well production is defined by the crossing point of the outflow and inflow curves (the working point).

Effort should be paid to in using all the available field data in order to define, in the best possible way, the inflow-outflow characteristics of the wells [27]. When this is not achieved, future well performances could be unrepresentative.

7.7.4 Running the Prediction Cases

Running the production forecast cases is, in general, a less difficult process than the history match phase. Nevertheless, the first trials always result in some problems, especially when working with VFP tables. The first step in the procedure of adjusting the prediction runs is therefore the PI calibration of the individual wells.

In fact, switching the model from history match to forecast very often results in discontinuities in the individual well rates, pressures and activities (Fig. 7.22). As mentioned in the previous section, this is related to the fact that the PI's calculated in the model are not generally calibrated with the actual field PI's. This difference is transparent in the history match phase, where wells work under imposed rate conditions, but becomes obvious in the prediction phase, where computed PI's actually determine well productivities.

In addition to the drainage radius, the differences between model and field PI's are related to the well skin factors, which are normally set to zero in the model at the start of the study. Therefore, the skin factor is usually the first parameter that is adjusted, in order to calibrate the performance of the wells and to obtain a smooth transition between the history match and the prediction phases. When this is not sufficient, more substantial adjustments of the model PI may be required.

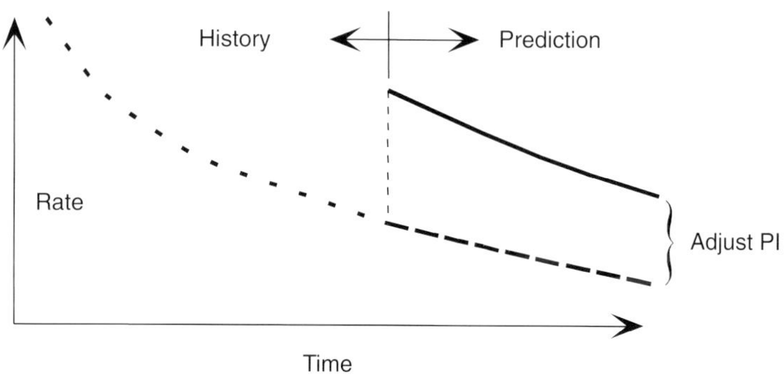

Figure 7.22 Well PI calibration.

Consistency of the Prediction Runs and Results Evaluation

Once the base case prediction run has been calibrated for the observed field conditions, a complete forward simulation can generally be attempted. The results of this run should be carefully checked for the presence of errors, oversight and numerical instabilities. In addition, a check should be made that the well management scheme has been correctly implemented and that no unexpected departures are observed in the resulting profiles.

Debugging and validating the base case is an essential step of the process. Once this is done, the simulation of all the subsequent cases will generally be straightforward.

As far as the results are concerned, the analysis of a production forecast can be made in a variety of ways, the most typical being tables and plots of oil rates and cumulative oil production vs. time. A comparison of the results of the various cases will show at a glance the most interesting (technical) exploitation options.

One important factor to consider in this respect is related to the accuracy of the expected results. Comparing the results of production forecasts performed in the eighties with the actual production performance, Saleri [28] reports a reasonable accuracy of the results concerning the global field profiles but, on the other hand, a poor match of the performance of the individual wells.

This behaviour is not unexpected, since individual well behaviour is largely governed by near-wellbore potentials, which are difficult to foresee with any certainty. Additionally, artificial factors like cementation and completion problems, stimulations etc., cannot be anticipated and properly accounted for in the model. At field scale, on the other hand, most factors compensate and the reservoir behaviour becomes predominant.

Again, it should be stressed that the results of the simulation study must be considered in a relative rather than in an absolute way, because of the many uncertainties existing in the reservoir characterisation and simulation phases of the study.

Likewise, the quality of the model should not be measured (or not only) by the difference between expected and observed production profiles. Accurate models may produce apparently poor forecasts when the actual field management strategies are different from those foreseen at the time of the study [5].

7.8 UNCERTAINTY ASSESSMENT

In many studies an assessment of the overall uncertainty related to the calculated reserves and the associated production profiles is required. From a purely technical point of view, this reduces to generating a number of simulations varying some of the input parameters, by considering for each one, for example, a reference as well as optimistic and pessimistic values. This methodology would produce a set of production profiles of the type shown in Fig. 7.23, from which, in turn, the overall uncertainty can be evaluated.

However, the generation of a statistically significant set of prediction cases is not straightforward and in all cases it is dependent on the exploitation stage of the field:

- **Undeveloped fields**. In the case of new fields, without historical production to match, the problem is related to the prohibitively large number of cases that should be run, in order to generate a significant set of prediction curves. Theoretically, one should evaluate the uncertainty existing in most of the parameters involved in the process, concerning both reservoir data and production and facility constraints. Moreover, the analysis should not be based on the simple approach of varying one parameter at a time, since some dependency among the parameters always exists [29]. In general, such sensitivity studies would require a very large and often unpractical number of simulation runs. A simplification can be obtained by concentrating on those parameters that are deemed to have a significant impact on the reservoir behaviour. Further reductions in the total number of cases to run can be obtained by applying methods based on experimental design [30].
- **Developed fields**. The evaluation of the uncertainty related to the future performance of fields with a production history is an even more difficult problem. Since the history match phase has fixed most of the static and dynamic reservoir parameters in some way, the only sensitivities that can be performed concern the wellbore dynamics and

5. Even in such cases, the ranking of the various cases is frequently preserved, which eventually guarantees the usefulness of the simulation study.

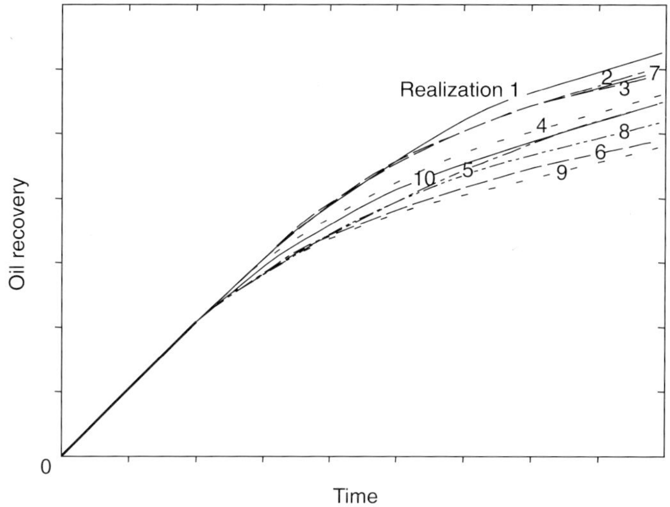

Figure 7.23 Set of production profiles for uncertainty assessment.

the surface production constraints. While this can be seen as a simplification of the problem, the non-uniqueness aspect of the history match phase (paragraph 7.6.1) casts a shadow over the actual representativeness of the uncertainty assessment performed in the forecast phase.

Another important point to consider is that the above general procedure for uncertainty assessment is based on the assumption of unbiased parameter estimates, i.e., free of systematic errors. Unfortunately, the available literature information reveals that a degree of bias always affects engineering estimates [31]. Interestingly, in the large majority this bias generates optimistic estimates.

In both the cases of undeveloped and developed fields, a proper uncertainty assessment is therefore a very complex, often prohibitive task. In the best cases, only a partial assessment can realistically be determined, within the practical constraints of a standard reservoir study.

When talking about uncertainty in the available geologic and dynamic data in what could be considered the Bible of reservoir engineering, Muskat stated that *the uniqueness of such specific data and their applicability to the actual producing intervals are assumptions that must at best be evaluated as necessary evils* [32]. Fifty years later, we are still faced with the same necessary evils, and at the same time the estimation of the related uncertainty still remains an uncertain issue.

References

1 Dogru AH (2000) Megacell reservoir simulation. JPT, May.

2 Coats KH, Use and misuse of reservoir simulation. SPE Reprint Series 11.

3 Arfonovsky JS, Cull GWL, Cox TF, Gaffney PD (1984) Use and abuse of reservoir simulation (3 parts). Oil and Gas Journal, Nov. 5 and 19, Dec. 3.

4 Odeh AS (1969) Reservoir simulation ... What is it? JPT, Nov.

5 Mattax CC, Dalton RL (1990) Reservoir simulation. SPE Monograph Series.

6 Pelgrain de Lestang A, Cosentino L, Lopez D, Gonzalez JE, A large scale geostatistical study: The Bachaquero 2 field. SPE paper 56657.

7 Brinkman FP, Kane TV, McCullogh RR, Miertschin JW (1999) Use of full field simulation to design a miscible CO_2 flood. SPE-REE, June.

8 Crichlow HB (1977) Modern Reservoir Engineering: A Simulation Approach. Prentice-Hall Inc.

9 Aziz K, Settari A (1979) Petroleum Reservoir Simulation. Applied Science Publishers Ltd, London.

10 Hagedorn KD, Coleman DR, Frank KJ, Janes RW, Pospisil G. Integrated reservoir management via full field modelling, Pt. McIntyre Field, Alaska. SPE paper 38851.

11 Kocberber S. An automatic, unstructured control volume generation system for geologically complex reservoirs. SPE paper 38001.

12 Gunasekera D, Cox J, Lindsay P. The generation and application of K-orthogonal grid systems. SPE paper 37998.

13 Aziz K (1993) Reservoir simulation grids: Opportunities and problems. JPT, July.

14 Tchelepi HA, Durlofsky LJ, Chen WH, Bernath A, Chien MCH (1999) Practical use of scale up and parallel simulation technologies in field studies. SPE-REE, August.

15 Christie MA (1996) Upscaling for reservoir simulation. JPT, Nov.

16 Cardwell WT, Parsons RL (1945) Average permeability of heterogeneous oil sands. Trans. AIME.

17 Deutsch CV (1989) Calculating effective absolute permeability in sandstones-shale sequences. SPE-FE, Sept.

18 Noetinger B, Haas A, Reservoir Helios Group. Permeability averaging for well tests in 3D stochastic reservoir models. SPE paper 36653.

19 Begg SH, Carter RR, Dranfield P (1989) Assigning effective values to simulator gridblocks parameters for heterogeneous reservoirs. SPE-RE, Nov.

20 King PR (1989) The use of renormalisation for calculating effective permeability. Transport in Porous Media 4, 37.

21 Killough JE (1976) Reservoir simulation with history dependent saturation functions. SPE Journal, Febr.

22 Leverett MC (1941) Capillary behaviour in porous solids. Trans. AIME.

23 Pederson JM, Moon MS, Al-Ajeel HY (1998) Data validation: Key to development of an integrated reservoir model for the Wara Formation, Greater Burgan Field. SPE-REE, August.

24 Vincent G, Corre B, Thore P (1999) Managing structural uncertainty in a mature field for optimal well placement. SPE-REE, August.

25 Toronyi RM, Saleri NG. Engineering control on reservoir simulation, Part 2. SPE paper 17937.

26 Saleri NG, Toronyi RM. Engineering control on reservoir simulation, Part 1. SPE paper 18305.

27 Nind TEW (1981) Principles of Oil Well Production. McGraw-Hill Books Co., New York City.

28 Saleri NG (1993) Reservoir performance forecasting: Acceleration by parallel planning. JPT, July.

29 Ovreberg O, Damsleth E, Haldorsen HH (1992) Putting error bars on reservoir engineering forecasts. JPT, June.

30 Damsleth E, Hage A, Volden R (1992) Maximum information at minimum cost: A North Sea development study with experimental design. JPT, Dec.

31 Brush RM, Marsden SS. Bias in engineering estimation: A case study. SPE paper 9569.

32 Muskat M (1949) Physical Principles Of Oil Production. McGraw-Hill Books Co., New York.

Chapter 8

Planning a Study

Every project requires a plan, and integrated studies are no exceptions. Planning a study is the work of understanding the resources needed to perform a given task, in terms of technical, economic and software constraints, and allocating these resources in time, with the global objective of optimising the synergy of the various phases of the project and minimising the associated costs.

In general, the art of planning a sequence of activities can be a relatively straightforward exercise. When for example an elevator is to be built, a rapid assessment of the condition of the site and some previous experience will be sufficient to design a realistic work schedule of the project. Even moving in the more complex petroleum world, most of the projects related to the building or revamping of surface facilities and infrastructures, can be planned with a relatively narrow margin of error. Deviations are always possible, but allowance can be foreseen for the perceived risks, in terms of delays or additional costs. Previously gained experience will provide a solid basis for quantifying those allowances.

Unfortunately, when reservoir studies are concerned, planning a realistic workflow and correctly estimating resources may become a very difficult exercise. Project managers know very well the pain of justifying to management unexpected delays in the final results or the embarrassment of asking for more time and more money to complete an ongoing study.

Reservoir project managers are not worse than other managers, of course. Simply, their planning task is more difficult, because of some particular characteristics that differentiate it from other types of projects. In particular, 2 aspects should be mentioned:

- **Every study is different from all others**. From a project management point of view, this means that each study will result in a different composition of work activities, depending on a number of internal (technical) and external (time and costs) constraints. Experience is useful, of course, but the variables involved are too many and, consequently, the range of unforeseen outcomes is just too wide to be taken into account properly.
- **There is an underlying technical uncertainty in most phases of the study**. The impact of this uncertainty in the project management phase can be appreciated when we consider that we cannot really say in advance whether the resources allocated to a particular phase of the study will be sufficient to obtain results within the expected range of accuracy.

Despite of these problems, the reservoir project manager is in general confronted with a task that is not different from the management of the elevator company or the surface facilities department. In other words, he needs to establish a plan of the study and to keep to the deadlines, since in most cases the operational activity is related to the results of the study and important decisions, e.g., building a water injection plant, must be taken in a given time frame. In fact, the activity related to the reservoir study is very often one of the many items in a higher level management plan and any delay in the study would ultimately result in a delay of the global project.

Being framed by such external conditions, the reservoir project manager must be able to set a flexible working schedule, where allowance is given for expected and unexpected factors that may influence the planned workflow.

In the next sections, we will analyse the technical issues that must be considered in order to take advantage of the integrated approach of the study. Later, we will review the traditional, sequential approach to reservoir studies planning and we also will see how this approach can be modified, in order to improve the synergy and to provide a more flexible framework for conducting the study.

8.1 PLANNING VS. INTEGRATION

In the traditional way of planning and performing a reservoir study, the project is generally divided into three broad phases, which could be called static model, dynamic model (or basic reservoir engineering, as it has been referred to in this text) and simulation model. These three phases are usually performed in sequential order, by different professionals or groups of professionals.

However, this way of performing the study has several limitations, the main one being the reduced possibility of technical exchange between the three phases. In fact, the degree of integration that can be reached in a particular study is related, among other factors, to the timing of the activities. Each phase of the study generates a piece of information that can be used in the following phases but, in general, cannot be utilised in previous, already completed parts of the study.

Of course, in the execution of an integrated reservoir study, it is important that the information generated within each discipline be available to the others. At the scale of the project schedule, this means that allowance must be made for the information to be exchanged among the three main modelling phases of the reservoir study. From a technical viewpoint, the following points are relevant:

- **Static vs. dynamic model**. Traditionally, the geological modelling of the reservoir ends with the computation of the OHIP, while subsequent modifications based on reservoir engineering evidence are only loosely integrated (or not considered at all). A better approach, as discussed throughout the text, is to make direct use of the dynamic information while building the geological model, which can be achieved through the direct co-operation of the geologists and the reservoir engineers in each phase of the study. This implies that the geological model can be considered concluded only when the reservoir engineering data have been fully reviewed and integrated.

- **Simulation vs. static and dynamic models**. In the simulation study, additional information can be obtained which is relevant to the static and dynamic models of the reservoir, and can therefore be used as a validation feedback. This in turn guarantees the global consistency of the study. From this point of view, it should be noted that most of the information needed to build the simulation model is available early in the study, therefore the numerical simulation phase can be started well before the other phases end. Furthermore, this allows for an early exploration in the study workflow of possible problems related to the numerical simulation approach.

It is therefore essential that the project workflow be planned in such a way to allow for the technical exchange among the various phases of the study. This is the base condition for attaining a true synergistic approach.

8.2 ESTIMATION OF INDIVIDUAL WORK PHASES

The number of work phases to be performed in a typical study and their duration obviously depend on the particular reservoir and the available resources. While this observation would preclude any generalisation, it is possibly useful to take a glance at the individual activities that would make up a typical reservoir study. We may refer in this case to the hypothetical example of an oil reservoir with a significant degree of geological heterogeneity, 30–50 producing wells, few injection wells and 20–30 years of exploitation.

For such a reservoir, Table 8.1 illustrates the activities that could be foreseen and their individual duration, on the basis of total man-weeks. This is also expressed in percentage of the total study duration, which should represent a more general reference value.

Needless to say, these estimations are subject to large variations, depending on the particular project. The study of a North Sea reservoir, for example, would typically need much more time in the analysis of the available data and less time in the database building and data pre-processing, since these fields have been developed in relatively recent years and a large amount of high quality data are usually available on existing and reliable databases. Also a smaller number of exploration and development wells are usually involved.

In other geographical contexts, on the other hand, the exploitation started much earlier and has been performed through the drilling of a large number of wells, where usually little quality information is available. This is typically the case of old producing basins, like some fields in the US, in Africa or Venezuela. For these studies, the time allocated to the data collection and database building, as well as the analysis of the production information, can be much higher than indicated in Table 8.1.

8.3 SEQUENTIAL PLANNING

Sequential planning is the type of work scheduling that is traditionally applied in most reservoir studies. Simply stated, sequential planning represents a logical approach to the organi-

Table 8.1 Typical phases and duration of a reservoir study.

Phase	Individual tasks	Duration (weeks)	Duration (%)
Data collection	Raw seismic, log and core data collection Water and oil PVT analyses data Well testing data Pressure data Field production and Injection data Existing studies and reports collection Database construction	9 to 20	6 to 14
Data pre-processing	Log data correction and normalisation Core-log depth matching Pressure data correction and cleaning Production data validation	3 to 6	2 to 4
Well data analysis	Petrophysical interpretation Synthetic seismograms generation Facies analysis and classification Production log analysis Well tests interpretation PVT study	20 to 28	14 to 18
Spatial distributions analysis	Sedimentological study Seismic interpretation and modelling Geological correlation Facies distribution analysis Petrophysical distribution study Pressure analysis Water/gas advance with time	30 to 44	20 to 28
Production analysis	Production/Injection performance analysis Material balance	9 to 12	6 to 8
Simulation model	Model building History match Production forecast	40 to 60	30 to 40
Final report	Report writing and editing	6 to 9	4 to 6
Total, average		**148 weeks**	**100%**

sation of the activities that compose a reservoir study, where all the tasks are related in a cascading progression, from data collection to reservoir performance forecasting (Fig. 8.1).

Actual studies planning will be much more complex, of course, and some overlap among the various tasks is always present even in the sequential approach, simply because more professionals are generally involved at the same time in the same study, each one performing a different work. However, from a conceptual point of view, in sequential planning the main parts of the integrated study are chronologically separated.

The most relevant factor in sequential planning is that each step can be performed only when the preceding steps have been completed. This in turn implies important consequences:

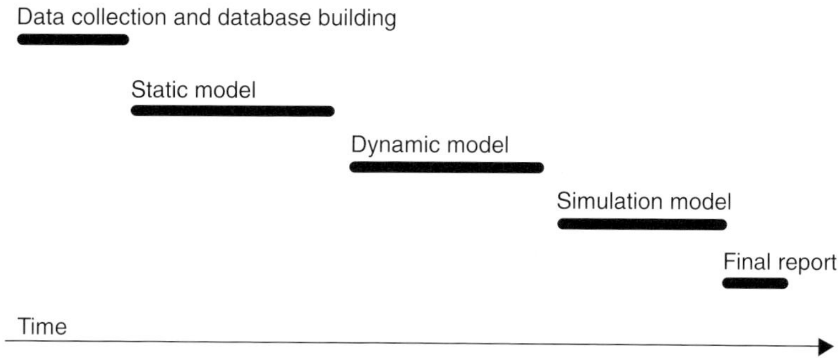

Figure 8.1 Sequential approach to reservoir studies planning.

- **Limited integration**. Each individual activity can integrate the information generated in previous phases, but cannot take advantage of the information that will be generated from subsequent phases, unless a later revision takes place.
- **Delays accumulate**. Each delay that is generated in a particular phase of the study, will be translated in a correspondent delay of the total project.
- **Limited flexibility**. Sequential planning offers little flexibility in terms of external and unforeseen factors. When for example, new data become available during the study or when the professional team changes, some parts of the study will be revised, thus leading to additional delays.

To overcome these problems, the reservoir project manager must be able to set up a more flexible planning study, which allows for the maximum team synergy, in terms of exchange of information, while limiting the possible overall delay.

8.4 INTEGRATED PLANNING

As discussed throughout the text and in paragraph 8.1, in the framework of an integrated study a considerable amount of information can be exchanged between the static, the dynamic and the simulation models of the reservoir. Therefore, an open planning system should be considered, in order to exploit all these integration opportunities. This simple observation sheds a new light over the type of approach that could be undertaken, when the study workflow is to be planned.

The basis for what could be referred to as **integrated planning** are the following:

- **Integration opportunities**. Each phase must be able to take advantage of the work being performed in the context of other disciplines.
- **Reduced delay**. A minimum tolerance should be considered for the total delay of the project, in order to comply with higher level planning strategies.
- **Reduced time frame**. The project should be completed in the shortest time frame. The longer the execution period, the more likely that internal and external factors will

result in changes and deviations with respect to the original objectives and budget. At the same global cost, a larger team will be able to complete the project in a shorter time frame.

To comply with these objectives, all the various phases of the study should ideally be performed simultaneously. Practically, however, this becomes impossible to do, because some dependency among the distinct disciplines always exists. It is impossible, for example, to start computing the spatial distribution of any reservoir parameter, e.g., porosity, before having completed at least a good part of the petrophysical evaluation.

Nevertheless, the basic idea of an integrated planning is to keep, as much as possible, simultaneous activities running. At the beginning of the study, for example, several steps could be undertaken in addition to the database building. When resources are available, a number of activities could be performed from the beginning, including for example the sedimentological study, the seismic interpretation, the well correlation study, the petrophysical evaluation and the conventional reservoir engineering. Fig. 8.2 shows a simplified image of a possible integrated planning.

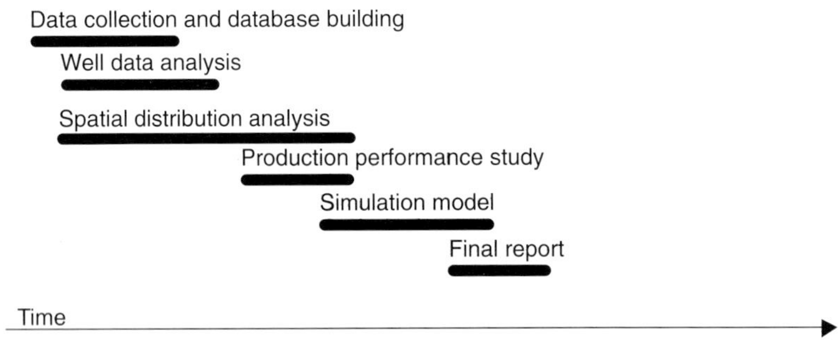

Figure 8.2 Integrated approach to reservoir studies planning.

Several features should be noted in such a planning. First, the items invoked are different from the traditional ones, based on the definition of the static, dynamic and simulation models (Fig. 8.1). The items referenced in here, consistently with the classification of Table 8.1, show that different groups of activities could be defined which better characterise the concept of integrated reservoir study. These items include both static and dynamic tasks.

This kind of classification is based on the objective of each item (e.g., spatial distribution analysis), rather than the nature of the data involved (e.g., dynamic data) and it is used here to stress that data coming from several different sources can be used and integrated to achieve the objectives of each individual task.

Note also that, when this planning is considered from a traditional point of view, the static and dynamic models of the reservoir will be completed about at the same time. This allows for the finalisation of the geological model taking into account the engineering information contribution, which, as discussed in paragraph 8.1, is one of the main technical requisites for a consistent scheduling of the study.

Likewise, the simulation model should be started before the completion of the other phases. Even when the definitive input data are not available, the model-building phase can be initiated, using provisional sets of data. This will allow for the generation of feedback information for the previous phases of the study and the early identification of problems in the simulation model. As long as definitive data become available, they can be input in the model, while the global behaviour and stability of the simulation can be tested in a stepwise fashion. This approach will provide a much tighter control over the numerical simulation, compared to the traditional approach of building and initialising the model with a final set of data that could give numerical problems that may be difficult to identify and solve. Interesting variations of this integrated planning approach, called **parallel planning**, have been presented by Saleri [1, 2].

Finally, it should be noted that the global execution time of the integrated approach is much shorter than the traditional, sequential approach. Although this is obtained through the involvement of a larger team, the global cost of the project should not be affected.

8.5 CONCLUSIONS

When integration is concerned, the traditional way of planning a reservoir study often represents too rigid a framework. The integration of the various disciplines that are involved in the study calls for the definition of a more flexible planning system, which gives professionals the opportunity to exchange their views and individual conclusions.

The recognition of such a need is critical in the proper execution of an integrated reservoir study, to the same degree as other external or non-technical factors, like the physical integration of the people and the computing environment. The project manager must schedule the study keeping in mind all the possible information exchange paths between the various disciplines and, at the same time, he must watch the development of the project closely and maintain the flexibility to impose any necessary corrections.

Integrated planning, as has been defined here, is the reference framework for reservoir studies. Not only is it a necessary condition for integrated studies: it may also generate integration, by giving the different professionals the chance to work in the same environment, at the same time and with the same global objective.

Its implementation requires a larger number of professionals to be involved in the study but, on the other hand, the global time frame will be shorter. Whenever this condition can be met and an integrated planning can be achieved, the chances are good that a more coherent and consistent study will be produced.

References

1 Saleri NG (1993) Reservoir performance forecasting: Acceleration by parallel planning. JPT, July.

2 Saleri NG (1998) Re-engineering simulation: Managing complexity and complexification in reservoir projects. SPE-REE, Febr.

APPENDIX

Material Balance

A.1 GENERAL FORMULATION OF THE MATERIAL BALANCE EQUATION

The mathematical derivation of the material balance equation is relatively simple and can be found in any basic reservoir engineering textbook [1, 2]. However, the general formulation will be reviewed here, since it will help in the understanding of the parameters that need to be considered in its application.

Let's consider the general case of a reservoir with a gas cap and with an underlying active aquifer, as depicted in Fig. A.1. Note that this represents the most general configuration, since all drive mechanisms are acting simultaneously.

Some time after the start of production, the reservoir pressure will be declined by a given amount $\Delta p = (p_i - p)$ from the initial pressure p_i. The general material balance equation can be expressed as follows:

$$F = N\,(E_o + mE_g + E_{f,w}) + W_e B_w \qquad \text{(A.1)}$$

This is the general formulation by Havlena and Odeh [3]. The terms that appear in this equation will be described hereafter, but it is useful to note at this stage that this formulation reflects precisely the qualitative statement given in paragraph 6.6.2. In fact, the left-hand side of Eq. A.1 represents the total underground withdrawal due to production, and can be expressed as:

$$F = N_p\,(B_o + (Rp - Rs)B_g) + W_p B_w \qquad \text{(A.2)}$$

In this equation, the withdrawal components related to oil ($N_p\,B_o$), gas ($N_p\,(Rp - Rs)B_g$) and water ($W_p B_w$) can be recognised.

At the right-hand side of Eq. A.1, we find the expansion terms of the reservoir. In particular, the term E_o represents the expansion of the oil plus the originally dissolved gas:

$$E_o = (B_o - B_{o_i}) + (Rs_i - Rs)B_g \qquad \text{(A.3)}$$

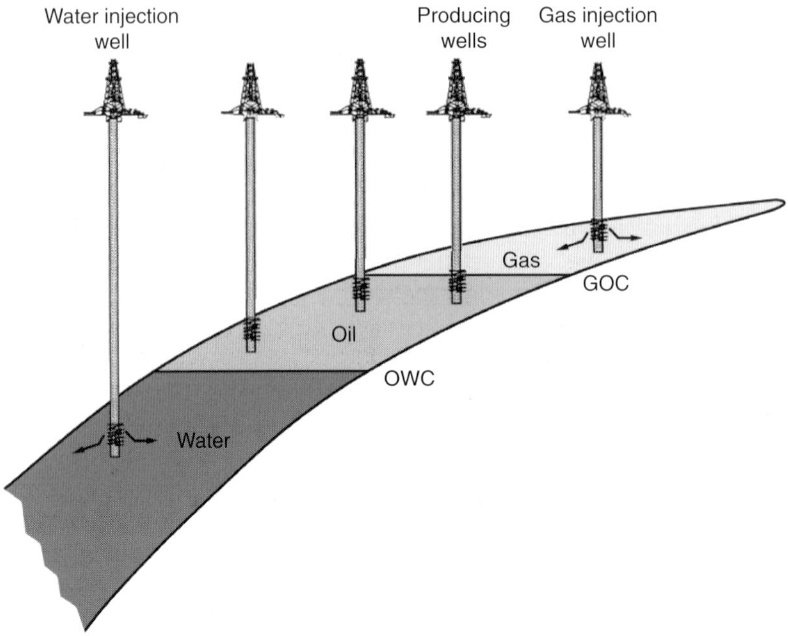

Figure A.1 Example of a combination drive reservoir.

The term E_g represents the expansion of the gas of the gas cap, and can be expressed as:

$$E_g = B_{o_i}\left(\frac{B_g}{B_{g_i}} - 1\right) \tag{A.4}$$

Finally, the term $E_{f,w}$ includes the expansion of the irreducible water saturation and the effect of compaction of the pore volume related to the compressibility of the formation:

$$E_{f,w} = (1+m)B_{o_i}\left(\frac{(c_w Sw_i + c_f)}{(1 - Sw_i)}\right)\Delta p \tag{A.5}$$

The factors appearing in the above equations are as follows:

N STOOIP (stb)

N_p cumulative oil production (stb)

m gas cap volume/oil volume ratio at initial conditions (vol/vol)

W_p cumulative water production (stb)

B_o, B_g, B_w oil, gas and water formation volume factors at the reduced pressure p (rb/stb)

B_{o_i}, B_{g_i}, B_{w_i} initial oil, gas and water formation volume factors (rb/stb)

Rs_i initial gas solubility ratio (scf/stb)

Rs gas solubility ratio at the reduced pressure p (scf/stb)

Rp cumulative GOR for the pressure drop Δp (scf/stb)

c_w water compressibility (1/psi)

c_f formation (pore volume) compressibility (1/psi)
Sw_i irreducible water saturation (fraction)
W_e water influx (rb)
$\Delta p = (p_i - p)$ pressure drop due to production (psi)

Therefore, a maximum of 18 parameters enter the general formulation of the material balance equation. This number can raise to 20 when water and gas injection are considered (Fig. A.1). In this case the cumulative gas and water injected (W_i and G_i) will be subtracted from the left-hand side of Eq. A.1, thus resulting in a net underground withdrawal, or added to the right-hand side as an extra component of the general expansion.

While these equations may appear a bit intimidating at first, it should be noted that most of these parameters are normally available to the reservoir engineer and their estimation in most cases does not require a particular effort. Furthermore, most reservoirs are less complex that that depicted in Fig. A.1, therefore the material balance formulation becomes simpler.

A.2 CHARACTERISTICS OF THE MATERIAL BALANCE APPROACH

In the way it has been formulated, the material balance equation exhibits some peculiar characteristics, which can be summarised as follows:

- The material balance is a comparison of voidage to expansion and mostly concentrates on evaluating fluid expansion. It is a volumetric approach, which does not specifically take into account fluid dynamics and therefore mobilities.
- The equation is zero-dimensional, because all the parameters are evaluated at a single point in the reservoir, typically the centre of gravity of the fluid distribution.
- The formulation expresses the dynamic behaviour of a tank-like reservoir, i.e., a reservoir whose pressure data show a uniform decline when referred to a common datum depth.
- The equation is not explicitly time dependent (even though the water influx often has a time dependence).

These features provide the main guidelines for the application of the material balance equation to a reservoir study. They represent the strong points and, at the same time, the limitation of the approach.

In the next section, we will analyse the conditions that underlie a correct application of the material balance method and the possible associated pitfalls. We will also review all the parameters that enter the general equation, trying to highlight their role in the calculation, as well as the uncertainty that is typically associated to each of them.

A.3 CONDITIONS FOR THE CORRECT APPLICATION OF MATERIAL BALANCE

To compute a material balance for a given reservoir is nearly always possible, but of course, this does not imply that the results will always be reliable. As in any other method, the accuracy of the results is primarily related to the reliability of the input information and it is therefore important, in each case, to carefully evaluate what we are introducing into the computation.

The inspection of the parameters that appear in the general formulation of the material balance equation allows for the identification of 5 main groups, that refer to production and injection parameters, PVT characteristics, rock properties, volumetric parameters and pressure data.

A.3.1 Production and Injection Parameters

Cumulative produced oil (N_p) water (W_p) and gas (R_p) are the production parameters that appear in the equation. When an injection project is active, the injection terms W_i and G_i also belong to this group. All these parameters are input to the material balance equation and are therefore considered as *knowns*. In fact, for commercial reasons, only the cumulative oil is practically always known with a good degree of confidence. Unless a sales contract exists, gas production is a much more uncertain measurement, especially when gas is flared or when dealing with old fields, where production measurements reports are scarce or unreliable. Also, bad management of gas lift procedures often impairs the reliability of gas measurement data.

As far as water production is concerned, the situation is not dissimilar, exacerbated by the fact that water never has a commercial interest. Unfortunately, in many operating companies and especially in the past, routine production procedures did not necessarily accommodate the needs of reservoir engineers, therefore these data should be regarded with a critical eye. The reservoir engineer is not responsible for the quality of the data he deals with, but he should always dedicate a particular effort in trying to evaluate the reliability of these data, through the direct assessment of the production measurement conditions of the field under study.

A.3.2 PVT Characteristics

In the general case, 8 PVT parameters enter the material balance equation. These are the initial and reduced oil, gas and water formation volume factors (B_{o_i}, B_{g_i}, B_{w_i} and B_o, B_g, B_w), as well as the initial and reduced gas solubility ratio (Rs_i and Rs). In fact, what is actually needed in practical application is a complete description of the PVT properties of oil, gas and water for a given range of pressures, obtained by means of laboratory measurements or correlations (see paragraph 6.2).

The PVT properties are an input to the material balance computation and are normally considered as a *hard* information, i.e., without important associated uncertainty. However, due to the impact of these parameters in the final results, it is always wise to investigate the quality of the available PVT data and to run sensitivity cases, by varying the PVT description within the range of the perceived uncertainty.

A.3.3 Rock Properties

Similarly to the parameters described above, rock properties are commonly an input to the material balance equation. Connate water saturation, Sw_i, represents the amount of water in the reservoir which is allowed to expand as a consequence of the pressure drop Δp. It is important that this parameter be consistent with the findings of the geological model: an average value should be back-calculated starting from the volumetric OOIP and the bulk reservoir pore volume.

As far as compressibility is concerned, in the majority of cases such parameters (water and formation compressibility, c_w and c_f) do not have a great impact in the final results, with the notable exception of overpressured reservoirs and reservoirs subject to compaction, where the formation compressibility may become important and provide a notable part of the energy to the system. In these cases, as already noted in paragraph 6.1.5, care must be taken in the selection of the formation compressibility value to enter in the equation, especially if the OOIP or the water influx are to be computed. Alternatively, when it is considered that a greater uncertainty is attached to the formation compressibility than to the OOIP, the material balance equation can be solved by fixing the OOIP and computing the compressibility value compatible with the observed field withdrawal. This is of particular importance when a simulation study is to be performed, since it provides an independent and often very reliable estimation for one of the most important dynamic parameters in compaction drive reservoirs.

A.3.4 Volumetric Parameters

Three volumetric parameters enter the material balance equation, i.e., the oil in place (N), the gas cap volume (expressed as a ratio between the gas cap and the oil volumes at initial conditions, m), and the cumulative water influx (W_e).

The first of these parameters, N, is one of the typical output of a material balance study. For many years, reservoir engineers have computed the OOIP through different formulations of the general material balance equation. One important point is that, being an estimate based on dynamic parameters, the OOIP figure provided by material balance refers to a connected or *active* OOIP, which is the OOIP actually drained by the wells used in the calculation. This simple matter accounts for most of the discrepancies that are often observed in the comparison between material balance and volumetric estimates of the OOIP. Any hydrocarbon accumulation not connected with the drained area, for example an isolated fault block, will not contribute to the pressure distribution and cannot be accounted for.

The computation of N can be straightforward in a number of basic material balance applications. In the relatively simple case of a solution gas drive reservoir, for which water and formation compressibility can be neglected, Eq. A.1 reduces to:

$$F = NE_o$$

In this case, the underground withdrawal is a linear function of the expansion of the oil plus its dissolved gas. From this relationship, the OOIP can be easily calculated.

When an additional source of energy is acting in the reservoir, for example a waterdrive, the material balance equation can be simultaneously solved for N and the aquifer influx, W_e.

In this case, the above relationship will deviate from linearity and in itself this behaviour is a diagnostic of the presence of an active water drive. A particular re-arrangement of Eq. A.1, obtained dividing both sides by E_o, still provides a linear behaviour:

$$\frac{F}{E_o} = N + \frac{W_e}{E_o}$$

When the corrected value of W_e has been selected, a plot of F/E_o vs. W_e/E_o gives a straight line whose intercept on the ordinates again gives an estimate of the OOIP, N. One of these plots is shown in Fig. 5. 4. From a practical standpoint, the value of W_e is estimated by choosing an analytical model for the aquifer (steady state or unsteady state, e.g., the Hurst and Van Everdingen model) and tuning the value of some parametric groups (e.g., the aquifer constant) until a satisfactory straight line is found. Note that this technique allows for a reliable estimation of the water influx, which is independent from the evaluation of the many single variables that actually appear in the fully analytical expressions of the aquifer influx of Hurst and Van Everdingen.

It is also important to note that, as mentioned in paragraph 5.2.2, the computed values of OOIP and water influx should be compared with independent estimations. Typically, the OOIP coming from a material balance estimation should be compared with the available volumetric figures, while the total aquifer influx can be checked with volumetric assessments made on maps of water advance as a function of time. This simple comparison often provides a good insight into the credibility of the reservoir model which is being built.

As far as the gas cap is concerned, a similar re-arrangement of Eq. A.1, can still provide a graphical solution for the OOIP and the gas cap volume, m. However, in this case, a much greater accuracy of the pressure data is required in order to perform a reliable computation. In fact, this need for greater accuracy severely limits the practical application of the method in the presence of a large gas cap. To further complicate the matter, pressure measurements are difficult and often questionable in reservoirs with a gas cap, since the presence of gas considerably complicates the interpretation of the build-up plots.

However, if the OOIP is known with a reasonable degree of confidence from volumetric computations, than the determination of the gas cap volume is often possible.

A.3.5 Pressure Data

Reservoir pressure, expressed as depletion ΔP, is the most important parameter that appears in the material balance equation. Interestingly, the inspection of Eq. A1 reveals that the quantity ΔP explicitly appears only in the last term $E_{f,w}$, the water and formation expansion term, but in fact the pressure is implicitly present in all the PVT parameters, which accounts for the volumetric expansion of the system.

Pressure enters the material balance expression as a depletion term or, from a practical standpoint, as a pressure decline trend. In fact, the reliability of a material balance calculation depends in large measure on the possibility of deriving a representative pressure decline trend for the reservoir under study, which is not a trivial task.

Actually, pressure data collected in the wells do not often display a coherent behaviour, but instead appear rather scattered in a pressure-time plot. There may be many reasons for this behaviour, but in fact, when reservoir compartmentalisation is excluded, the cause is often a lack of pressure re-equilibration due to low transmissibility of the system. Pressure gradients related to production travel through the reservoir with a rapidity that depends on the diffusivity constant of the system, $k/\phi\mu c$. Therefore the higher the porosity ϕ, the fluid viscosity μ and the effective compressibility c, and the lower the permeability k, the slower a pressure equilibrium condition will be attained.

When the pressure data for different wells show a degree of dispersion, an attempt can be made to compute a weighted average value, at fixed time intervals, following the procedure that has been discussed in paragraph 6.4.3. However, in low permeability reservoirs or in the presence of heavy, viscous oils, the identification of a common pressure trend may be difficult, if not impossible. In these cases, the results of the material balance may prove to be unreliable. Fig. A.2 shows an example for a heavy oil reservoir (13 API), where the pressure measurements are probably too sparse to allow for an unambiguous recognition of a common pressure trend.

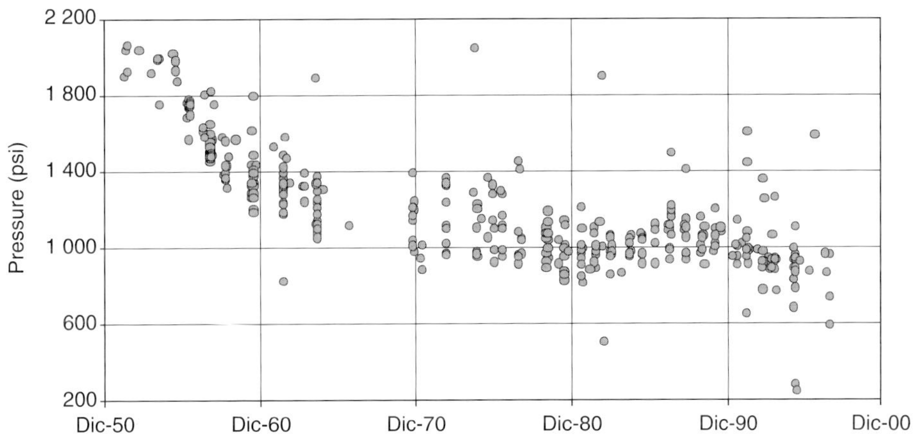

Figure A.2 Pressure data in a heavy oil reservoir.

Another important point to consider is that a significant pressure drop is usually required for a reliable material balance computation. Large pressure declines produce large expansions, making inaccuracies in the production volumes relatively less significant. Similarly, in the case of large depletion, the uncertainty in the pressure measurements become less important and a pressure trend is more easily identified. It is generally assumed that a pressure decline of at least 10% of the original formation pressure is required for a reliable computation to be performed. Below this threshold, high quality data would be required, since the expansion terms become very small. For this reason, strong water or gas drive systems are not often modelled with material balance.

A.4 CONCLUSION

Material balance has a number of merits, which justify its survival in these years of high technology, dominated by complex numerical techniques. Therefore, when performing an integrated reservoir study, material balance should be an essential step of the workflow.

The following points can be remembered here:

- The material balance approach can always be attempted. Whenever reliable basic reservoir engineering data are available, the method can provide useful results. It is also worth noting that if these basic requirements do not exist, than any other reservoir engineering technique is likely to fail in its objectives, especially in more complex simulation studies.
- The technique offers an invaluable tool for investigating the coherence and the consistency of the available dynamic data. The results of this stage provide the engineer with significant additional information to be used in the following phases of the study.
- Material balance is the only dynamic technique that provides reliable estimates of the OOIP, as well as the aquifer and/or gas cap volumes. These estimates should be compared with the available volumetric figures, while differences should be reconciled or explained. From this point of view, the method offers the possibility of checking assumptions made in the static study concerning the actual compartmentalisation of the reservoir.
- The possibility of evaluating the reservoir drive mechanisms and some of the main dynamic parameters, like water influx or compressibility, justifies the use of material balance as a preliminary investigation prior to a more detailed numerical simulation study.

References

1 Dake LP (1987) Fundamentals of Reservoir Engineering. Elsevier, Amsterdam.
2 Craft BC, Hawkins MF (1959) Applied Reservoir Engineering. Prentice-Hall.
3 Havlena D, Odeh AS (1963) The material balance as the equation of a straight line. JPT, August.

Index

R

S

X

Achevé d'imprimer
Impression : EUROPE MEDIA DUPLICATION SA
F 53110 Lassay-les-Châteaux
N° 8215 - Dépôt légal : février 2001
N° d'éditeur : 1055